Cotton Fibers

Developmental Biology,
Quality Improvement,
and Textile Processing

THE FOOD PRODUCTS PRESS
Crop Science
Amarjit S. Basra, PhD
Senior Editor

New, Recent, and Forthcoming Titles of Related Interest:

Dictionary of Plant Genetics and Molecular Biology by Gurbachan S. Miglani

Advances in Hemp Research by Paolo Ranalli

Wheat: Ecology and Physiology of Yield Determination by Emilio H. Satorre and Gustavo A. Slafer

Mineral Nutrition of Crops: Fundamental Mechanisms and Implications by Zdenko Rengel

Conservation Tillage in U.S. Agriculture: Environmental, Economic, and Policy Issues by Noel D. Uri

Cotton Fibers: Developmental Biology, Quality Improvement, and Textile Processing edited by Amarjit S. Basra

Heterosis and Hybrid Seed Production in Agronomic Crops edited by Amarjit S. Basra

Intensive Cropping: Efficient Use of Water, Nutrients, and Tillage by S. S. Prihar, P. R. Gajri, D. K. Benbi, and V. K. Arora

Plant Growth Regulators in Agriculture and Horticulture: Role and Commercial Uses edited by Amarjit A. Basra

Crop Responses and Adaptations to Temperature Stress: New Insights and Approaches edited by Amarjit A. Basra

Physiological Bases for Maize Improvement edited by Maria Elena Otegui and Gustavo A. Slafer

Cotton Fibers
Developmental Biology, Quality Improvement, and Textile Processing

Amarjit S. Basra
Editor

Routledge
Taylor & Francis Group
NEW YORK LONDON

First published 1999 by The Haworth Press, Inc.

Published 2020 by Routledge
52 Vanderbilt Avenue, New York, NY 10017
2 Park Square, Milton Park, Abingdon, Oxon OX14 4RN

Routledge is an imprint of the Taylor & Francis Group, an informa business

Softcover edition published 2000.

Cover design by Jennifer M. Gaska.

The Library of Congress has cataloged the hardcover edition of this book as:

Basra, Amarjit S.
   Cotton fibers : developmental biology, quality improvement, and textile processing / Amarjit S. Basra.
      p.   cm.
   Includes bibliographical references and index.
   ISBN 1-56022-867-9 (alk. paper)
      1. Cotton. 2. Cotton—Quality. 3. Cotton manufacture—Quality control. I. Title.
SB249.B35 1999
633.5′1233—DC21   99-20218
                                                                                        CIP

ISBN 13: 978-1-56022-898-1 (pbk)

CONTENTS

ABOUT THE EDITOR

Amarjit S. Basra, PhD, is an eminent Botanist (Associate Professor) at Punjab Agricultural University in Ludhiana, India. His outstanding research on mechanisms of cotton fiber development has been internationally recognized and has significantly contributed to cotton fiber quality improvement programs around the world. He is a member of the Crop Science Society of America, the Australian Society of Plant Physiologists, the Indian Society of Developmental Biologists, and the International Association for Plant Molecular Biology. Dr. Basra has more than seventy research and professional publications to his credit, including serving as editor of *Seed Quality: Basic Mechanisms and Agricultural Implications, Crop Sciences: Recent Advances,* and the forthcoming book *Heterosis and Hybrid Seed Production in Agronomic Crops,* all published by The Haworth Press, Inc. He is also the Founding Editor of the *Journal of Crop Production* and *Journal of New Seeds* (The Haworth Press, Inc.). A recipient of several coveted awards and honors, Dr. Basra has made scientific visits to several countries, fostering cooperation in crop science research at the international level.

CONTRIBUTORS

W. Stanley Anthony is Supervisory Agricultural Engineer, U.S. Cotton Ginning Laboratory, USDA-ARS, Stoneville, Mississippi.

Antony J. Buchala, PhD, is Lecturer, Institute of Plant Biology, University of Fribourg, Switzerland.

Gayle H. Davidonis, PhD, is Plant Physiologist, USDA-ARS, Southern Regional Research Center, New Orleans, Louisiana.

Deborah P. Delmer, PhD, is Professor and Chair, Section of Plant Biology, University of California at Davis.

Yehia E. El Mogahzy, PhD, is Professor, Department of Textile Engineering, Auburn University, Auburn, Alabama.

You-Lo Hsieh, PhD, is Professor of Fiber and Polymer Science, Division of Textiles and Clothing, University of California at Davis.

Judith A. Jernstedt, PhD, is Associate Professor, Department of Agronomy and Range Science, University of California at Davis.

Maliyakal E. John is affiliated with Agracetus, a Unit of Monsanto, Middleton, Wisconsin.

Russell J. Kohel, PhD, is Research Geneticist, USDA-ARS, Crop Germplasm, Research Unit, College Station, Texas.

O. Lloyd May, PhD, is Research Geneticist, USDA-ARS, Pee Dee Research and Education Center, Florence, South Carolina.

Ulrich Ryser, PhD, is affiliated with the Institute of Plant Biology, University of Fribourg, Fribourg, Switzerland.

Sukumar Saha, PhD, is Research Geneticist, USDA-ARS, Crop Science Laboratory, Stoneville, Mississippi.

Thea A. Wilkins, PhD, is Associate Professor, Department of Agronomy and Range Science, University of California at Davis.

Preface

Cotton, an oilseed and fiber crop, is grown in more than seventy countries of the world, and plays an important role in the global economy. No crop competes with it in the potential of value-added processing. Valued at over $20 billion, it produces the premier natural fiber for the textile industry. Cotton textile products have existed for over two and one-half millennia, and despite competition from man-made fibers, cotton fiber has maintained its importance and utility to this day.

The commercial cotton fiber is a product of four cultivated species of Gossypium, that is *G. arboreum, G. hirsutum, G. barbadense,* and *G. herbaceum.* Botanically, the fiber is a single-celled hair or trichome developing from individual epidermal cells on the outer integument of cotton ovules. Cotton fiber development has been divided into four phases: (1) initiation, (2) elongation, (3) secondary wall thickening, and (4) maturation. The fibers elongate considerably during development and may end up over 1,000 to 3,000 times longer than their diameter. In the form of a single-celled epidermal hair, the cotton plant produces one of the purest forms of cellulose known to man.

Cotton fiber quality is important to the spinning and weaving industry and determines the use to which it is put, as well as influencing the price paid for the crop. Several characters are used to assess quality, some requiring sophisticated measuring and testing devices. Fiber length, fineness and length distribution, fiber strength, and maturity are the most important quality factors for textile processing. The challenge facing cotton breeders and biotechnologists is to produce a cultivar that meets the needs of the textile industry and also produces enough lint for growers to make a profit. Of more fundamental concern is an understanding of how the fiber quality traits are biologically regulated.

As we enter the new millennium, innovations in textile technology demand increasingly better fiber quality to meet the processing needs and the quality of the end product. The specialty fibers under development using genetic engineering techniques are creating new opportunities for novel product concepts in woven and nonwoven markets, in addition to improving the economics of cotton fiber processing.

The explosion of information on cotton fibers generated in recent years certainly warrants the publication of a fully integrated and comprehensive book dealing with all aspects of fiber development, quality improvement, and processing. In the present volume, I have endeavored to achieve this with the cooperation of foremost cotton scientists in the field to provide an authoritative text and reference source for everyone who uses, produces, markets, or researches cotton fibers. Furthermore, it will enable students, researchers, and teachers in many agricultural and botanical disciplines to become better informed about the interesting and unique vistas that cotton fibers provide for studies on cell and developmental biology, and in particular the pathway of cellulose biosynthesis. This is the only book that combines basic information on the biology of cotton fibers with a wealth of new approaches to produce increasingly better quality and specialty fibers.

I hope that this book will serve to stimulate further research and technological development in the field that will help in meeting the fiber needs of an ever-growing population as well as satisfying new consumer preferences.

Amarjit S. Basra

Acknowledgments

I wish to express my appreciation and gratitude to the following people:

Professor C. P. Malik for introducing me to the world of cotton fibers.

All the contributing authors of this book for contributing their special knowledge and splendid cooperation.

Bill Palmer and Bill Cohen for undertaking this project under the Food Products Press Crop Science book series.

All the staff at the Haworth Press Book Division for making this book a reality, particularly the meticulous editing of Andy Roy, Peg Marr, and Amy Rentner; the creative cover design by Jennifer Gaska, and the administrative support of Melissa Devendorf.

My alma mater, Punjab Agricultural University, for fostering my scientific career and creative growth.

My wife Ranjit, daughter Sukhmani, and son Nishchayjit for their patience, understanding, and inspiring support.

Chapter 1

Cotton Fiber Initiation and Histodifferentiation

Ulrich Ryser

INTRODUCTION

Shorter or longer seed fibers are characteristic for the genus *Gossypium*. The functions of the fibers in wild cotton species are unknown, but may be related to seed dispersal, protection against herbivores, and extreme climatic conditions. The fibers are single, elongated cells of the seed coat epidermis. Fiber development can be divided into four phases: initiation, elongation, secondary wall formation, and maturation (Basra and Malik, 1984). Fiber initials elongate by a factor of 1000 to 3000 and then deposit a thick cellulosic secondary wall. The following material provides a short overview of cotton fiber development.

About 10 percent of the epidermal cells of cotton ovules develop into fibers. Fiber initials first become visible at the day of flowering (anthesis) through bulging and spherical expansion above the epidermal surface. Spherical expansion is followed by cell elongation, and the elongating cells orient their tips initially against the micro-

The author would like to thank Nikolaus Amrhein, Anthony Buchala, Titus Jenny, Jean-Pierre Métraux, Laurence Moire, and Alain Schmutz for their collaboration in several projects; Tony Buchala, Brigitte Mauch-Mani, and Heinz Müller-Schärer for critical reading of the manuscript; Martine Schorderet for skillful technical assistance and help with the figures; and Ruth Bosshard for cultivating the cotton plants. This work was supported in part by the Roche Research Foundation (Grant No. 93-226), the Sandoz Foundation, and the Swiss National Science Foundation (Grant No. 31-39648.93).

pylar end of the ovule. After 2 to 3 days, the tips become tapered. Longitudinal growth apparently exceeds the rate of diametric expansion. At this stage, the fibers start to grow in spirals and are no longer oriented exclusively toward the micropylar end. Elongation continues for about 20 to 30 days, depending on the genotype and the culture conditions, until the fibers reach their final length of 20 to 60 mm (millimeters). Elongating cells contain a large central vacuole and a thin layer of cytoplasm. During elongation, the thin (0.2-0.4 μm [micrometer]) and extensible primary wall is deposited. The cellulose content of the primary wall is low (about 25 percent), and the amount of cellulose per unit fiber length remains constant at about 1 ng mm (nanogram per millimeter)$^{-1}$ during primary wall formation, increasing sharply at the onset of secondary wall formation to reach about 130 ng mm^{-1} at maturity (Beasley, 1979). The cellulosic secondary wall is deposited inside the primary wall and pushes the layer of cytoplasm toward the interior of the cell at the expense of the vacuole. At maturation, 50 to 60 days postanthesis (dpa), the fruit capsules open, and the cylindrical fibers dehydrate and collapse to ribbonlike, twisted structures. The direction of the twist changes at frequent intervals along the fibers. These changes, the "reversals," were detected more than a century ago and are a diagnostic feature used to distinguish cotton and other textile plant fibers. Twist and reversals reflect the orientation of the cellulose microfibrils in the secondary wall. The cellulosic secondary wall of mature fibers has a final thickness of up to 8 to 10 μm and essentially defines the textile properties of the fiber.

Mature fibers can be easily detached from the seeds. After detachment of the longer fibers (lint fibers), the seeds of many cultivars remain covered by very short fibers, the so-called fuzz. Some cultivars have naked seeds; that is, fuzz fibers are missing.

The development of cotton fibers has been described many times. Reviews that include the ultrastructure of the differentiating fibers were published some years ago by Ryser (1985) and Berlin (1986). This chapter gives a short overview of cotton fiber differentiation, concentrating mainly on recent progress made in the following domains: (1) fiber initiation, (2) the cytoskeleton of differentiating fibers and its function in cellulose deposition, (3) the chemical composition and function of suberin and waxes, associated with the fibers of the green lint mutant and the seed coat epidermal cells of the genus *Gossypium* in general, and (4) the ultrastructure and function of the fiber base.

Fertilized and unfertilized cotton ovules can be cultured in vitro on a defined liquid medium (Beasley and Ting, 1973, 1974; Beasley et al., 1974). These methods were of invaluable help in many of the studies reported here.

FIBER INITIATION

Structural Analysis, Cytochemistry, and Autoradiography

The emergence of the fiber initials at the day of anthesis was beautifully illustrated by scanning and transmission electron microscopy (Beasley, 1975; Stewart, 1975; Berlin, 1986). In *Gossypium hirsutum*, fiber initials first appear at the crest of the funiculus near the chalazal end and then around the lateral circumference of the ovule. The emergence of the fiber initials is delayed for a few hours at the chalazal cap and for a day or more at the micropylar region.

The nucleoli of the fiber initials undergo dramatic changes in size and structure (Ramsey and Berlin, 1976; Ryser, 1985; Berlin, 1986; Kosmidou-Dimitropoulou, 1986). The nucleoli enlarge, and ring-shaped nucleoli are typical for the later stages of this process (see Figure 1.1). The incorporation of tritiated uridine into nucleoli and cytoplasm of the fiber initials of ovules cultured in vitro (Berlin, 1986) indicates that the nucleoli are active in the formation of preribosomal particles necessary for ribosome formation and protein synthesis. Intense incorporation of uridine was observed from 1 to 6 dpa (Berlin, 1986). During this time period, a large proportion of the ribosomes used for the rapid elongation of the fibers is probably produced. In this context, it is important to remember that the developmental time scale of the in vitro cultures is considerably compressed as compared to the in vivo situation (Meinert and Delmer, 1977; Ryser, 1979). In fibers from plants grown in vivo, the nucleolar volume increased rapidly after anthesis, reached a maximum at 6 to 10 dpa, and declined slowly to a constant size at 20 dpa

FIGURE 1.1. Ovule Epidermis 4 dpa (*G. arboreum* L.)

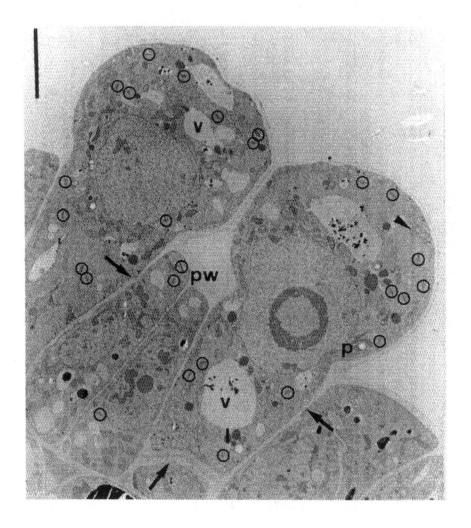

Source: Reprinted with permission from Ryser (1985, p. 248).

Note: Notice the enlarged nucleus and ring-shaped nucleoli in fiber initials. Numerous plasmodesmata are present in this section, mainly in the side walls of epidermal cells (arrows). Golgi complexes (encircled); mitochondria (arrowheads); l, lipid droplets; p, leucoplasts, containing small starch grains; pw, primary wall; v, vacuoles. Bar = 5 μm.

(DeLanghe, 1986; Peeters et al., 1987, 1988; Peeters, Dillemans, and Voets, 1991). A review on nucleolar ultrastructure and function was published recently (Thiry and Goessens, 1996).

The volume of the nucleus was also reported to increase in fiber initials, but not in other epidermal cells (Berlin, 1986; Kosmidou-Dimitropoulou, 1986). However, no incorporation of tritiated thymidine was observed in the nuclei of the fiber initials during the critical time period (Berlin, 1986), suggesting that fiber initiation is not associated with endopolyploidization, as was reported for trichomes of *Arabidopsis* (Melaragno, Mehrotra, and Coleman, 1993). Interesting changes in the labeling frequency of the nuclei in the epidermis were observed before and after anthesis. At 12 days preanthesis, about 10 percent of the nuclei in the epidermis were labeled after a short pulse with tritiated thymidine (S-phase nuclei). The frequency of labeled nuclei continuously dropped to zero at 2 days preanthesis, then increased again to 7 percent at 2 dpa, before finally declining to zero at 6 dpa. These data indicate that cell cycling and mitotic activity in the epidermis of ovules cultured in vitro came to a rest 2 days preanthesis, resumed, and then stopped definitively at 6 dpa.

The incorporation of a mixture of four radioactively labeled amino acids (glu, gly, leu, lys) was compared with the incorporation of phenylalanine (phe) into macromolecules of the epidermis and the subepidermal layers of ovules cultured in vitro (Berlin, 1986). Phe labels, in addition to proteins, phenolic compounds derived from cinnamate, a compound formed from phe by the enzyme phenylalanine-ammonia-lyase. At times, near anthesis, the subepidermal cells incorporated a greater amount of the mixture of the radioactive amino acids than did the epidermis. Phe, however, labeled the epidermal cell layer more strongly than the subepidermal cells, indicating that phenolic compounds are synthesized in the epidermis. This observation correlated well with the accumulation of vacuolar phenolic compounds in the epidermal cells, as observed by cytochemical staining reactions in the light microscope and by the accumulation of electron-dense vacuolar compounds observed in the transmission electron microscope. However, no difference was detected between fiber initials and other epidermal cells in the incorporation of phe. The bulk of the proteins in fibers of ovules cultured in vitro was synthesized in the first 6 days following anthesis, as already reported for rRNA synthesis (Berlin, 1986).

According to Berlin (1986), fiber initials can be recognized in electron micrographs already 1 day preanthesis by the leakage of electron-dense, phenolic compounds from the vacuoles into the cytoplasm, leading to a darkening of the fiber initials. However, the leakage of the phenols may be a fixation artifact. A critical examination of the different electron micrographs published by Berlin (1986) raises serious doubts about the generality of the hypothesis that the leakage of the phenols differentiates nonelongating epidermal cells from fiber initials. Having examined numerous ovules of *G. arboreum* and *G. hirsutum* St. 406 before, at, and after anthesis by transmission electron microscopy, leakage of phenols was only occasionally observed and was not correlated with the type of epidermal cell (U. Ryser, unpublished observations).

At the day of anthesis, fiber initials are characterized by increased numbers of Golgi complexes, probably involved in the biosynthesis of cell wall polysaccharides (Berlin, 1986). By EM-cytochemistry, β-glycerophosphatase activity was demonstrated in the initials at anthesis and in the following days, but not in other epidermal cells or in the epidermal cells of a fiberless mutant (Joshi, Stewart, and Graham, 1985). The enzyme had a pH optimum of 6.0 and was localized in changing proportions in the cell wall, the cytoplasm, the nucleus, and the nucleolus. These multiple enzyme localizations may indicate the diffusion of enzyme or reaction product during the incubation of the tissue blocks with the substrates (Essner, 1973). Nevertheless, the increased enzyme activity observed specifically in the fiber initials clearly indicates a function of the phosphatase in elongating fibers, probably reflecting increased carbohydrate metabolism. Another enzyme of carbohydrate metabolism, sucrose synthase, was also specifically localized immunologically in initials and elongating fibers (Nolte et al., 1995). Surprisingly, the enzyme was unevenly distributed in the fiber initials, the growing part of the cells being labeled more intensely, suggesting that sucrose synthase has an integral role in the formation of cell wall constituents. At about the same time, a membrane-associated form of sucrose synthase, probably involved in callose or cellulose synthesis, was described (Amor et al., 1995).

Using the immunofluorescence technique, Jernstedt and colleagues (1993) observed a reorientation of the cortical microtubules in fiber initials, but not in other epidermal cells. No change in the orientation of the microtubules was observed in the epidermis of a fiberless cotton cultivar. Already 1 day preanthesis, a reorientation of the microtubules from longitudinal (perpendicular to the ovule surface) to a random or crisscross arrangement, typical for ballooning fiber initials at anthesis, was observed.

Plant Hormones and Fiber Initiation

Auxins and gibberellins, alone or in combination, can promote fiber initiation in unfertilized ovules, and their effects are additive (Beasley et al., 1974). However, unfertilized ovules cultured in the presence of an inhibitor of gibberellin biosynthesis, at 1 day preanthesis, were unable to initiate fibers even in the presence of an auxin (Sharma et al., 1995). This experiment indicates that auxins in the absence of endogenous giberellins are not able to promote fiber initiation.

The endogenous concentrations of IAA (indole-3-acetic acid)$_3$, GA (gibberellic acid), and ABA (abscisic acid) of whole cotton ovules were determined by GLC (gas-liquid chromatography) from 2 days preanthesis to 5 dpa (Nayyar et al., 1989). Complex variations in the amounts of the three hormones were observed.

Several oxidase activities were measured in cotton ovules before and after anthesis (Naithani, 1987). The activities of peroxidase, IAA oxidase, catechol oxidase, and DOPA oxidase were low preanthesis, reached a minimum at the day of anthesis, and then sharply increased. Interestingly, the drop in the oxidase activities at anthesis was more pronounced in a long-staple cultivar in comparison to a short-staple cultivar, and the enzyme activities were generally higher in the short-staple cultivar.

In a series of elegant experiments, Graves and Stewart (1988a) investigated the chronology of fiber initiation. Using excised ovules cultured in vitro, they showed that epidermal cells already become competent to differentiate into fibers a few days before anthesis in response to IAA and GA$_3$. The time window for a full response to the plant hormones (determined as percent of responding ovules) was 2 days before anthesis until the day of anthesis. During this time, the fiber cells are latent, awaiting appropriate stimulation, which, in the intact plant, is apparently associated with anthesis.

Initiation-Specific RNA and Protein Species

Fiber initiation is dependent on poly(A)-RNA synthesis, as shown by inhibition of fiber initiation with α-amanitin at the day of anthesis to 2 dpa (Triplett, 1996). Cool night temperatures retard fiber initiation (Xie, Trolinder, and Haigler, 1993). Interestingly, a chemical compound, a C16 terpenoid lactone, appears to be able to counteract the adverse effect of low temperature on fiber initiation in ovules cultured in vitro (Singh et al., 1995).

The patterns of soluble proteins of ovules and fibers were analyzed at different times pre- and postanthesis on two-dimensional gels (Graves and Stewart, 1988b). The complexity of the gels increased up to 3 days preanthesis and again at 2 dpa. From 3 days preanthesis to 2 dpa, the protein patterns remained relatively similar. The rise in the complexity of the protein patterns at 3 days preanthesis correlates well with the beginning of fiber initiation, and the rise at 2 dpa, with the beginning of the cell elongation phase (Graves and Stewart, 1988a).

By comparing the protein profiles of a fiberless and a normal cotton cultivar, five proteins were found to be unique to ovules at anthesis or postanthesis in the fiber-producing cultivar (Turley and Ferguson, 1996). Further analysis may provide clues to the cellular function of these proteins and their putative roles in fiber initiation and elongation.

Genes Involved in Fiber Initiation

Several lines of evidence indicate that cotton fibers are initiated at 3 to 2 days preanthesis. Further development of the fiber initials depends on RNA and protein synthesis. However, proteins, and their genes responsible for the critical switch leading to fiber initiation, have not been isolated so far in cotton. In *Arabidopsis,* the *GL1* and *GL2* genes encode regulatory proteins that are key components of the early events in trichome development. Mutations in *GL1* affect trichome initiation, whereas mutations in *GL2* affect trichome expansion. GL1 is a member of the MYB class of transcription factors (Larkin et al., 1993), and GL2 is a homeodomain protein (Rerie, Feldmann, and Marks, 1994). Recently, data were presented indicating that GL1 regulates GL2 expression (Szymansky, Jilk, and Marks, 1996).

FIBER ELONGATION AND THE TRANSITION TO SECONDARY WALL FORMATION

Specific information on the structure of elongating fibers is given in the following material and in the section on the cytoskeleton in this chapter. The mechanism of fiber elongation is discussed in greater detail by Basra and Saha, Chapter 2, this volume.

Cotton Fibers Grow by Diffuse Growth

The growth of a plant cell may occur by extension of the wall, all over its surface (diffuse growth), or by localized extensions, for example, at the apex of the cell (tip growth). Root hairs, pollen tubes, and fungal hyphae grow exclusively at their tips. Such cells are characterized by a typical zonation of their organelles and a steep Ca^{2+} gradient in the tip region (Heath, 1990).

Although, for technical reasons, the growth kinetics of cotton fibers have not been measured, several lines of evidence clearly indicate that cotton fibers elongate by diffuse growth. It was observed already in 1951, by transmission electron microscopy of shadowed primary wall fragments of cotton fibers, that the cellulose microfibrils at the inner surface of the primary wall are oriented perpendicularly to the long axis of the cell. However, the microfibrils at the outside of the wall are oriented longitudinally (Roelofson, 1951; Houwink and Roelofson, 1954; see Figure 1.2). This change in the orientation of the microfibrils from transverse on the inner side of the wall to longitudinal on the outer side was explained by a displacement of the newly deposited transverse microfibril layers to the exterior by the following microfibril layers and by a rearrangement of the outer layers in the direction of the cell axis due to continued cell extension (reviewed by Roelofson, 1965). After

FIGURE 1.2. Schematic Drawing of the Fibrillar Architecture of the Outer and Inner Faces of the Primary Cell Wall at the Tip of Elongating Cotton Fibers

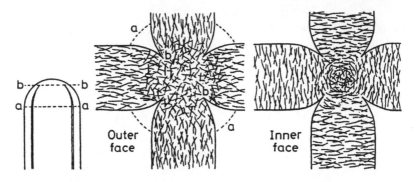

Source: Reprinted with permission from Houwink and Roelofson (1954, p. 392).

Note: Cell wall fragments, 15 dpa, were deincrusted with boiling 5 percent NaOH and shadowed before their examination in the transmission electron microscope. An exception is the inner surface of the extreme tip, where the orientation of the microfibrils was deduced from observations with the polarizing microscope.

incorporation of tritiated glucose, the cell wall of cotton fibers was labeled all over the length of the fibers, an extended tip region being labeled somewhat more intensely (Ryser, 1977; see Figure 1.3). Golgi complexes were observed throughout the length of the fibers and not only in the fiber tip (Itoh, 1974; Westafer and Brown, 1976), and it was shown that the Golgi complexes secrete noncellulosic polysaccharides (Ryser, 1979). Using rapid freeze-fixation and freeze-substitution protocols, the cytology of the tips of young cotton fibers (2 dpa) was studied. The fiber tips did not show any characteristics of other tip-growing plant cells and did not behave as tip-growing cells toward inhibitors interfering with the functions of the cytoskeleton (Tiwari and Wilkins, 1995).

It should be kept in mind that cells with a diffuse growth type grow at their tips also. Even if the growth rate may be somewhat higher in the tip region, adhering to a strict terminology, this phenomenon should not be called tip growth.

FIGURE 1.3. Autoradiograph of Cotton Fibers 4 dpa After Labeling with ³H- Glucose for 10 Hours (*G. arboreum* L.): Bright Field Illumination

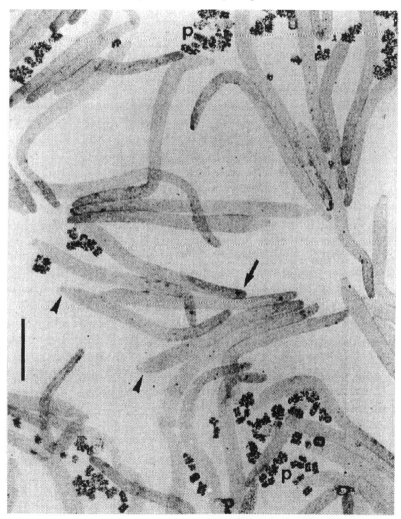

Source: Reprinted with permission from Ryser (1977, p. 79).

Note: Under these conditions, unlabeled cell walls are not visible, as demonstrated by the parenchyma cells of the outer integument (p). Only the labeled starch grains, but not the cell walls, are visible. The elongating cotton fibers are labeled all over their length, the tip region being labeled somewhat more intensely. Arrows indicate fiber tips, arrowheads, and fiber bases. Similar results were obtained with older fibers (11 dpa), having reached about one-third of their final length. Bar = 0.1 mm.

Kinetics of Elongation and Cellulose Accumulation

Considerable overlapping occurs between the elongation and secondary wall formation phases in the order of 5 to 10 days (Schubert et al., 1973, 1976; Meinert and Delmer, 1977; Ryser, 1979; Naithani, Rao, and Singh, 1982). The reported studies deal with mean fiber length and mean fiber dry weight, or cellulose accumulation. Therefore, it is not known whether the overlapping occurs between different seeds, between fibers with different positions on the seeds, or within single fibers. DeLanghe (1986) discussed the problem in detail and, on the basis of a detailed study of the elongation rate at nine different regions of the ovule surface, concluded that overlapping of elongation and secondary wall formation must occur at the level of individual fibers. Evidence, based on the swelling of secondary wall cellulose, indicated that cellulose was preferentially deposited in the middle of individual fibers at the beginning of secondary wall formation, extending toward both ends, until the whole fibers showed the typical swelling patterns over their entire length at about 50 dpa (Marx-Figini, 1967). Although growth in the nonswelling parts of the fibers was not demonstrated, it is possible that the fibers continue to grow at their tips after the beginning of secondary wall deposition. In this context, it may also be interesting to note that the reversal frequencies in developing cotton fibers increase rapidly between 3 and 4 weeks postanthesis and then more slowly up to 6 weeks postanthesis (Hebert and Boylston, 1984).

Golgi Complexes and Their Function in Elongating Fibers

In plant cells, the Golgi complexes are involved in the biogenesis of the cell wall and the vacuole, as well as in the formation of the plasma and vacuolar membranes. The targeting of membrane-bound and soluble proteins, as well as the correct delivery of the transport vesicles to their target membranes, is now intensely studied in animal and plant cells (for recent reviews, see Staehelin and Moore, 1995; Okita and Rogers, 1996). It was estimated that in elongating cotton fibers with a length of 2 cm (centimeters), as many as 75,000 Golgi complexes are present. This number decreases at the transition to secondary wall formation (Westafer and Brown, 1976).

Recently, the effects of tunicamycin, an inhibitor of lipid-linked glycosyl transfer, and monensin, an ionophore interfering with Golgi function, were examined in fertilized cotton ovules cultured in vitro (Davidonis, 1993). Both inhibitors interfered with cell elongation and secondary wall deposition. At lower concentrations, monensin interfered only with secondary wall formation.

In elongating cotton fibers, three types of vesicles are associated with the Golgi complexes, apparently smooth secretory vesicles and two types of vesicles with protein coats (Ryser, 1979). Neither the coat proteins (clathrin and COP) nor their genes have been studied in cotton fibers so far. In elongating fibers, noncellulosic polysaccharides are secreted by apparently smooth vesicles (Ryser, 1979). Immuno-localization of methylesterified pectins, xyloglucans, and carbohydrate moieties of arabinogalactan proteins and N-linked glycoproteins traced the likely sites of synthesis and assembly of these complex polysaccharides to individual Golgi cisternae. Double immunolabeling experiments demonstrated that a single Golgi complex is capable of synthesizing both pectins and hemicelluloses (T. Wilkins, personal communication, February 20, 1997).

Disappearance of Cytoplasmic Polysomes at the Transition to Secondary Wall Formation

The transition between cell elongation and secondary wall formation occurs 15 to 22 dpa. It is marked by an almost complete disappearance of the numerous free polyribosomes characteristic of the cytoplasm of elongating fibers. During secondary wall formation, only a few endoplasmic reticulum-associated polyribosomes persist (Westafer and Brown, 1976). A 20 percent reduction in the total number of protein bands was observed from 10 to 20 dpa on two-dimensional gels of total protein extracted from fibers at 10, 16, and 20 dpa (Graves and Stewart, 1988b).

SECONDARY WALL FORMATION

The main event during the period of secondary wall formation is the biosynthesis of secondary wall cellulose. Here, and in the section on the cytoskeleton, the oriented deposition of the cellulose microfibrils is discussed. An overview of the cross section of a fiber approaching the end of secondary wall formation is shown in Figure 1.4, and an enlarged view of the plasma membrane-associated cortical microtubule system is presented in Figure 1.5. Cellulose biosynthesis is discussed by Delmer (Chapter 4, this volume).

Cellulose-Synthesizing Complexes

Rosettes, the putative cellulose-synthesizing complexes of higher plants, have been localized with the freeze-fracture technique in cotton fibers by Willison (1983) and Herth (Ryser, 1985). However, during secondary wall formation, the frequency of the rosettes on the P-fracture face was about ten times lower than the frequency of the corresponding terminal globules on the E-fracture face. This observation led Willison to conclude that rosettes are not an integral part of the cellulose-synthesizing complex in higher plants. However, other explanations for the observed discrepancies are possible (Ryser, 1985), and today, the "rosette hypothesis" seems to be generally accepted (for reviews, see Giddings and Staehelin, 1991; Cyr, 1994; Brown, Saxena, and Kudlicka, 1996).

Callose

The fluorescent dyes aniline blue and sirofluor, which are fairly specific for (1-3)-β-glucans (callose), stain the innermost wall layer of cotton fibers intensely throughout fiber development, suggesting the presence of a thin layer of callose at the interface between plasma membrane and cell wall (Waterkeyn, 1981; Haigler et al., 1991). The function of this callose layer is unknown.

Wound callose deposits were observed in damaged or immature fibers. Interestingly, some callose was apparently also deposited in the reversals in the form of a fluorescent band oriented perpendicularly to the cellulose microfibrils (Waterkeyn, 1981). It would be worthwhile to test the presence of callose in these locations with specific antibodies against (1-4)-β- and (1-3)-β-glucans.

Growth Rings in the Cellulosic Secondary Walls

When cotton plants in the field or ovule cultures are subjected to cycling day and night temperatures, growth rings in the cellulosic

FIGURE 1.4. Cross Section of a Cotton Fiber 50 dpa (*G. arboreum* L.)

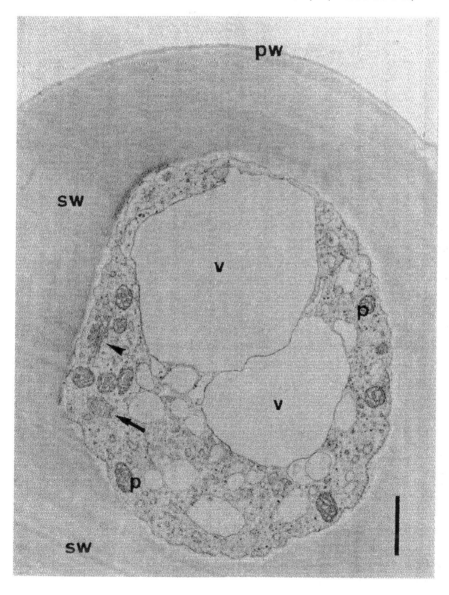

Note: Golgi complex (arrowhead); peroxisome (arrow); p, leucoplasts; pw, primary wall; sw, secondary wall; v, vacuole. Bar = 1 μm.

FIGURE 1.5. Enlarged View of the Plasma Membrane and Cortical Microtubules of the Fiber Shown in Figure 1.4

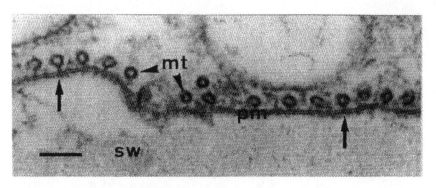

Note: Cross-links between microtubules and plasma membrane (arrows); mt, microtubules; pm, plasma membrane; sw, secondary wall. Bar = 0.1 μm.

secondary walls can be visualized in the light and scanning electron microscope after treatment with different swelling agents (Haigler et al., 1991). The rings correspond to regions of high and low cellulose accumulation. The experiments with ovules cultured in vitro demonstrated that cool temperature effects on fiber development are at least partly ovule- or fiber-specific events and do not depend on carbohydrate supply. Ovules cultured in vitro are, therefore, valid models by which to study the regulation of the response of field-grown fibers to cool temperature (Haigler et al., 1991; Xie, Trolinder, and Haigler, 1993).

The effects of temperature on rates of cellulose synthesis, respiration, and glucose uptake were investigated using cultured ovules (*G. hirsutum* L. cv. Acala SJ1). The optimal temperature for the accumulation of secondary wall cellulose versus respiration was determined to correspond to 28°C (celsius) (Roberts et al., 1992).

THE CYTOSKELETON OF DIFFERENTIATING FIBERS

Cortical Microtubules

With transmission electron microscopy, a cortical array of microtubules was observed in differentiating fibers. Most of these tubules are located at a rather constant, small distance to the plasma membrane. During secondary wall formation, electron-dense links between the microtubules and the plasma membrane can be seen (Westafer and Brown, 1976; see Figure 1.5). A parallel orientation of cortical microtubules and the innermost layer of cellulose microfibrils occurs in elongating and secondary wall-forming cotton fibers (reviewed in Ryser, 1985). During the elongation phase, cortical microtubules and cellulose microfibrils are widely spaced and are oriented roughly perpendicular to the long axis of the fiber, as observed in many other plant cells. At the transition to secondary wall formation, the density of the microtubules increases abruptly and their orientation becomes almost perfectly parallel. At the same time, microtubules and the cellulose microfibrils reorient and form first flat, and later on steeper, helical arrays. The parallel orientation of microtubules and cellulose microfibrils is maintained even through the reversals, that is, a change in the direction of the helically deposited secondary wall microfibrils (Yatsu and Jacks, 1981).

The alkaloid colchicine, which binds to α- and β-tubulin dimers and interferes with the polymerization of microtubules, did not inhibit cellulose biosynthesis, but did drastically alter the orientation of the deposited cellulose microfibrils (Yatsu, 1983). These observations indicate that cotton fibers employ a microtubule-based system for the orientation of cellulose microfibrils. These early results were fully confirmed, beautifully illustrated, and extended with the immunofluorescence methodology used by Seagull (1986, 1989, 1990a).

Due to the lower resolution of the fluorescence microscope, the fluorescent images have to be interpreted carefully (Seagull, 1986; Williamson, 1991). Single microtubules are not normally resolved in the fluorescence microscope. Bundles of microtubules are seen, creating the illusion of very long microtubular helices. Individual microtubules, as observed in the electron microscope, seem to be relatively short. The mean length of the tubules is often less than 10 µm (Hardham and Gunning, 1978; Gunning and Hardham, 1982). The relative length of the microtubules in cotton fibers was found to be more or less constant during the elongation phase and then to increase slowly during secondary wall formation (Seagull,

1992, 1993). Using a special technique, microtubules with a length of 7 to 20 μm were observed in differentiating cotton fibers (Quader et al., 1987) and of 20 to 35 μm in the seed hairs of *Cobaea* (Quader, Deichgräber, and Schnepf, 1986). The question of the real length of the cortical microtubules may, therefore, still be open to discussion.

Another effect of imaging in the fluorescence microscope is that the proportion of the space occupied by the microtubules is exaggerated and that fluorescent fibrils are "created" by physically unlike microtubules. Consequently, the divergence of microtubule orientation may be underestimated. In developing cotton fibers, the orientation of the microtubules has been documented in most developmental stages with both methods, fluorescence and electron microscopy.

In summary, two main stages of repatterning of the microtubule arrays have been identified during cotton fiber development: (1) the transition between fiber initiation and elongation, during which microtubules develop from a random to a roughly transverse orientation and (2) the transition between primary and secondary wall synthesis, during which microtubules abruptly shift orientation to a steeply pitched helical pattern (Seagull, 1992, 1993).

Tubulins

The concentrations of α- and β-tubulin rapidly increase in developing cotton fibers until the end of cell elongation when they reach a plateau (Kloth, 1989). Tubulin accumulation and cell elongation follow the same kinetics. Combining the results of Kloth (1989) and Seagull (1992), it must be concluded that the ratio of free and microtubule-bound tubulin monomers dramatically changes at the transition of the elongation and secondary wall deposition phases. As a percentage of total protein, cotton fibers at the beginning of secondary wall formation had greater amounts of tubulin than did hypocotyls, roots, leaves, or cotyledons (Dixon, Seagull, and Triplett, 1994).

Changes in the accumulation of specific α- and β-tubulin isotypes (Hussey et al., 1987; Hussey, Snustad, and Silflow, 1991) also occur during cotton fiber development (Dixon, Seagull, and Triplett, 1994). This observation showed for the first time differential usage of tubulin isotypes during the development of a single plant cell type. The functional significance of the heterogeneity of α- and β-tubulins is unknown. An interesting proposal is that tubulin iso-forms differentially modulate the dynamic behavior of microtubules (Luduena, 1993).

Recently, γ-tubulin has been identified in animal and plant cells. Gamma-tubulin is present only in small amounts in the cells, binds to the minus end of microtubules, and is probably involved in the nucleation of microtubules (Liu et al., 1993, 1994; Goddard et al., 1994; Lambert, 1995).

Turnover of Cortical Microtubules

After microinjection of a fluorescent derivative of rat brain tubulin into stamen hair cells of *Tradescantia*, microtubule turnover within a living plant cell could be estimated for the first time. The method employs laser photobleaching followed by the observation of fluorescence redistribution (Hush et al., 1994; Hepler and Hush, 1996). The turnover rate of cortical microtubules (half-time for fluorescence recovery = 67 ± 3 s [seconds]) was similar to the turnover of preprophase and phragmoplast microtubules, but was about two times lower than that of spindle microtubules in plant and mammalian cells (half-time = 31 ± 6 s). The turnover rate was, however, several times higher than that of the microtubules of mammalian cells in the interphase of the cell cycle (half-times = 200 ± 85 s and 270 s ± 73 s). Therefore, the plant cell cortical microtubules appear to be highly dynamic in comparison to their animal counterparts.

Actin Filaments

The presence of a system of actin filaments in cotton fibers was first shown by Quader and colleagues (1987) using the immuno-fluorescence technique. The orientation of the actin filaments was different from the orientation of the microtubules. Relatively thick actin bundles were oriented longitudinally in the cotton fibers, and it was proposed that they were involved in cytoplasmic streaming. Using improved fixation procedures, very fine actin filaments, branching from the main actin cables, and fine patches of an amor-phous actin-containing material were observed with the immunofluorescence method (Seagull, 1990a). After cryofixation, cryosubstitution, and observation by transmission electron microscopy, single microfilaments, probably actin, were found to be parallel to the cortical microtubules (Seagull, 1993).

The Effect of Cytoskeleton Inhibitors on the Orientation of Cellulose Microfibrils

Microtubule inhibitors did not affect the actin network. However, cytochalasin D, an inhibitor that binds to the plus end of actin filaments and interferes with their polymerization, changed the orientation of the microtubules, as well as the orientation of the cellulose microfibrils, the second being probably an indirect effect (Seagull, 1990b). Similar results were obtained with differentiating tracheary elements (Kobayashi, Fukuda, and Shibaoka, 1988). The actin filaments in the cortex of higher plant cells may therefore have a function in orienting the cortical microtubules. In this context, it is interesting to note that genes encoding small GTP-binding proteins, analogous to mammalian Rac, are preferentially expressed in cotton fibers at the transition from primary to secondary wall formation (Delmer et al., 1995). In animals, the small GTP-binding proteins Rac and Rho, of the *ras* gene superfamily, participate in the signal transduction pathway that regulates the actin cytoskeleton (Hall, 1994).

As proposed initially by Herth (1980) for chitin deposition in *Poteriochromonas* and by Staehelin and Giddings (1982) for cellulose deposition in algae and higher plants, the actual models of cellulose deposition assume that the driving force for the movement of the synthesizing complexes is the crystallization of the polysaccharide. An attachment of the synthesizing complexes to elements of the cytoskeleton is not considered to be necessary and has not been observed thus far. According to these models, the movement of the synthesizing complexes is channeled by the cortical microtubules, being linked via unknown proteins to the plasma membrane.

THE CUTICLE OF COTTON FIBERS AND THE GASEOUS COMPONENTS INSIDE COTTON BOLLS

Cotton fibers, similar to the other cells of the seed coat epidermis, are covered by an extremely thin, lamellated cuticle. For a review on the ultrastructure and chemical composition of cotton fiber cutin see Ryser (1985).

A full-length cDNA clone coding for a lipid transfer protein (LTP) has been isolated recently from a cotton fiber cDNA library. Northern analysis showed that the LTP-mRNA concentration steadily increased during fiber elongation and sharply decreased by the stage of secondary wall formation. This observation, and the presence of a transmembrane signal peptide at the N-terminal end of the protein, suggested an involvement of the LTP in cuticle biogenesis (Ma et al., 1995). Possible functions of lipid transfer proteins in plants and the biochemistry and molecular biology of cutin and associated wax have been reviewed recently (Hamilton, 1995; Kader, 1996; Kerstiens, 1996; Post-Beittenmiller, 1996).

Cutin monomers are recognized by a number of fungal plant pathogens as chemical signals leading to cutinase gene activation and strongly enhanced cutinase activity (Kolattukudy et al., 1995). Free cutin monomers can be perceived also by plant cells (Schweizer et al., 1996).

The gaseous composition inside cotton bolls was analyzed by GC-MS (gas chromatography-mass spectrometry) (Jacks et al., 1993). The concentration of the permanent gases in the boll was, by weight, 46 percent nitrogen, 29 percent oxygen, 4 percent argon, and 20 percent carbon dioxide, whereas outside air was assayed at 73 percent nitrogen, 25 percent oxygen, 2 percent argon, and 0.3 percent carbon dioxide. No significant differences in the gas composition existed between day and night nor among different fruit ages (15 to 44 dpa). Only aggressive infection by *Aspergillus flavus* raised carbon dioxide to 31 percent and lowered oxygen to 17 percent. The gas volume inside a locule of a mature fruit was about 0.3 ml (milliliter). The maintenance of a constant oxygen concentration in the boll during day- and nighttime is difficult to understand and cannot depend directly on the photosynthesis of carpels and bracts.

NATURALLY COLORED COTTON FIBERS

Mutants of cotton with green and differently shaded brown fibers have been known for a long time, and the genetics of many of them are well-established (Endrizzi, Turcotte, and Kohel, 1984, 1985;

Kohel, 1985). Increasingly, colored fibers are being used for the production of textiles, although the colors may change somewhat during use, especially if exposed to sunlight. The chemistry and the cellular localization of the colored compounds are discussed next.

Brown Fibers

The brown fiber color is determined by different genetic loci: *Lc1-6* and *Dw*. The *Dw* locus is considered to be homologous to the *Lc1* locus, the first being localized on chromosome 16 of the D genome, the second on the homologous chromosome 7 of the A genome (Rhyne, 1957). The chemistry of the colored compounds in brown cotton fibers has not yet been studied in detail. However, (+)-catechin and derived tannins have been identified by cytochemical methods combined with thin-layer chromatography in different layers of cotton seed coats and in the adhering fuzz fibers. The brown color develops only at the time of the dehiscence of the cotton bolls and depends on the presence of oxygen (Halloin, 1982). In the fuzz fibers, the tannin precursor (+)-catechin was localized in the lumen of the cells, and upon dehiscence, the fuzz fibers assumed a light brownish color. It may therefore be assumed that the different colors of brown lint cultivars are due to vacuolar tannins derived from (+)-catechin. The brown cotton fibers are not suberized (see next paragraph). The absence of suberin layers in the brown fibers was also verified in fibers removed from different textiles containing naturally colored fibers and in the mocha, brown, and red cultivars of BC-Cotton, Inc. (U. Ryser, unpublished results).

Green Fibers

So far, in all cotton lines displaying green lint or fuzz, the green color proved to be conditioned by alleles of the *Lg* locus. The green fuzz allele was designated Lg^f and conditions white lint and green fuzz lines, whereas the *Lg* allele conditions green lint and fuzz (Kohel, 1985). The fibers associated with the *Lg* genotype turn green during secondary wall formation, contain relatively high amounts of wax (Conrad, 1941; Conrad and Neely, 1943), and are suberized (Ryser, Meier, and Holloway, 1983; Yatsu, Espelie, and Kolattukudy,

1983; Ryser and Holloway, 1985). The *Lg* genotype is also expressed in ovules cultured in vitro. Ultrastructure and aliphatic monomer composition of cotton fiber suberin are markedly different from the suberin of the stem periderm of the same plants (Ryser and Holloway, 1985). Cotton fiber suberin is deposited in concentric, osmiophilic layers, alternating with layers of cellulose. Each concentric suberin layer is lamellated with a periodicity of about 4.2 nm (nanometer) (see Figure 1.6). The periodic change in cellulose and suberin deposition has, as yet, not been observed in any other plant genus. However, in the pits regularly occurring at the base of cotton fibers, a compact suberin layer is observed, as in the suberin deposits of other

FIGURE 1.6. Cross Section of a Green Cotton Fiber (*G. hirsutum* L.) at High Magnification, Showing the Lamellation of the Suberin Layers (s)

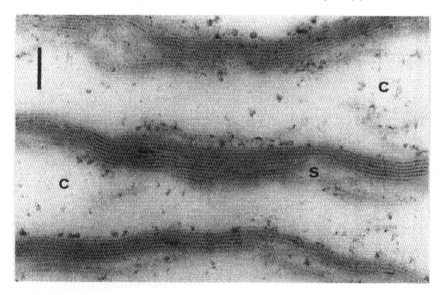

Source: Reprinted with permission from Ryser, Meier, and Holloway (1983, p. 201).

Note: The cellulose layers (c) are electron transparent. The periodicity of the suberin layers corresponds to 4.2 nm, the thickness of the electron translucent suberin lamellae to about 3.4 nm, and the length of extended C22 fatty acids to about 3 nm. Bar = 50 nm.

plants. Obviously, cellulose deposition is inhibited in the pits, but suberin deposition is not (Ryser, 1992).

Suberin As a Constituent of the Fibers of Wild Cotton Species and of the Seed Coat Epidermis in the Genus Gossypium

Suberin and associated wax are not only deposited in the green fibers but in all cells of the seed coat epidermis of the green lint cultivar, with the exception of the guard cells of stomata, which occur mainly in the chalazal region of cotton seed coats (Ryser, 1992).

The fibers of twelve wild cotton species were examined, and all contained secondary walls with suberin layers, whereas the fibers of white and brown lint cultivars are not suberized. Interestingly, suberin layers are also observed in the seed coat epidermis of white lint cotton cultivars (see Figure 1.7), where suberin and associated wax have also been identified with chemical methods (Ryser and Holloway, 1985; Schmutz, Jenny, and Ryser, 1994). In fact, suberin layers have been found in the seed coat epidermis of all cotton species and cultivars examined so far (see Table 1.1). The green fibers are therefore a useful material to study a genetically selected group of chemical compounds occurring in the cell wall of the seed coat epidermis. The results also suggest that the cotton cultivars with white fibers are the mutants; that is, they have a fiber-specific defect at the *Lg* locus or at some regulatory genes associated with it.

The Wax of Green Cotton Fibers

Green cotton fibers are characterized by a high wax content of 14 to 17 percent of their dry weight, whereas white fibers contain only 0.4 to 0.7 percent of wax (Conrad, 1941; Conrad and Neely, 1943). During imbibition of whole delinted seeds, white cotton seeds took up water at a rate three times that of green seeds with the weight gains after 10 h (hours), being 29 percent and 9.8 percent, respectively (Yatsu, Espelie, and Kolattukudy, 1983).

A quantitative assessment of the wax of white cotton fibers was provided by Ryser and Holloway (1985), and the major components

FIGURE 1.7. Seed Coat Epidermis of White Lint Cotton Cultivar (*G. hirsutum* L., cv. St 406)

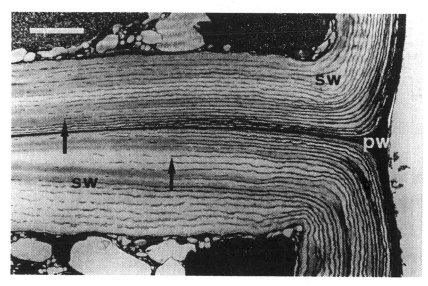

Source: Reprinted with permission from Ryser, Meier, and Holloway (1983, p. 201).

Note: Regular suberin layers are observed in the outer tangential wall. In the radial walls, the suberin layers progressively disappear (arrows), and the inner tangential walls do not contain suberin layers (not visible on this micrograph), sw, secondary wall; pw, primary wall. Bar = 3 μm.

were free 1-alkanols and free alkanoic acids. The wax fraction of the green fibers contained similar compounds, but in much lower proportion, and large amounts of unidentified materials. Further analysis of the wax fraction of the green cotton fibers showed that a large proportion is composed of oligomeric materials, consisting mainly of a colorless, but fluorescent, compound and several yellow compounds. After acid-catalyzed transesterification, mainly long chain fatty acids, glycerol, caffeate, and ferulate were liberated (Schmutz et al., 1993; Schmutz, Jenny, and Ryser, 1994a; Schmutz, Buchala, and Ryser, 1996).

TABLE 1.1. Occurrence of Suberin in the Fibers and the Seed Coat Epidermis of Different Cotton Species and in Colored Fibers of *G. hirsutum*

Genome	Species	Fibers	Seed Coat Epidermis
Diploid			
A1	*G. herbaceum* L.	−	+
A2	*G. arboreum* L.	−	+
B1	*G. anomalum* Wawr. et Peyr.	+	+
C1	*G. sturtianum* Willis	+	+
C3	*G. australe* F. Muell.	+	+
D1	*G. thurberi* Tod.	+	+
D2-2	*G. harknessii* Brandeg.	+	+
D3-d	*G. davidsonii* Kell.	+	+
D4	*G. aridum* (R. et S.) Skovsted	+	+
D5	*G. raimondii* Ulbr.	+	+
E1	*G. stocksii* Mast.	+	+
F1	*G. longicalyx* Hutch. et Lee	+	+
G1	*G. bickii* Prokh.	+	+
Tetraploid			
AD1	*G. hirsutum* L.		
	(cv. ST 406)	−	+
	(cv. green lint: *Lg*)	+	+
	(cv. brown lint: *Lc1*)	−	+
	(cv. brown lint: *Lc1/Lc2*)	−	+
	(cv. dirty white: *Dw*)	−	+
AD2	*G. barbadense* L.	−	+
AD3	*G. tomentosum* Nutt.	+	+

Source: Reprinted with permission from Ryser and Holloway (1985, p. 157).

The fluorescent compound, a caffeoyl–fatty acid–glycerol ester, was isolated from the wax of green cotton fibers and characterized chemically (Schmutz, Jenny, and Ryser, 1994). The function of this major wax component is unknown. However, it was proposed that the caffeate moiety protects cotton seeds against ultraviolet (UV)-induced damage, either by scavenging primary photoproducts or by absorbing UV light, leading to *trans-cis* isomerization (Schmutz et al., 1994). Another possible function is protection against microorganisms. Hydrophilic caffeic acid-containing compounds are the most widespread hydroxycinnamic acid conjugates in plants, and their biological activities include inhibition of insect growth, and antiviral, antibacterial, and antifungal properties (Hohlfeld, Veit, and Strack, 1996).

The Monomer Composition of the Suberin of Green Cotton Fibers

The suberin of green cotton fibers is characterized by a high content of C_{22} fatty acids (about 80 percent of the total monomeric mixture), the dominant components being 22-hydroxy-docosanoic acid and 1,22-docosandioic acid (Ryser, Meicr, and Holloway, 1983; Yatsu, Espelie, and Kolattukudy, 1983; Ryser and Holloway, 1985; Schmutz, Buchala, and Ryser, 1996). In addition to the aliphatic suberin domain, an aromatic domain was postulated (Kolattukudy, 1980; Kolattukudy and Espelie, 1989; Davin and Lewis, 1992). For potato suberin, this aromatic domain was recently shown to be composed mainly of cinnamic acids, not of the corresponding alcohols, as in lignins (Bernards et al., 1995). The covalent linkages of the cinnamic acids in suberin still remain to be determined.

Recently, we obtained experimental evidence that glycerol is a monomer of cotton fiber suberin and maybe of suberins in general (Schmutz et al., 1993; Schmutz, Buchala, and Ryser, 1996). In analogy to the situation in biological membranes, it was proposed that the thickness of the electron-translucent part of the suberin layers (3 nm) is determined by the length of the C_{22} ω-hydroxyalkanoic and α,ω-alkanedioic fatty acids of the suberin polyester (Schmutz et al., 1993; Schmutz, Buchala, and Ryser, 1996). Significantly, the dimensions of the electron-translucent suberin lamellae could be changed with a specific inhibitor of the endoplasmic reticulum-associated fatty acid elongases. These results have led to a tentative model for the aliphatic part of cotton fiber suberin (see Figure 1.8). This model is currently being tested in our laboratory by different approaches.

FIGURE 1.8. Hypothetical Model of the Aliphatic Domain of Cotton Fiber Suberin

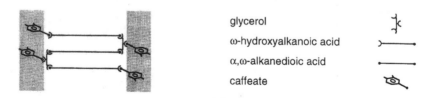

glycerol

ω-hydroxyalkanoic acid

α,ω-alkanedioic acid

caffeate

Note: The dark shaded areas correspond to the electron dense lamellae, containing the putative aromatic suberin domain. In this model, glycerol has a structural role, covalently linking the aromatic and the aliphatic suberin domain by ester bonds.

Suberin Biosynthesis

The biosynthesis of the aromatic and aliphatic suberin monomers is relatively well understood (for reviews, see Von Wettstein-Knowles, 1993; Cassagne et al., 1994; Kader and Mazliak, 1995; Whetten and Sederoff, 1995; Schmid and Amrhein, 1995; Boudet, Goffner, and Grima-Pettenati, 1996; Harwood, 1996). However, essentially nothing is known concerning the assembly of the suberin monomers or oligomers on a preexisting cell wall and the final cross-linking of the polymer (Kolattukudy and Espelie, 1989; Davin and Lewis, 1992). Recently, an acyltransferase, which transfers hydroxycinnamic acids from hydroxycinnamoyl-CoA thioesters to several hydroxylated fatty acid derivatives, was characterized in wound-healing potato tuber discs (Lotfy, Negrel, and Javelle, 1994). The time course of induction and the tissue distribution of the enzyme in the wound periderm coincides with the deposition of suberin. The enzyme has been isolated and purified to near homogeneity from tobacco cell suspension cultures, where it is expressed constitutively (Lotfy, Javelle, and Negrel, 1996). Possible functions of the enzyme are the formation of wax esters and of suberin oligomers. At present, Laurence Moire, in our lab, is involved in the isolation of suberization specific cDNAs of green cotton fibers, using RT-PCR (reverse transcription-polymerase chain reaction) differential mRNA display (Song, Yamamoto, and Allen, 1995) and related methods.

THE BASE OF COTTON FIBERS

The structure and mechanical properties of the basal region of cotton fibers are of interest for the separation of fibers and seed in the ginning process. A long-term objective is to develop methods to remove fibers from seed at reduced levels of fiber breakage (Fryxell, 1963; Vigil et al., 1996).

Pits and Plasmodesmata in the Base of the Fibers

From a physiological viewpoint, of view, the fiber base plays a vital role in fiber development. The low molecular weight carbohydrates necessary for fiber elongation and secondary wall formation have to be imported through the fiber base. The morphology of the base of mature fibers was thoroughly investigated by Fryxell (1963). The base of young cotton fibers is not readily distinguished from other epidermal cells, whereas older fuzz and lint fibers have an extended base, the so-called "foot," and are somewhat constricted in the zone of contact with the neighboring epidermal cells. Plasmodesmata are present in the epidermal cells during fiber initiation and elongation, but their frequency in periclinal and anticlinal walls has not been determined. Concomitant with secondary wall formation, pits develop in the base of the fibers and in ordinary epidermal cells (see Figures 1.9 and 1.10). About 25 percent of the length of the inner periclinal wall of the seed coat epidermis of a white lint cultivar of *G. hirsutum* was occupied by pits, but only 2 percent of the anticlinal walls was pitted, mainly in their proximal part (Ryser, 1992). The pits contained numerous plasmodesmata, $22 \pm 2 \cdot 3$ in the periclinal and $27 \pm 3.3 \cdot \mu m^{-2}$ in the anticlinal walls (see Figure 1.11). It was estimated that the transport capacity of the plasmodesmata should be sufficient for the transport of the assimilates necessary for secondary wall formation. For a recent review on the structure and function of plasmodesmata, see Wolf and Lucas (1994). Although technically difficult, it would be interesting to determine the size exclusion limit and the transport capacity of the plasmodesmata at the fiber base during the different stages of fiber development.

FIGURE 1.9. The Different Layers of the Seed Coat of *G. hirsutum* L., Median Cross Section, 69 dpa, Shortly Before Opening of the Fruit Capsule

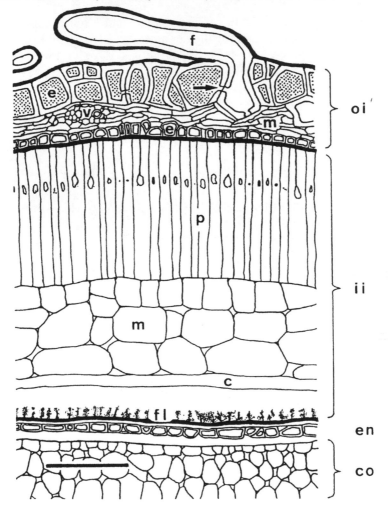

Source: Reprinted with permission from Ryser and colleagues (1988, p. 89).

Note: oi, outer integument; e, outer and inner epidermis—note pits in the outer epidermis (arrow); f, fiber; v, vascular bundle; ii, inner integument; c, collapsed cells of the mesophyll (and outer tangential wall of the fringe layer); fl, fringe layer (= inner epidermis of the inner integument, without outer tangential wall); m, mesophyll; p, palisade layer (outer epidermis of the inner integument); en, endosperm; co, cotyledons. Bar = 0.1 mm.

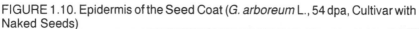

FIGURE 1.10. Epidermis of the Seed Coat (*G. arboreum* L., 54 dpa, Cultivar with Naked Seeds)

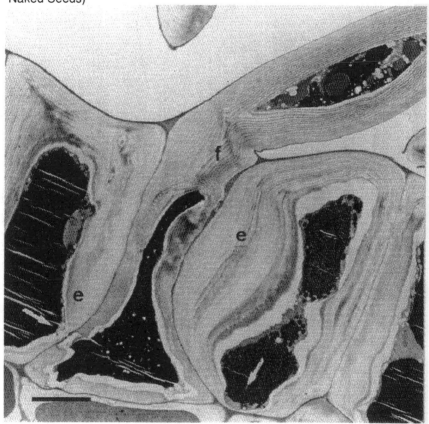

Source: Reprinted with permission from Ryser (1992, p. 17).

Note: Several pits are visible in the fiber base and in the adjacent epidermal cells (arrows). The distal part of the fiber (not visible on this micrograph) is not suber-ized, but the whole fiber base contains concentric suberin layers. f, fiber; e, epidermal cells not forming fibers. Bar = 10 μm.

Suberin in the Base of Cotton Fibers and Possible Implications for Assimilate Transport

In wild cotton species and in green lint cultivars, the fibers are suberized, and the uptake of sugars via the cell wall space, by

FIGURE 1.11. Tangential Section of a Pit Localized Between Epidermal Cells in the Chalazal Region of the Seed Coat (*G. hirsutum* L., White Link Cultivar, 50 dpa)

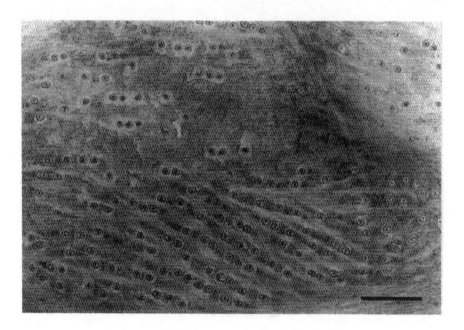

Source: Reprinted with permission from Ryser (1992, p. 20).

Note: Numerous cross-sectioned plasmodesmata are visible. It is evident that in 50 nm ultrathin cross sections of the same pit, only a few plasmodesmata may be seen, depending on the direction of sectioning. Bar = 0.5 μm.

plasma membrane-bound sugar transporters, is probably not possible during secondary wall formation. Therefore, in the genus *Gossypium*, the transport via the symplast appears to be the evolutionary original mode of assimilate transport into fibers and epidermal cells during secondary wall formation. White and brown cotton fibers are not suberized, but such fibers still contain various amounts of suberin in the fiber base (Ryser, 1992). Whether apoplastic sugar transport occurs in elongating fibers or in secondary wall-forming white or brown cotton fibers is still an open question.

An unsolved problem is also the deposition of suberin layers in the base of white or brown cotton fibers. From genetic studies, we know that cell type-specific expression of the *Lg* gene is possible (Kohel, 1985). But, how is the apparent leakiness of the *Lg* gene explained? It may be assumed that a factor, missing or defective in white and brown fibers, is imported from the neighboring epidermal cells through the plasmodesmata in the anticlinal walls of the fiber base. The factors may be small molecules or proteins or protein-mRNA complexes. In fact, evidence is accumulating that not only viral movement proteins and nucleic acids (Wolf and Lucas, 1994) but also plant-encoded macromolecules move from cell to cell via plasmodesmata. This has been shown for KNOTTED1, a homeo-box-containing transcription factor and its mRNA (Lucas et al., 1995; Mezitt and Lucas, 1996).

FIXATION OF FIBERS FOR ULTRASTRUCTURAL AND IMMUNOLOCALIZATION STUDIES (APPENDIX)

The fixation of differentiating cotton fibers, especially at the transition between elongation and secondary wall formation, is a difficult task. First, an intact locule with seeds and adhering fibers has to be prepared from a capsule, as gently as possible. A single section through the locule probably already destroys most of the fibers (see Figure 1.12). Therefore, intact locules have to be fixed and then sectioned very carefully into smaller pieces in the fixative in a well-ventilated fume hood (Westafer and Brown, 1976). At the initial step, a gentle degassing and the addition of nonionic deter-gents at low concentrations may be helpful (Ryser, 1979; Yatsu and Jacks, 1979).

However, the method of choice is certainly the rapid cryofixation of fibers of in vitro cultured ovules by plunge freezing (Seagull, 1993; Tiwari and Wilkins, 1995). Precautions have to be taken against the extremely dangerous splashing of the cryogen, due to the relatively large size of the samples. The volume of the sample should be as small as possible to minimize heat transfer to the cryogen, but large enough to contain a sufficient number of undam-aged fiber cells. The bulk of the ovule tissue should be removed, therefore, without touching the fibers to be examined. Cryofixation

FIGURE 1.12. Cross Sections of Cotton Locules

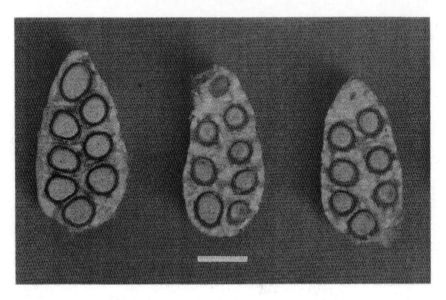

Note: Notice that the space between the seeds is only a few mm wide, whereas the fibers are much longer and must be entangled around their seed of origin in a very complicated manner. A single section through a seed and the adhering fibers destroys the cytoplasmic fine structure of most of the fibers. Bar = 1 cm.

is then followed by cryosubstitution, and finally, the samples are embedded into epoxy or acrylic resins. For immunocytochemistry, often a low temperature embedding procedure may be preferred (Griffiths, 1993; Newman and Hobot, 1993).

CONCLUSION AND PERSPECTIVES

The microscopic structure and the chemistry of differentiating cotton fibers have been relatively well documented in recent decades. White cotton fibers have emerged as a classical model system for the study of cellulose biosynthesis, the involvement of the cytoskeleton in the oriented deposition of cellulose microfibrils, and cell elongation. Green cotton fibers are a new model system for the biosynthesis and deposition of suberin and associated wax. A further promising field of research awaits the brown fiber mutants, having received little attention by the scientific community thus far, and programmed cell death, possibly occurring during fiber maturation.

The expression of heterologous genes coding for colored compounds is feasible in white cotton fibers, but further problems have to be overcome to arrive at products satisfying the demands of the market. Another interesting approach may be the genetic engineering of naturally colored cotton fibers, interesting goals being the stabilization or modification of the existing green and brown colors. For this approach, however, a better knowledge of the chemistry of the colored compounds will be essential.

Using PCR technology and other methods of molecular biology, rapid progress can be expected in the next few years in our understanding of the key events in cotton fiber differentiation: initiation, cell elongation, secondary wall formation and maturation, and the regulatory mechanisms involved in the transitions between the different developmental stages.

An important goal will be a molecular understanding of cell- and tissue-specific gene expression. How is the biosynthesis of suberin and wax specifically suppressed in the white fibers or in white fibers and fuzz, but not in other epidermal cells, or how is the number of fiber initials regulated? These problems can be solved with the standard methodology of current molecular biology and molecular genetics.

Probably a more difficult goal will be the understanding of localized events within a single cell type: (1) the expansion of the outer tangential wall, but not of the other walls of fiber initials at anthesis; (2) the reorientation of the cortical microtubule system; (3) the coexistence of secondary wall formation and cell elongation in the fibers at the transition between primary and secondary wall formation; (4) the localized deposition of suberin layers in the base of white cotton fibers in some cotton cultivars; and, finally, (5) the suppression of cellulose, but not of suberin biosynthesis, in the pits at the base of maturing fibers. The identification of plasmodesmata in the fiber base is a first step toward an integrated understanding of cotton fiber differentiation, taking into account possible interactions between the fibers and epidermal and subepidermal cell types.

The cortical microtubules of developing cotton fibers are morphogenetically important elements that are involved in the orientation of the cellulose microfibrils. How the microtubules acquire the necessary spatial information remains unknown, but the reported results indicate that cortical microtubules do not function in isolation, and future efforts will probably focus on the interaction of microtubules with other cytoskeletal elements, the plasma membrane, and the cell wall. According to Cyr (1994), biophysical factors, for example, in the form of stresses in the wall, are likely elements of the signaling chains involved in the reorientation of microtubules. Mechanosensory microtubule reorientation has been demonstrated in the epidermis of maize coleoptiles subjected to bending stress (Zandomeni and Schöpfer, 1994), and it was postulated that growth changes, elicited by auxin, light, and mechanical stress, affect microtubule orientation through a common signal perception and transduction pathway (Fischer and Schöpfer, 1997).

The importance of microscopy in cotton fiber research is rapidly changing from a descriptive to a more experimental approach, involving immunolocalization of newly described proteins, reporter gene experiments, in situ hybridization, and dye coupling experiments to probe the function of plasmodesmata.

As in the past, it will also be a pleasure in the future to be open to the unexpected, leading to new concepts and further progress in our understanding of plant cell development.

REFERENCES

Amor, Y., C.H. Haigler, S. Johnson, M. Wainscott, and D.P. Delmer (1995). A membrane-associated form of sucrose synthase and its potential role in synthesis of cellulose and callose in plants. *Proceedings of the National Academy of Sciences USA* 92: 9353-9357.

Basra, A.S. and C.P. Malik (1984). Development of the cotton fiber. *International Review of Cytology* 87: 65-113.

Beasley, C.A. (1975). Developmental morphology of cotton flowers and seed as seen with the scanning electron microscope. *American Journal of Botany* 62: 584-592.

Beasley, C.A. (1979). Cellulose content in fibers of cotton which differ in their lint lengths and extent of fuzz. *Physiologia Plantarum* 45: 77-82.

Beasley, C.A. and I.P. Ting (1973). The effects of plant growth substances on *in vitro* fiber development from fertilized cotton ovules. *American Journal of Botany* 60: 130-139.

Beasley, C.A. and I.P. Ting (1974). Effects of plant growth substances on *in vitro* fiber development from unfertilized cotton ovules. *American Journal of Botany* 61: 188-194.

Beasley, C.A., I.P. Ting, A.E. Linkins, E.H. Birnbaum, and D.P. Delmer (1974). Cotton ovule culture: A review of progress and a preview of potential. In *Plant Cell, Tissue and Organ Culture*, Ed. H.E. Street. New York: Academic Press, pp. 169-192.

Berlin, J.D. (1986). The outer epidermis of the cotton seed. In *Cotton Physiology*, Eds. J.R. Mauney and J. McD. Stewart. Memphis, TN: The Cotton Foundation, pp. 375-414.

Bernards, M.A., M.L. Lopez, J. Zajicek, and N.G. Lewis (1995). Hydroxycinnamic acid-derived polymers constitute the polyaromatic domain of suberin. *Journal of Biological Chemistry* 270: 7382-7386.

Boudet, A.M., D.P. Goffner, and J. Grima-Pettenati (1996). Lignins and lignification: Recent biochemical and biotechnological developments. *Comptes Rendus de l'Académie des Sciences*, Paris, *Life Sciences* 319: 317-331 (review).

Brown, R.M., Jr., I.M. Saxena, and K. Kudlicka (1996). Cellulose biosynthesis in higher plants. *Trends in Plant Science* 1: 149-156.

Cassagne, C., J.J. Lessire, P. Bessoule, A. Moreau, A. Creach, and F. Schneider (1994). Biosynthesis of very long chain fatty acids in higher plants. *Progress in Lipid Research* 33: 55-69.

Conrad, C.M. (1941). The high wax content of green lint cotton. *Science* 94: 113.

Conrad, C.M. and J.W. Neely (1943). Heritable relation of wax content and green pigmentation of lint in upland cotton. *Journal of Agricultural Research* 66: 307-312.

Cyr, R.J. (1994). Microtubules in plant morphogenesis: Role of the cortical array. *Annual Review of Cell Biology* 10: 153-180.

Davidonis, G. (1993). Cotton fiber growth and development *in vitro*: Effects of tunicamycin and monensin. *Plant Science* 88: 229-236.

Davin, L.B. and N.G. Lewis (1992). Phenylpropanoid metabolism: Biosynthesis and monolignols, lignans and neolignans, lignins and suberins. In *Phenolic Metabolism in Plants*, Eds. H.A. Stafford and R.K. Ibrahim. New York: Plenum Press, pp. 325-375.

DeLanghe, E.A.L. (1986). Lint development. In *Cotton Physiology*, Eds. J.R. Mauney and J. McD. Stewart. Memphis, TN: The Cotton Foundation, pp. 325-349.

Delmer, D.P., J.R. Pear, A. Andrawis, and D.M. Stalker (1995). Genes encoding small GTP-binding proteins analogous to mammalian rac are preferentially expressed in developing cotton fibers. *Molecular and General Genetics* 248: 43-51.

Dixon, D.C., R.W. Seagull, and B.A. Triplett (1994). Changes in the accumulation of α- and β-tubulin isotypes during cotton fiber development. *Plant Physiology* 105: 1347-1353.

Endrizzi, J.E., E.L. Turcotte, and R.J. Kohel (1984). Qualitative genetics, cytology, and cytogenetics. In *Cotton*, Eds. R.J. Kohel and C.F. Lewis. Madison, WI: American Society of Agronomy-Crop Science Society of America/Soil Science Society of America, pp. 81-129.

Endrizzi, J.E., E.L. Turcotte, and R.J. Kohel (1985). Genetics, cytology and evolution in *Gossypium*. *Advances in Genetics* 23: 271-375.

Essner, E. (1973). Phosphatases. In *Electron Microscopy of Enzymes*, Volume 1, Ed. M.A. Hayat. New York: Van Nostrand Reinhold Company, pp. 44-76.

Fischer, K. and P. Schöpfer (1997). Interaction of auxin, light, and mechanical stress in orienting microtubules in relation to tropic curvature in the epidermis of maize coleoptiles. *Protoplasma* 196: 108-116.

Fryxell, P.A. (1963). Morphology of the base of seed hairs of *Gossypium*. I. Gross morphology. *Botanical Gazette* 124: 196-199.

Giddings, T.H. and L.A. Staehelin (1991). Microtubule-mediated control of microfibril deposition: A reexamination of the hypothesis. In *The Cytoskeletal Basis of Plant Growth and Form*, Ed. C.W. Lloyd. London: Academic Press, pp. 85-99.

Goddard, R.H., S.M. Wick, C.D. Silflow, and D.P. Snustad (1994). Microtubule components of the plant cytoskeleton. *Plant Physiology* 104: 1-6 (review).

Graves, D. A. and J.M. Stewart (1988a). Chronology of the the differentiation of cotton (*Gossipium hirsutum* L.) fiber cells. *Planta* 175: 254-258.

Graves, D. A. and J.M. Stewart (1988b). Analysis of the protein constituency of developing cotton fibers. *Journal of Experimental Botany* 39: 59-69.

Griffiths, G., Ed. (1993). *Fine Structure Immunocytochemistry*. Berlin: Springer-Verlag.

Gunning, B.E.S. and A.R. Hardham (1982). Microtubules. *Annual Review of Plant Physiology* 33: 651-698.

Haigler, C.H., N.R. Rao, E.M. Roberts, J.Y. Huang, D.R. Upchurch, and N.L. Trolinder (1991). Cultured ovules as models for cotton fiber development under low temperatures. *Plant Physiology* 95: 88-96.

Hall, A. (1994). Small GTP-binding proteins and the regulation of the actin cytoskeleton. *Annual Review of Cell Biology* 10: 31-54.

Halloin, J.M. (1982). Localization and changes in catechin and tannins during development and ripening of cottonseed. *New Phytologist* 90: 651-657.

Hamilton, R.J., Ed. (1995). *Waxes: Chemistry, Molecular Biology and Functions*. Dundee, Scotland: The Oily Press Ltd.

Hardham, A.R. and B.E.S. Gunning (1978). Structure of cortical microtubule arrays in plant cells. *Journal of Cell Biology* 77: 14-34.

Harwood, J.L. (1996). Recent advances in the biosynthesis of plant fatty acids. *Biochimica et Biophysica Acta* 1301: 7-56 (review).

Heath, I.B., Ed. (1990). *Tip Growth in Plant and Fungal Cells*. San Diego, CA: Academic Press.

Hebert, J.J. and E.K. Boylston (1984). Reversal frequencies in developing cotton fibers. *Textile Research Journal* 54: 23-26.

Hepler, P.K. and Hush, J.M. (1996). Behavior of microtubules in living cells. *Plant Physiology* 112: 455-461.

Herth, W. (1980). Calcofluor white and congo red inhibit chitin microfibril assembly of *Poteriochromonas*: Evidence for a gap between polymerization and microfibril formation. *Journal of Cell Biology* 87: 442-450.

Hohlfeld, M., M. Veit, and D. Strack (1996). Hydroxycinnamoyl-transferases involved in the accumulation of caffeic acid esters in gametophytes and sporophytes of *Equisetum arvense*. *Plant Physiology* 111: 1153-1159.

Houwink, A.L. and P.A. Roelofson (1954). Fibrillar architecture of growing plant cell walls. *Acta Botanica Neerlandica* 3: 385-395.

Hush, J.M., P. Wadsworth, D.A. Callaham, and P.K. Hepler (1994). Quantification of microtubule dynamics in living plant cells using fluorescence redistribution after photobleaching. *Journal of Cell Science* 107: 775-784.

Hussey, P.J., D.P. Snustad, and C.D. Silflow (1991). Tubulin gene expression in higher plants. In *The Cytoskeletal Basis of Plant Growth and Form*, Ed. C.W. Lloyd. London: Academic Press, pp. 15-27.

Hussey, P.J., J.A. Traas, K. Gull, and C.W. Lloyd (1987). Isolation of cytoskeletons from isolated plant cells: The interphase microtubule array utilizes multiple tubulin isotypes. *Journal of Cell Science* 88: 225-230.

Itoh, T. (1974). Fine structure and formation of cell wall of developing cotton fiber. *Wood Research* 56: 49-61.

Jacks, T.J., T.P. Hensarling, M.G. Legendre, and S.M. Buco (1993). Permanent gases inside healthy and microbially infected cotton fruit during development. *Biochemical and Biophysical Research Communications* 191: 1284-1287.

Jernstedt, J.A., K. Sugimoto, P. Lu, and T.A. Wilkins (1993). Microtubule involvement in cotton fiber initiation. *American Journal of Botany* 80: 28 (abstract).

Joshi, P.A., J.M. Stewart, and E.T. Graham (1985). Localization of β-glycerophosphatase activity in cotton fiber during differentiation. *Protoplasma* 125: 75-85.

Kader, J.C. (1996). Lipid transfer proteins in plants. *Annual Review of Plant Physiology and Plant Molecular Biology* 47: 627-654.

Kader, J.C. and P. Mazliak, Eds. (1995). *Plant Lipid Metabolism*. Dordrecht, Netherlands: Kluwer Academic Publishers.

Kerstiens, G., Ed. (1996). *Plant Cuticles*. Oxford, UK: BIOS Scientific Publishers Ltd.

Kloth, R.H. (1989). Changes in the level of tubulin subunits during development of cotton (*Gossypium*) fiber. *Physiologia Plantarum* 76: 37-41.

Kobayashi, H., H. Fukuda, and H. Shibaoka (1988). Interrelation between the spatial disposition of actin filaments and microtubules during the differentiation of tracheary elements in cultured *Zinnia elegans*. *Protoplasma* 143: 29-37.

Kohel, R.J. (1985). Genetic analysis of fiber color variants in cotton. *Crop Science* 25: 793-797.

Kolattukudy, P.E. (1980). Biopolyester membranes of plants. *Science* 208: 990-1000.

Kolattukudy, P.E. and K.E. Espelie (1989). Chemistry, biochemistry, and function of suberin and associated waxes. In *Natural Products of Woody Plants*, Volume 1, Ed. J.W. Rowe. Berlin: Springer-Verlag, pp. 304-367.

Kolattukudy, P.E., L.M. Rogers, D. Li, C.S. Hwang, and M.A. Flaishman (1995). Surface signaling in pathogenesis. *Proceedings of the National Academy of Sciences USA* 92: 4080-4087.

Kosmidou-Dimitropoulou, K. (1986). Hormonal influences on fiber development. In *Cotton Physiology*, Eds. J.R. Mauney, and J. McD. Stewart. Memphis, TN: The Cotton Foundation, pp. 361-373.

Lambert, A.M. (1995). Microtubule-organizing centers in plants: Evolving concepts. *Botanica Acta* 108: 535-537.

Larkin, J.C., D.G. Oppenheimer, S. Pollock, and M.D. Marks (1993). Arabidopsis *glabrous1* gene requires downstream sequences for function. *Plant Cell* 5: 1739-1748.

Liu, B., H.C. Joshi, T.J. Wilson, C.D. Silflow, and B.A. Palevitz (1994). γ-tubulin in *Arabidopsis*: Gene sequence, immunoblot, and immunofluorescence studies. *Plant Cell* 6: 303-314.

Liu, B., J. Marc, H.C. Joshi, and B.A. Palevitz (1993). A γ-tubulin related protein associated with microtubule arrays of higher plants in a cell-cycle-dependent manner. *Journal of Cell Science* 104: 1217-1228.

Lotfy, S., Javelle, F., and J. Negrel (1996). Purification and characterization of hydroxycinnamoyl-CoA:ω-hydroxypalmitic acid O-hydroxycinnamoyltransferase from tobacco (*Nicotiana tabacum* L.) cell suspension cultures. *Planta* 199: 475-480.

Lotfy, S., J. Negrel, and F. Javelle (1994). Formation of ω-feruloyl-palmitic acid by an enzyme from wound-healing potato tuber discs. *Phytochemistry* 35: 1419-1424.

Lucas, W.J., S. Bouche-Pillon, D.P. Jackson, L. Nguyen, L. Baker, B. Ding, and S. Hake (1995). Selective trafficking of KNOTTED1 homeodomain protein and its mRNA through plasmodesmata. *Science* 270: 1980-1983.

Luduena, R.F. (1993). Are tubulin isotypes functionally significant? *Molecular Biology of the Cell* 4: 445-457.

Ma, D.P., H. Tan, Y. Si, R.G. Creech, and J.N. Jenkins (1995). Differential expression of a lipid transfer protein gene in cotton fiber. *Biochimica et Biophysica Acta* 1257: 81-84.

Marx-Figini, M. (1967). Zur Biosynthese der Zellulose: Bevorzugte Ablagerung der Sekundärwand in den mittleren Partien des Samenhaares von *Gossypium*. *Zeitschrift für Pflanzenphysiologie* 57: 235-242 (abstract in English).

Meinert, M.C. and D. Delmer (1977). Changes in biochemical composition of the cell wall of the cotton fiber during development. *Plant Physiology* 59: 1088-1097.

Melaragno, J.E., B. Mehrotra, and A.W. Coleman (1993). Relationship between endopolyploidy and cell size in epidermal tissue of *Arabidopsis*. *Plant Cell* 5: 1661-1668.

Mezitt, L.A. and W.J. Lucas (1996). Plasmodesmal cell-to-cell transport of proteins and nucleic acids. *Plant Molecular Biology* 32: 251-273.

Naithani, S.C. (1987). The role of IAA oxidase, peroxidase and polyphenoloxidase in the fiber initiation of the cotton ovule. *Beiträge zur Biologie der Pflanzen* 62: 79-90.

Naithani, S.C., N.R. Rao, and Y.D. Singh (1982). Physiological and biochemical changes associated with cotton fiber development. *Physiologia Plantarum* 54: 225-229.

Nayyar, H., K. Kaur, C.P. Malik, and A.S. Basra (1989). Regulation of differential fiber development in cotton by endogenous plant growth regulators. *Proceedings of the Indian National Science Academy* B55: 463-468.

Newman, G.R. and J.A. Hobot, Eds. (1993). *Resin Microscopy and On-Section Immunocytochemistry*. Berlin: Springer-Verlag.

Nolte, K.D., D.L. Hendrix, J.W. Radin, and K.E. Koch (1995). Sucrose synthase localization during initiation of seed development and trichome differentiation in cotton ovules. *Plant Physiology* 109: 1258-1293.

Okita, T.W. and J.C. Rogers (1996). Compartmentation of proteins in the endomembrane system of plant cells. *Annual Review of Plant Physiology and Plant Molecular Biology* 47: 327-350.

Peeters, M.C., W. Dillemans, and S. Voets (1991). Nucleolar activity in differentiating cotton fibres is related to the position of the boll on the plant. *Journal of Experimental Botany* 42: 353-357.

Peeters, M.C., S. Voets, G. Dayatilake, and E. DeLanghe (1987). Nucleolar size at early stages of cotton fiber development in relation to final fiber dimension. *Physiologia Plantarum* 71: 436-440.

Peeters, M.C., S. Voets, J. Wijsmans, and E. DeLanghe (1988). Pattern of nucleolar growth in differentiating cotton fibres (*Gossypium hirsutum* L.). *Annals of Botany* 62: 377-382.

Post-Beittenmiller, D. (1996). Biochemistry and molecular biology of wax production in plants. *Annual Review of Plant Physiology and Plant Molecular Biology* 47: 405-430.

Quader, H., G. Deichgräber, and E. Schnepf (1986). The cytoskeleton of *Cobaea* seed hairs: Patterning during cell wall differentiation. *Planta* 168: 1-10.

Quader, H., W. Herth, U. Ryser, and E. Schnepf (1987). Cytoskeletal elements in cotton seed hair development *in vitro*: Their possible regulatory role in cell wall organization. *Protoplasma* 137: 56-62.

Ramsey, J.C. and J.D. Berlin (1976). Ultrastructure of early stages of cotton fiber differentiation. *Botanical Gazette* 137: 11-19.

Rerie, W.G., K.A. Feldmann, and M.D. Marks (1994). The *glabra2* gene encodes a homeo domain protein required for normal trichome development in *Arabidopsis. Genes and Development* 8: 1388-1399.

Rhyne, C.L. (1957). Duplicated linkage groups in cotton. *Journal of Heredity* 48: 59-62.

Roberts, E.M., N.R. Rao, J.Y. Huang, N.L. Trolinder, and C.H. Haigler (1992). Effects of cycling temperatures on fiber metabolism in cultured cotton ovules. *Plant Physiology* 100: 979-986.

Roelofson, P.A. (1951). Orientation of cellulose fibrils in the cell wall of growing cotton hairs and its bearing on the physiology of cell wall growth. *Biochimica et Biophysica Acta* 7: 43-53.

Roelofson, P.A. (1965). Ultrastructure of the wall in growing cells and its relation to the direction of growth. In *Advances in Botanical Research*, Volume 2, Ed. R.D. Preston. London: Academic Press, pp. 69-149.

Ryser, U. (1977). Cell wall growth in elongating cotton fibres: An autoradiographic study. *Cytobiologie* 15: 78-84.

Ryser, U. (1979). Cotton fibre differentiation: Occurrence and distribution of coated and smooth vesicles during primary and secondary wall formation. *Protoplasma*: 98: 223-239.

Ryser, U. (1985). Cell wall biosynthesis in differentiating cotton fibres. *European Journal of Cell Biology* 39: 236-256 (review).

Ryser, U. (1992). Ultrastructure of the epidermis of developing cotton (*Gossypium*) seeds: Suberin, pits, plasmodesmata, and their implication for assimilate transport into cotton fibers. *American Journal of Botany* 79: 14-22.

Ryser, U. and P.J. Holloway (1985). Ultrastructure and chemistry of soluble and polymeric lipids in cell walls from seed coats and fibres of *Gossypium* species. *Planta* 163: 151-163.

Ryser, U., H. Meier, and P.J. Holloway (1983). Identification and localization of suberin in the cell walls of green cotton fibres (*Gossypium hirsutum* L., var. green lint). *Protoplasma* 117: 196-205.

Ryser, U., M. Schorderet, U. Jauch, and H. Meier (1988). Ultrastructure of the "fringe-layer," the innermost epidermis of cotton seed coats. *Protoplasma* 147: 81-90.

Schmid, J. and N. Amrhein (1995). Molecular organization of the shikimate pathway in higher plants. *Phytochemistry* 39: 737-749 (review no. 104).

Schmutz, A., A. Buchala, T. Jenny, and U. Ryser (1994). The phenols in the wax and in the suberin polymer of green cotton fibres and their functions. In *International Symposium on Natural Phenols in Plant Resistance*, Volume 1, Eds. M. Geibel, D. Treutter, and W. Feucht. *Acta Horticulturae* 381: 269-275.

Schmutz, A., A.J. Buchala, and U. Ryser (1996). Changing the dimensions of suberin lamellae of green cotton fibers with a specific inhibitor of the endoplasmic reticulum-associated fatty acid elongases. *Plant Physiology* 110: 403-411.

Schmutz, A., T. Jenny, N. Amrhein, and U. Ryser (1993). Caffeic acid and glycerol are constituents of the suberin layers in green cotton fibres. *Planta* 189: 453-460.

Schmutz, A., T. Jenny, and U. Ryser (1994). A caffeoyl-fatty acid-glycerol ester from wax associated with green cotton fibre suberin. *Phytochemistry* 36: 1343-1346.

Schubert, A.M., C.R. Benedict, J.D. Berlin, and R.J. Kohel (1973). Cotton fiber development—Kinetics of cell elongation and secondary wall thickening. *Crop Science* 13: 704-709.

Schubert, A.M., C.R. Benedict, C.E. Gates, and R.J. Kohel (1976). Growth and development of the lint fibers of Pima S-4 cotton. *Crop Science* 16: 539-543.

Schweizer, P., G. Felix, A. Buchala, C. Müller, and J.P. Métraux (1996). Perception of free cutin monomers by plant cells. *Plant Journal* 10: 331-341.

Seagull, R.W. (1986). Changes in microtubule organization and microfibril orientation during *in vitro* cotton fiber development: An immunofluorescent study. *Canadian Journal of Botany* 64: 1373-1381.

Seagull, R.W. (1989). The role of the cytoskeleton during oriented microfibril deposition. II. Microfibril disposition in cells with disrupted cytoskeletons. In *Cellulose and Wood—Chemistry and Technology*, Proceedings of the Tenth Cellulose Conference, Ed. C. Schuerch. New York: J. Wiley and Sons, pp. 811-825.

Seagull, R.W. (1990a). Tip growth and transition to secondary wall synthesis in developing cotton hairs. In *Tip Growth in Plant and Fungal Cells*, Ed. I.B. Heath. San Diego, CA: Academic Press, pp. 261-284.

Seagull, R.W. (1990b). The effect of microtubule and microfilament disrupting agents on cytoskeletal arrays and wall deposition in developing cotton fibers. *Protoplasma* 159: 44-59.

Seagull, R.W. (1992). A quantitative electron microscopic study of changes in microtubule arrays and wall microfibril orientation during *in vitro* fiber development. *Journal of Cell Science* 101: 561-577.

Seagull, R.W. (1993). Cytoskeletal involvement in cotton fiber growth and development. *Micron* 24: 643-660 (review).

Sharma, P., B. Singh, C.P. Malik, and K.L. Bajaj (1995). Effect of a plant growth retardant (E)-1-(4-chlorophenyl)-4,4-dimethyl-2(1,2,4-triazol-1-yl)-1-penten-3-ol (S-3307) on the fiber initiation and enzyme activities of cotton ovules (*Gossypium arboreum*). *Phytomorphology* 45: 79-86.

Singh, B., P. Sharma, C.P. Malik, P.S. Kalsi, and Sanju (1995). Effects of guaianolide derivative I on fiber initiation and elongation in *Gossypium arboreum* L. in relation to adverse effects of cool temperature. *Plant Growth Regulation* 17: 101-107.

Song, P., E. Yamamoto, and R.D. Allen (1995). Improved procedure for differential display of transcripts from cotton tissues. *Plant Molecular Biology Reporter* 13: 174-181.

Staehelin, L.A. and T.H. Giddings (1982). Membrane-mediated control of cell wall microfibrillar order. In *Developmental Order: Its Origins and Regulation*, Eds. S. Subtelny and P.B. Green. New York: Liss, pp. 133-147.

Staehelin, L.A. and I. Moore (1995). The plant Golgi apparatus: Structure, functional organization and trafficking mechanisms. *Annual Review of Plant Physiology and Plant Molecular Biology* 46: 261-288.

Stewart, J.M. (1975). Fiber initiation on the cotton ovule (*Gossypium hirsutum*). *American Journal of Botany* 62: 723-730.

Szymanski, D.B., R. Jilk, and M.D. Marks (1996). Immunolocalization of GL1 and GL2 proteins in wild type and mutant *Arabidopsis*. *Plant Physiology* 111: 144 (abstract).

Thiry, M. and G. Goessens, Eds. (1996). *The Nucleolus During the Cell Cycle*. New York: Springer-Verlag.

Tiwari, S.C. and T.A. Wilkins (1995). Cotton (*Gossypium hirsutum*) seed trichomes expand via diffuse growing mechanisms. *Canadian Journal of Botany* 73: 746-757.

Triplett, B.A. (1996). Inhibition of fiber development in cotton ovule cultures by α-amanitin. *Plant Physiology* 111: 145 (abstract).

Turley, R.B. and D.L. Ferguson (1996). Changes of ovule proteins during early fiber development in a normal and a fiberless line of cotton (*Gossypium hirsutum* L.). *Journal of Plant Physiology* 149: 695-702.

Vigil, E.L., W.S. Anthony, E. Columbus, E. Erbe, and W.P. Wergin (1996). Fine structural aspects of cotton fiber attachment to the seed coat: Morphological factors affecting saw ginning of lint cotton. *International Journal of Plant Sciences* 157: 92-102.

Von Wettstein-Knowles, P.M. (1993). Waxes, cutin, and suberin. In *Lipid Metabolism in Plants*, Ed. T.S. Moore Jr. Boca Raton, FL: CRC Press, pp. 127-166.

Waterkeyn, L. (1981). Cytochemical localization and function of the 3-linked glucan callose in the developing cotton fibre wall. *Protoplasma* 106: 49-67.

Westafer, J.M. and M.R. Brown Jr. (1976). Electron microscopy of the cotton fiber: New observations on cell wall formation. *Cytobios* 15: 111-138.

Whetten, R.W. and R.R. Sederoff (1995). Lignin biosynthesis. *Plant Cell* 7: 1001-1013.

Williamson, R.E. (1991). Orientation of cortical microtubules in interphase plant cells. *International Review of Cytology* 129: 135-206.

Willison, J.H.M. (1983). The morphology of supposed cellulose-synthesizing structures in higher plants. *Journal of Applied Polymer Science: Applied Polymer Symposium* 37: 91-105.

Wolf, S. and W.J. Lucas (1994). Virus movement proteins and other molecular probes of plasmodesmal function. *Plant, Cell and Environment* 17: 573-585 (review).

Xie, W., N.L. Trolinder, and C.H. Haigler (1993). Cool temperature effects on cotton fiber initiation and elongation clarified using *in vitro* cultures. *Crop Science* 33: 1258-1264.

Yatsu, L.Y. (1983). Morphological and physical effects of colchicine treatment on cotton (*Gossypium hirsutum* L.) fibers. *Textile Research Journal* 53: 515-519.

Yatsu, L.Y., K.E. Espelie, and P.E. Kolattukudy (1983). Ultrastructural and chemical evidence that the cell wall of green cotton fiber is suberized. *Plant Physiology* 73: 521-524.

Yatsu, L.Y. and T.J. Jacks (1979). Fixation of cotton (*Gossypium hirsutum* L.) hair cells (fibers). *Microscope* 27: 113-116.

Yatsu, L.Y. and T.J. Jacks (1981). An ultrastructural study of the relationship between microtubules and microfibrils in cotton (*Gossypium hirsutum* L.) cell wall reversals. *American Journal of Botany* 68: 771-777.

Zandomeni, K. and P. Schöpfer (1994). Mechanosensory microtubule orientation in the epidermis of maize coleoptiles subjected to bending stress. *Protoplasma* 182: 96-101.

Chapter 2

Growth Regulation of Cotton Fibers

Amarjit S. Basra
Sukumar Saha

INTRODUCTION

The mature fiber is an elongated epidermal cell of the cotton ovule with a thickened secondary cell wall composed mainly of cellulose ($\sim$95 percent). A developing fiber passes through four discrete yet overlapping phases: (1) initiation, (2) elongation, (3) secondary cell wall synthesis, and (4) maturation (Basra and Malik, 1984).

Primordial fiber cells, destined to become lint fibers, initiate elongation on the day of anthesis and continue to do so for a period of $\sim$16 to 25 days, reaching variable fiber lengths in different *Gossypium* species. Lint fibers refer to those spinnable fibers which are removed from the seed coat during the first pass through the cotton gin. Other epidermal cells forming fuzz fibers are initiated 4 to 10 days postanthesis (dpa), but never attain lengths greater than 10 mm (millimeters). Typically, the fuzz fibers are 1.5 to 3.3 mm long. These fibers usually remain attached to the seed coat during the first pass through the gin, thus giving the seed a fuzzy appearance. Why fuzz fibers do not elongate as much as lint fibers poses an intriguing question.

Secondary cell wall synthesis begins slightly before the cessation of fiber elongation ($\sim$16 to 18 dpa) and continues for several weeks thereafter (Meinert and Delmer, 1977; Beasley, 1979; Delmer, Chapter 4, this volume). The composition of the fiber secondary cell wall is highly crystalline, nearly pure cellulose. Fiber maturation begins $\sim$50 to 60 dpa when the bolls open and the fibers desiccate into a flattened twisted ribbon.

In addition to its economic importance as a natural textile fiber, the developing cotton fiber is an excellent model system for unraveling the fundamental processes of plant cell growth, differentiation, wall biogenesis, and programmed cell death. Cultured cotton ovules undergo normal fiber morphogenesis, facilitating the study of growth regulatory factors (see Davidonis, Chapter 3, this volume). In addition, a number of fiber mutants are available for dissecting the control mechanisms of fiber growth and developmental processes.

Normally, growth in a plant is a combination of cell division and cell elongation. The conventional idea is that fibers originate and elongate as single epidermal cells, without any cell division. However, it has been shown recently that a large fraction of fully differentiated fiber cells are able to divide up to 4 dpa (Van't Hof and Saha, 1997). An understanding of growth regulation of cotton fibers is a prerequisite for the manipulation of fiber quality traits.

FIBER GROWTH

Cell Division

Normally, the cotton fiber originates and ends as a single cell without any complications from cell division. This is especially unique because the epidermal cells surrounding the fiber-forming cells in the cotton ovule undergo continuous cell division as the seed develops. A comparison of dividing versus nondividing fiber cells and their morphogenetic processes at a very young stage can provide an opportunity to identify some of the substances or factors involved in cell division, growth, and differentiation.

Van't Hof and Saha (1997) have made the exciting discovery that cotton fibers can undergo cell division in vitro to form multicellular fibers. It was observed that fibers from ovules cultured at 2 dpa, in Beasley's medium, without any hormone, at 34°C (celsius), and in constant darkness undergo cell division within 24 h (hours); lengthening the time of culture to 72 h produces about 30 percent multicellular fibers (see Figure 2.1). The addition of IAA and GA_3 to the culture medium lowers the frequency of fiber cell division. The

FIGURE 2.1. Multicellular Fibers

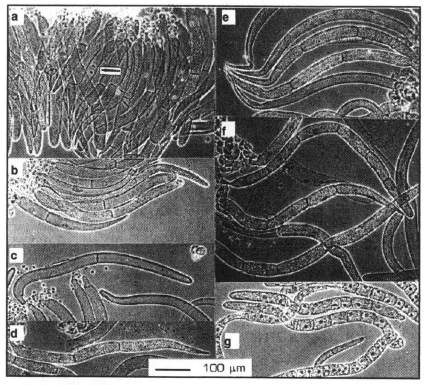

a Two-celled multicellular fibers; the arrow indicating mitosis in a fiber cell.
b Two-celled fibers.
c A two-celled fiber showing the cell wall formation almost in the middle region.
d A three-celled fiber.
e A four-celled fiber.
f A six-celled fiber.
g Multicelled fibers.

basal and tip cells of multicellular fibers possess the same characteristics as the polar ends of unicellular fibers.

The age of the ovule is a critical factor in fiber cell division. Ovules cultured at 1 dpa produce about 2 percent multicellular fibers; ovules at 2 and 3 dpa, more than 35 percent; ovules at 4 dpa, about 5 percent; and ovules at 7 dpa, none (Van't Hof and Saha,

1997). The dividing fibers are part of the normal lint fiber population because the multicellular fibers develop at the same time as the lint fibers (0 to 4 dpa) and because the number of fiber cells per ovule is nearly the same whether measured on ovules from the plant at 4 dpa or from cultured ovules. The number of fiber cells per cultured ovule ranges between 30,000 and 35,000 and is not significantly different from the 32,000 seen in the plant at 4 dpa.

This discovery indicates that genes controlling fiber cell division exist and can thus be identified. Ultimately, their expression could be modified to benefit fiber properties that contribute to textile performance. These genes appear to respond to fertilization because fertilization is needed to develop fibers in vitro (Beasley and Ting, 1974) and because the frequency of multicellular fibers coincides with the time when fertilization occurs (Beasley and Ting, 1973). The role of sugars, phytohormones, and polyamines in the regulation of cell division genes is worthy of investigation.

Finally, the cell cycle phase in which fiber cells are arrested during the very initial stage of development is also of significance. Work of Saha and Van't Hof, using the S-phase inhibitor 5-aminouracil (5-AU) has shown that 98 percent of the cotton fiber cells in ovules 2 dpa are arrested at G_1 and so do not divide (unpublished data). Thus, only 2 percent of the cells had passed through the S phase as 5-AU was unable to stop them from undergoing cell division. However, the fiber cells undergo cell division in the control medium without 5-AU. This is in accordance with a recent report that the most critical step in the regulation of cell cycle occurs late in G_1. This step is control at the cytokinesis stage. Hence, multicellular fibers provide a unique opportunity for understanding this critical control point in cell cycle regulation.

The multicellular fiber in ovule culture may also provide a quick assay system to evaluate transgene expression and nuclear incorporation in cotton fiber cells. Genetic engineering strategies for fiber modification require a suitable tool to quickly evaluate fiber-specific expression (see John, Chapter 10, this volume). If the transformation is successful, the transgene expression in all of the daughter cells of the multicellular fiber will indicate that the foreign gene has successfully gone through the DNA replication and cell division processes from the mother cell in the fiber. Accordingly, the transgene incorporation and expression can be easily determined without waiting for harvest of bolls.

Fiber Elongation

The fiber cells undergo enormous elongation during development, increasing their length 1000 to 3000 times the diameter of the cell. Fiber cell diameter is typically 12 to 20 μm (micrometer). The maximal rate of fiber elongation occurs between 6 and 12 dpa. (Meinert and Delmer, 1977). The log phase of fiber elongation occurs during the first 10 dpa (DeLanghe, 1986). The rate of elongation and ultimate length are regulated by both environmental and genetic factors.

Based on final fiber length, the cultivated *Gossypium* spp. can be arranged into five staple-length classes (Basra and Malik, 1984). Staple length is the average of lint fibers on a seed. It is an inherited characteristic of cultivars, which are classified into five groups from short-staple Asian cottons (less than 1 mm) to extra-long-staple Egyptian and Sea Island cottons (35 mm and longer). The majority of the cotton fiber production in the world is of medium (22 to 25 mm) and medium-long (25 to 28 mm) staple from Upland cultivars.

During most of the elongation phase, the fibers are encased in a thin (0.2 to 0.4 μm) primary cell wall that is plastic enough to accommodate the incorporation of new cell wall material. As elongation slows, the rate of secondary wall synthesis increases to accomplish fiber thickening. There is significant overlap between these two phases, of the order of 5 to 10 days (Schubert et al., 1973; Meinert and Delmer, 1977; Beasley, 1979), depending upon the cultivar and environmental conditions. Overlap tends to be greater in long-fibered cultivars than in short-fibered cultivars (Naithani, Rama Rao, and Singh, 1981).

Controlling Factors

Fiber elongation is subject to temporal and spatial control imposed both from within (its developmental program) and from the external environment. Plant hormones play an important role in mediating both the internal and external control of fiber growth. Fiber elongation is also sensitive to environmental factors, especially the temperature.

Temperature

Temperature is an important environmental factor influencing the final fiber length within a species or cultivar (Gipson and Joham, 1969; Gipson and Ray, 1969, 1970; Gipson, 1986). It is well documented that developing fibers exposed to cool temperatures (generally at night) have prolonged periods and reduced overall rates of elongation and thickening (Hawkins, 1930; Gipson and Joham, 1969; Gipson and Ray, 1970; Thaker et al., 1989) and growth rings in their secondary walls (Balls, 1919; Kerr, 1936; Haigler et al., 1991; Xie, Trolinder, and Haigler, 1993). Cell wall development slows during cool nights, giving rise to rings that appear similar to the growth rings of trees. These rings may be weak points in the fibers, as slippage at the rings has been observed.

Even though the duration of the elongation period is increased at lower temperatures, the fibers attain the same final length as those grown under more favorable conditions (Hawkins, 1930; Gipson and Joham, 1969; Gipson and Ray, 1969; Thaker et al., 1989). Fibers cultured in vitro under 34°C/22°C and 34°C/15°C cycling experienced delayed onset of substantial elongation and required prolonged periods to attain fiber length equal to controls (Haigler et al., 1991). After an initial delay of variable length, the maximum rates of elongation were similar under all conditions, suggesting that the prolonged elongation period under cool temperatures is more attributable to the delay in its onset than to its reduced rate.

Secondary wall synthesis may be relatively more sensitive to cool temperatures than primary wall synthesis (Ryser, 1985), possibly due to differences in the diversity of proteins and noncellulosic carbohydrates in the primary wall (see Buchala, Chapter 5, this volume), compared to the almost exclusively cellulosic composition of the secondary cell wall (see Delmer, Chapter 4, this volume). Cool temperatures significantly decrease the rate of cellulose synthesis (Gipson, 1986) and cause perturbations in fiber sugar metabolism (Roberts et al., 1992; Malik et al., 1995).

Comparatively little is known about the regulation of fiber growth and metabolism by other environmental factors such as drought (Bradow et al., 1997), and this work needs to be accelerated.

Phytohormones

Plant hormones have a decisive role in fiber elongation as revealed by effects of exogenous plant hormones on fiber elongation in vitro or by determination of their endogenous levels. It has been found that exogenously applied indole-3-acetic acid (IAA) and gibberellic acid (GA_3) could enhance the differentiation of fiber and promote its elongation, whereas abscisic acid (ABA) is inhibitory for fiber growth (Beasley and Ting, 1973, 1974; Zheng and Xu, 1982; Wang, Shen, and Zhang, 1985; Chen et al., 1988). Although fiber cells have the ability to initiate growth 2 days preanthesis, the elongation does not take place until the day of anthesis (Graves and Stewart, 1988), coinciding perhaps with the synthesis and/or sensitivity to hormonal factors promoting elongation.

The importance of exogenously applied hormones is dependent on the fertilization status of ovules before culturing. Unfertilized ovules produce no fibers, but the addition of IAA and GA_3 permitted ovule growth and fiber elongation (Beasley and Ting, 1974). Using the antiauxin, p-chlorophenoxyisobutyric acid, Dhindsa (1978a) showed that IAA was essential for the elongation of fiber initials, whereas GA_3 was involved in ovule growth.

With fertilized ovules, however, exogenous GA_3 appears to be essential for fiber elongation, whereas IAA only showed a weak effect, indicating that fertilization triggers IAA production (Beasley and Ting, 1973; Shen, Chang, and Yhn, 1978; Zhang, 1982; Chen et al., 1988). Endogenous levels of GA_{1+3} in cotton ovules increased after flowering and maintained a high level until 8 dpa, but the level of GA_{4+7} was rather low (Chen et al., 1996). The presence of embryo-endosperm complex in cultured ovules was linked to the production of auxin needed for fiber growth (Baert, DeLanghe, and Waterkeyn, 1975).

In vivo studies of auxin-metabolizing enzymes, for example, IAA oxidase and peroxidase, showed that IAA catabolism was low during the elongation phase, whereas it increased significantly during secondary thickening (Jasdanwala, Singh, and Chinoy, 1977, 1980; Rama Rao, Naithani,and Singh, 1982). However, no positive correlation was found between endogenous IAA levels and the rate of fiber elongation in different *Gossypium* spp. (Naithani, Rama Rao, and Singh, 1981; Nayyar et al., 1989a,b). Similarly, no discernible effect of increased free IAA content on fiber length was seen in transgenic cotton transformed with two auxin biosynthetic genes, *iaaM* and *iaaH* from *Agrobacterium tumefaciens*, linked to the fiber-specific E6 promoter (see John, Chapter 10, this volume).

Exogenous cytokinins and ABA inhibit fiber growth in both fertilized and unfertilized ovules in vitro (Beasley, 1973; Beasley and Ting, 1973, 1974; Zhang, 1982; Chen et al., 1988). Increased cytokinin content in the ovules of a mutant cotton was suggested to be the main factor for inhibition of fiber elongation (Chen et al., 1996). However, overexpression of the cytokinin biosynthetic gene isopentenyl transferase into cotton did not influence either elongation or other fiber traits (see John, Chapter 10, this volume).

In vitro, ABA reduced the capacity of the ovule to produce fibers in the presence of IAA or GA_3 (Beasley and Ting, 1974). Inhibition of ABA biosynthesis by addition of fluridone to the culture medium caused a marked promotion of fiber production and elongation (Nayyar et al., 1989a). Concomitant with a reduction in ABA levels, fluridone treatment also caused increased IAA levels. Chen and colleagues (1996) reported a gradual decline in the ABA level of cotton ovules during the period of fiber elongation. An inverse correlation between fiber elongation and ABA content was also recently reported by Gokani, Kumar, and Thaker (1998). A low level of endogenous polyamines seems to promote fiber elongation, as reflected in the declining levels of polyamines observed during the active period of fiber elongation (Davidonis, 1995).

Physiology of Fiber Elongation

Fiber elongation is the net result of a complex interplay between cell turgor and cell wall extensibility. The genes expressed during various stages of fiber expansion are discussed in detail by Wilkins and Jernstedt (see Chapter 9, this volume).

In conjunction with increased water uptake through aquaporin activity (Ferguson, Turley, and Kloth, 1997; Wilkins and Jernstedt,

Chapter 9, this volume), the sequestration of osmotically active solutes (osmolytes) by the expanding vacuole is an important aspect of turgor-driven wall extension. In elongating fibers, K^+ and malate are the predominant osmolytes, and their levels fluctuate in correlation with the growth rate (Dhindsa, Beasley, and Ting, 1975; Basra and Malik, 1983a). Plant K^+ channels reside in both the plasma membrane and the tonoplast (Chéryl et al., 1996), as do the two membrane-bound electrogenic proton pumps that hydrolyze ATP (adenosine triphosphate) to pump H^+ ions against a concentration gradient, thereby generating both an electrochemical and a pH gradient across the membrane (see Wilkins and Jernstedt, Chapter 9, this volume). The electrochemical potential provides the energy to activate osmolyte transport. Joshi, Stewart, and Graham (1985) demonstrated a positive relationship of ATPase activity to the process of fiber elongation. No ATP-specific enzyme activity was seen in nonelongating epidermal cells. As the fibers started elongating, the enzyme activity gradually increased in the tonoplast of enlarging vacuoles. The enzyme inhibitor N, N-dicyclohexylcarbodiimide inhibited, while KCl (potassium chloride) stimulated tonoplast ATPase activity. This provides a strong indication of active transport of osmolytes into the vacuoles.

Recent research has led to the conclusion that auxin-induced growth and proton secretion strictly depend on extracellular K^+ ions and the uptake of K^+ mediated by K^+ channels at the plasma membrane (Claussen et al., 1997). Whether a similar regulation operates in elongating cotton fibers is worthy of investigation.

Malate is derived by dark metabolism of CO_2 via PEP (phosphoenolpyrurate) carboxylase (Dhindsa, Beasley, and Ting, 1975; Basra and Malik, 1983a). Fiber growth is inhibited in vitro in the absence of CO_2 and K^+ in the culture medium (Dhindsa, Beasley, and Ting, 1975). CO_2 concentration in the cotton bolls is high enough to support high rates of CO_2 fixation. Besides malate production, dark metabolism of CO_2 is involved in linking other pathways of fiber metabolism (Basra and Malik, 1985).

ABA, which inhibits fiber elongation, also inhibits the enzyme activities associated with dark metabolism of CO_2 (Dhindsa, 1978b). Conversely, cotton ovules cultured in the presence of fluridone, an inhibitor of ABA biosynthesis, respond with a marked increase in these enzyme activities (Kaur et al., 1990). Ca^{2+} also modulates the fiber capacity for CO_2 fixation and is possibly involved in ABA action (Basra et al., 1993).

Apart from K^+ and malate, the contribution of other osmolytes, such as sugars, amino acids, and their derivatives, is also important, which needs to be quantified. There exist separate "storage" and "metabolic" pools of sugars in the fiber showing mutual exchange (Carpita and Delmer, 1981). Since the fibers are highly vacuolated, the storage pool is probably the vacuole. This is supported by the fact that only about 11 percent of the total of reducing sugars is released by treatment of the fibers with 7.5 percent dimethyl sulfoxide (Carpita and Delmer, 1981). This treatment alters permeability of the plasma membrane, while having much less effect on the vaculoar membrane.

In addition to the generation of the fiber turgor, the primary cell wall must be able to undergo plastic extension and incorporate new cell wall material. The incorporation of new wall and membrane material can occur primarily at the fiber tip (tip growth) or over its entire surface (diffuse growth) (Seagull, 1990). Recent evidence is in favor of random growth over the entire fiber surface (Tiwari and Wilkins, 1995). Sustaining rapid and prolonged growth of developing fibers requires the extensive involvement of numerous metabolic and biosynthetic pathways (Basra and Malik, 1982, 1983a,b,c; Kumar et al., 1987; Nayyar, Basra, and Malik, 1988; Nayyar et al., 1990; Basra et al., 1990; Basra, Sarlach, Nayyar, and Malik, 1992; Kaur et al., 1992a,b).

Increase or decrease in the susceptibility of walls to turgor-driven stretching can be described in general terms as wall "loosening" and "tightening," respectively. The underlying mechanisms causing these physical changes in the fiber cell walls are not well understood. Orford and Timmis (1997) isolated five fiber-specific cDNA clones by differential screening of a cDNA library. One cDNA encoded a lipid transfer protein (LTP), and a second encoded a member of a group of well-characterized proline-rich proteins (PRP) from plants. The presence of signal peptide-encoding sequences indicated that both the LTP and the PRP are targeted to the extracellular matrix of the fiber, possibly playing a role in cell elongation. Shimizu and colleagues (1997) studied changes in the levels of mRNAs for cell wall-related enzymes in growing cotton fibers by reverse transcription-PCR analysis. Both endo-1, 4-β-glucanase and expansin mRNA levels were high during cell elongation, but decreased when cell elongation ceased and xyloglucan decreased. The endo-1, 4-β glucanase may be responsible for solubilization of xyloglucan in the primary wall, which may be associated with the auxin-induced cell elongation in plants (Hayashi, 1989). Expansins appear to disrupt hydrogen bonds not only between cellulose microfibrils but also between xyloglucan and cellulose (Cosgrove, 1993; McQueen-Mason and Cosgrove, 1994; McQueen-Mason, 1997), allowing cell walls to "relax," but they require acid conditions to do so. Pertinently, functional cell wall proteins in plant stems that induce cell wall extension were found to be activated by acidic pH (McQueen-Mason, Durachko, and Cosgrove, 1992; Okamoto and Okatomoto, 1994, 1995; McQueen-Mason and Cosgrove, 1995). IAA activates the plasma membrane-bound H^+ pump and is considered to participate in wall extension by extruding H^+ into the surface cell wall (Kitamura, Mizuno, and Katou, 1998). Progress is needed in hormonal regulation of wall-loosening and wall-tightening processes.

The period between 16 to 21 dpa represents a switch in emphasis from primary to secondary cell wall synthesis during fiber development, exemplified by the transition in cell wall metabolism (Shimizu et al., 1997). A dramatic shift in the IAA/ABA ratio coincident with the decrease in the rate of fiber elongation is indicative of the signal-transducing role of the phytohormones (Nayyar et al., 1989a,b). A precise understanding of this regulation is crucial for devising strategies for fiber modification that could affect ultimate fiber properties.

During the fiber elongation phase, the ionically bound wall peroxidase activity remained low, but it increased significantly during the secondary thickening phase (Rama Rao, Naithani, and Singh, 1982). The role of wall peroxidase in cessation of elongation growth was postulated. MacAdam, Nelson, and Sharp (1992) also established that ionically bound cell wall peroxidase plays a key role in determining when cell elongation ceases in growing leaves, both temporally and spatially. Few reports in the literature have suggested the involvement of ABA in control of cell wall-associated peroxidase activity, possibly through controlling the cellular redox state (Lee and Lin, 1996; Takahama, 1994). Evidence suggests a relationship of ABA and CA^{2+} to peroxidase activity in cotton fibers (Basra, Sarlach, Dhillon-Grewal, and Malik, 1992). H_2O_2 and redox regulation of plant metabolism and gene expression is extensively documented (Foyer et al., 1997). Interestingly, H_2O_2 has been proposed as a signal molecule that is initiated exactly coincident with the onset of secondary wall formation and the termination of fiber elongation (Potikha et al., 1999). Furthermore, H_2O_2 causes an increase in cytosolic Ca^{2+}, indicating a role for calcium in the signal transduction process (Price et al., 1994).

CLOSING REMARKS

If we are to understand the regulation of the growth and functioning of cotton fibers, and ultimately perhaps to manipulate fiber properties, then we need more fundamental information on the signals and mechanisms associated with signal transduction, leading to changes in fiber gene expression.

REFERENCES

Baert, T., E. DeLanghe, and L. Waterkeyn (1975). In vitro culture of cotton ovules. III. Influence of hormones upon fiber development. *La Cellule* 71: 55-63.

Balls, W.L. (1919). The existence of daily growth rings in the cell wall of cotton hairs. *Proceedings of the Royal Society of London, B. Biological Sciences* 90: 542-555.

Basra, A.S., R. Dhillon-Grewal, R.S. Sarlach, and C.P. Malik (1993). Influence of ABA and calcium on dark fixation of carbon dioxide in cotton fibres. *Phyton* 33: 1-6.

Basra, A.S. and C.P. Malik (1982). Adenine nucleotides and energy charge during fibre elongation of two cottons differing in their lint lengths. *Proceedings of Indian National Science Academy* 46: 229-233.

Basra, A.S. and C.P. Malik (1983a). Dark metabolism of CO_2 during elongation of two cottons differing in fibre lengths. *Journal of Experimental Botany* 34: 1-9.

Basra, A.S. and C.P. Malik (1983b). Glucose oxidation of elongating fibres of two species of cotton differing in fibre lengths. *Tropical Plant Science Research* 1: 329-333.

Basra, A.S. and C.P. Malik (1983c). A comparison of lipid synthesizing capacity in elongating fibres of two cottons differing in fibre lengths. *Current Science* 52: 30-31.

Basra, A.S. and C.P. Malik (1984). Development of the cotton fiber. *International Review of Cytology* 89: 65-113.

Basra, A.S. and C.P. Malik (1985). Non-photosynthetic fixation of carbon dioxide and possible biological roles in higher plants. *Biological Reviews* 60: 357-401.

Basra, A.S., R.S. Sarlach, R. Dhillon-Grewal, and C.P. Malik (1992). Calcium-mediated changes in peroxidase and O-diphenol oxidase activities of cotton fibres (*Gossypium* spp.) and its possible relationship to ABA. *Plant Growth Regulation,* 11: 159-164.

Basra, A.S., R.S. Sarlach, H. Nayyar, and C.P. Malik (1990). Source hydrolysis in relation to development of cotton (*Gossypium* spp.) fibres. *Indian Journal of Experimental Biology* 28: 985-988.

Basra, A.S., R.S. Sarlach, H. Nayyar, and C.P. Malik (1992). Hormonal effects on partitioning of ^{14}C-sucrose in cotton fibres. *Acta Physiologiae Plantarum* 14: 137-142.

Beasley, C.A. (1973). Hormonal regulation of growth in unfertilized cotton ovules. *Science* 179: 1003-1005.

Beasley, C.A. (1979). Cellulose content in fibers of cotton which differ in their lint lengths and extent of fuzz. *Physiologia Plantarum* 45: 77-82.

Beasley, C.A. and I.P. Ting (1973). The effects of plant growth substances on in vitro development from fertilized cotton ovules. *American Journal of Botany* 60: 130-139.

Beasley, C.A. and I.P. Ting (1974). Effects of plant growth substances on in vitro development from unfertilized cotton ovules. *American Journal of Botany* 61: 188-194.

Bradow, J.M., P.J. Bauer, I. Hinojosa, and G. Sassenrath-Cole (1997). Quantitation of cotton fibre-quality variations arising from boll and plant growth environments. *European Journal of Agronomy* 6: 191-204.

Carpita, N.C. and D.P. Delmer (1981). Concentration and metabolic turnover of UDP glucose in developing cotton fibers. *Journal of Biological Chemistry* 256: 308-315.

Chen, J.G., X.M. Du, H.Y. Zhao, and X. Zhou (1996). Fluctuation in levels of endogenous plant hormones in ovules of normal and mutant cotton during flowering and their relation to fiber development. *Journal of Plant Growth Regulation* 15: 173-177.

Chen, Y.N., C.Y. Shen, Z.L. Zhang, and J.Q. Yan (1988). Study of the fiber development of cotton ovules. *Acta Biol Exp Sinica* 21: 417-421.

Chéryl, I., P. Daram, F. Gaymard, C. Horeau, B. Thibaud, and H. Sentenac (1996). Plant K[+] channels: Structure, activity, and function. *Biochemical Society Transactions* 24: 961-971.

Claussen, M., H. Luthen, M. Blatt, and M. Bottger (1997). Auxin-induced growth and its linkage to potassium channels. *Planta* 201: 227-234.

Cosgrove, D.J. (1993). How do plant cell walls extend? *Plant Physiology* 102: 1-6.

Davidonis, G. (1995). Changes in polyamine distribution during cotton fiber and seed development. *Journal of Plant Physiology* 145: 105-112.

DeLanghe, E.A.L. (1986). Lint development. In *Cotton Physiology*, Eds. J.R. Mauney and J.M. Stewart. Memphis, TN: The Cotton Foundation, pp. 325-350.

Dhindsa, R.S. (1978a). Hormonal regulation of cotton ovule and fiber growth. Effects of bromodeoxyuridine, AMO-1618, and p-chlorophenoxyisobutyric acid. *Planta* 141: 269-273.

Dhindsa, R.S. (1978b). Hormonal regulation of enzymes of non-autotrophic CO_2 fixation in unfertilized cotton ovules. *Zeitschrift für Pflanzenphysiologie* 89: 355-362.

Dhindsa, R.S., C.A. Beasley, and I.P. Ting (1975). Osmoregulation in cotton fiber: Accumulation of potassium and malate during growth. *Plant Physiology* 56: 394-398.

Ferguson, D.L., R.B. Turley, and R.H. Kloth (1997). Identification of a δ-TIP cDNA clone and determination of related A and D genome subfamilies in *Gossypium* species. *Plant Molecular Biology* 34: 111-118.

Foyer, C.H., H. Lopez-Delgado, J.F. Dat, and I.M. Scott (1997). Hydrogen peroxide- and glutathione-associated mechanisms of acclimatory stress tolerance and signalling. *Physiologia Plantarum* 100: 241-254.

Gipson, J.R. (1986). Temperature effects on growth, development and fiber properties. In *Cotton Physiology*, Eds. J.R. Mauney and J.M. Stewart. Memphis, TN: The Cotton Foundation, pp. 47-56.

Gipson, J.R. and H.E. Joham (1969). Influence of night temperature on growth and development of cotton (*Gossypium hirsutum* L.). III. Fiber elongation. *Crop Science* 9: 127-129.

Gipson, J.R. and L.L. Ray (1969). Fiber elongation rates in five varieties of cotton (*Gossypium hirsutum* L.) as influenced by night temperature. *Crop Science* 9: 339-341.

Gipson, J.R. and L.L. Ray (1970). Temperature-variety interrelationships in cotton. I. Boll and fiber development. *Cotton Growers Review* 47: 257-271.

Gokani, S.J., R. Kumar, and V.S. Thaker (1998). Potential role of abscisic acid in cotton fiber and ovule development. *Journal of Plant Growth Regulation* 17: 1-5.

Graves, D.A. and J.M. Stewart (1988). Chronology of the differentiation of cotton (*Gossypium hirsutum* L.) fiber cells. *Planta* 175: 254-258.

Haigler, C.H., N. Ramo Rao, E.M. Roberts, J.Y. Huang, D.R. Upchurch, and N. Trolinder (1991). Cultured ovules as models for cotton fiber development under low temperatures. *Plant Physiology* 95: 88-96.

Hawkins, R.S. (1930). Development of cotton fibers in Pima and Acala varieties. *Journal of Agricultural Research* 40: 1017-1029.

Hayashi, T. (1989). Xyloglucans in the primary cell wall. *Annual Reviews of Plant Physiology and Plant Molecular Biology* 40: 139-168.

Jasdanwala, R.T., Y.D. Singh, and J.J. Chinoy (1977). Auxin metabolism in developing cotton hairs. *Journal of Experimental Botany* 28: 1111-1116.

Jasdanwala, R.T., Y.D. Singh, and J.J. Chinoy (1980). Changes in components related to auxin turnover during cotton fibre development. *Beitrage Biologie Planzenphysiologie* 55: 23-36.

Joshi, P.C., J.M. Stewart, and E.T. Graham (1985). Localization of β-glycerophosphatase activity in cotton fiber differentiation. *Protoplasma* 143: 1-10.

Kaur, K., H. Nayyar, A.S. Basra, and C.P. Malik (1990). Stimulation of enzymes of non-photosynthetic C_4 metabolism in cultured cotton ovules by fluridone. *Acta Physiologiae Plantarum* 12: 3-6.

Kaur, K., H. Nayyar, A.S. Basra, and C.P. Malik (1992a). Cytochemical activity of dehydrogenases during early ovule and fibre development in cotton. *Journal of Plant Science Research* 8: 17-20.

Kaur, K., H. Nayyar, A.S. Basra, and C.P. Malik (1992b). Activity profiles of some transaminases during early ovule and fibre development in cotton. *Journal of Plant Science Research* 8: 56-58.

Kerr, T. (1936). The structure of the growth rings in the secondary wall of the cotton hair. *Protoplasma* 27: 229-243.

Kitamura, S., A. Mizuno, and K. Katou (1998). IAA-dependent adjustment of the in vivo wall-yielding properties of hypocotyl segments of *Vigna unguiculata* during adaptive growth recovery from osmotic stress. *Plant and Cell Physiology* 39: 627-631.

Kumar, H., K. Kaur, A.S. Basra, and C.P. Malik (1987). The relationship of hydrolases to fibre development in *Gossypium hirsutum* L. *Proceedings of Indian National Science Academy* 53: 259-261.

Lee, M. and Y.H. Lin (1996). Peroxidase activity in ethylene-treated, ABA-treated or Meja-treated rice (*Oryza sativa*) roots. *Botanical Bulletin of Academia Sinica* 37: 201-207.

MacAdam, J.W., C.J. Nelson, and R.E. Sharp (1992). Peroxidase activity in the leaf elongation zone of tall fescue. I. Spatial distribution of ionically bound peroxidase activity in genotypes differing in length of the elongation zone. *Plant Physiology* 99: 872-878.

Malik, C.P., B. Singh, P. Sharma, and A.S. Basra (1995). Fibre initiation and early elongation in vitro of cotton ovules as influenced by low temperature: Scanning electron microscopic comparison and biochemical changes. *Physiology and Molecular Biology of Plants* 1: 129-134.

McQueen-Mason, S. (1997). Plant cell walls and the control of growth. *Biochemical Society Transactions* 25: 204-214.

McQueen-Mason, S.J. and D.J. Cosgrove (1994). Disruption of hydrogen bonding between plant cell wall polymers by proteins that induce wall extension. *Proceedings of the National Academy of Sciences USA* 91: 6574-6578.

McQueen-Mason, S.J. and D.J. Cosgrove (1995). Expansin mode of action on cell walls: Analysis of wall hydrolysis, stress relaxation and bending. *Plant Physiology* 107: 87-100.

McQueen-Mason, S.J., D.M. Durachko, and D.J. Cosgrove (1992). Two exogenous proteins that induce cell wall extension in plants. *Plant Cell* 4: 1425-1433.

Meinert, M.C. and D.P. Delmer (1977). Changes in biochemical composition of the cell wall of the cotton fiber during development. *Plant Physiology* 59: 1088-1097.

Naithani, S.C., N. Rama Rao, and Y.D. Singh (1981). Physiological and biochemical changes associated with cotton fibre development. I. Growth kinetics and auxin content. *Physiologia Plantarum* 54: 225-229.

Nayyar, H., A.S. Basra, and C.P. Malik (1988). $H^{14}CO_3^-$ fixation into lipids of elongating cotton fibres in two *Gossypium* species. *Journal of Nuclear Agriculture and Biology* 17: 243-245.

Nayyar, H., K. Kaur, A.S. Basra, and C.P. Malik (1989a). Hormonal regulation of cotton fibre elongation in *Gossypium arboreum* L. in vitro and in vivo. *Biochemie und Physiologie der Pflanzen* 185: 415-421.

Nayyar, H., K. Kaur, A.S. Basra, and C.P. Malik (1989b). Regulation of differential fibre development in cotton by endogenous plant growth regulators. *Proceedings of Indian National Science Academy* 55: 463-468.

Nayyar, H., K. Kaur, A.S. Basra, and C.P. Malik (1990). Activity profiles of pyruvate orthophosphate dikinase and phosphoenolpyruvate carboxylase in developing cotton fibres. *Beitrage Biologie Pflanzenphyiologie* 65: 319-324.

Okamoto, A. and H. Okamoto (1995). Two proteins regulate the cell wall extensibility and the yield threshold in glycerinated hollow cylinders of cowpea hypocotyl. *Plant and Cell Physiology* 18: 827-830.

Okamoto, H. and A. Okamoto (1994). The pH-dependent yield threshold of the cell wall in a glycerinated hollow cylinder (in vitro system) of cowpea hypocotyl. *Plant and Cell Physiology* 17: 979-983.

Orford, S.J. and J.N. Timmis (1997). Abundant mRNAs specific to the developing cotton fibre. *Theoretical and Applied Genetics* 94: 909-918.

Potikha, T., C. Collins, D.I. Johnson, D.P. Delmer, and A. Levine (1999). The involvement of hydrogen peroxide in the differentiation of secondary walls in cotton fibers. *Plant Physiology* 119: 849-858.

Price, A.H., A. Taylor, S.J. Ripley, A. Griffiths, A.J. Trewavas, and M.R. Knight (1994). Oxidative signals in tobacco increase cytosolic calcium. *Plant Cell* 6: 1301-1310.

Rama Rao, N., S.C. Naithani, and Y..D. Singh (1982). Physiological and biochemical changes associated with cotton fibre development II. Auxin oxidising system. *Physiologia Plantarum* 55: 204-208.

Roberts, E.M., R.R. Nunna, J.Y. Huang, N.L. Trolinder, and C.H. Haigler (1992). Effects of cycling temperatures on fiber metabolism in cultured cotton ovules. *Plant Physiology* 10: 979-986.

Ryser, U. (1985). Cell wall biosynthesis in differentiating cotton fibres. *European Journal of Cell Biology* 39: 236-256.

Schubert, A.M., C.R. Benedict, J.D. Berlin, and R.J. Kohel (1973). Cotton fiber development—Kinetics of cell elongation and secondary wall thickening. *Crop Science* 13: 704-709.

Seagull, R.W. (1990). Tip growth and transition to secondary wall synthesis in developing cotton hairs. In *Tip Growth in Plant and Fungal Cells*, Ed. I.S. Heath. San Diego, CA: Academic Press, pp. 261-284.

Shen, T.Y., S.C. Chang, and C.C. Yhn (1978). The growth of fibers on excised cotton ovules and the formation of seedlings. *Acta Phytophysiol Sinica* 4: 183-187.

Shimizu, Y., S. Aotsuka, O. Hasegawa, T. Kawada, T. Sakuno, R. Sakai, and T. Hayashi (1997). Changes in levels of mRNAs for cell wall-related enzymes in growing cotton fiber cells. *Plant and Cell Physiology* 38: 375-378.

Takahama, U. (1994). Changes induced by abscisic acid and light in the redox state of ascorbate in the apoplast of epicotyls of *Vigna argularis. Plant and Cell Physiology* 35: 975-978.

Thaker, V.S., S. Saroop, P.P. Vaishnav, and Y.D. Singh (1989). Genotypic variations and influence of diurnal temperature on cotton fibre development. *Field Crops Research* 22: 1-13.

Tiwari, S.C. and T.A. Wilkins (1995). Cotton (*Gossypium hirsutum* L.) seed trichomes expand via a diffuse growing mechanism. *Canadian Journal of Botany* 73: 746-757.

Van't Hof, J. and S. Saha (1997). Cotton fibers can undergo cell division. *American Journal of Botany* 84: 1231-1235.

Wang, S., Z.Y. Shen, and Z.L. Zhang (1985). A study of elongation of the cotton fiber cell. *Acta Phytophysiol Sinica* 11: 409-417.

Xie, W., N.L. Trolinder, and C.H. Haigler (1993). Cool temperature effects on cotton fiber initiation and elongation clarified using an in vitro culture. *Crop Science* 33: 1258-1264.

Zhang, S.Z. (1982). Advances in cotton fiber cell development. *Plant Physiology Communications* 6: 1-7.

Zheng, Z.R. and D.W. Xu (1982). The role of plant hormones in the reproductive growth of cotton plant. *Sci. Agric. Sinica* 5: 40-47.

Chapter 3

Cotton Fibers in Vitro

Gayle H. Davidonis

INTRODUCTION

Cotton ovule cultures and ovule-derived cotton suspension cultures can be used to address questions about fiber growth and development that cannot easily be answered using ovules in situ. Researchers have proposed that fiber grown in culture could serve as a valid model for plant-grown fiber. Cotton ovule culture has been employed in many aspects of fiber development, including metabolic studies of cell wall glucans, fiber ultrastructure, and cytoskeletal components, discussed elsewhere in this volume (see Chapters 1, 2, 4, 5, and 9). The purpose of this chapter is to present evidence that fibers developing in vitro are valid models for cotton fiber development (ovule culture) and fiber cell elongation (suspension culture). Ovule culture and suspension culture procedures involve controlled conditions and changes in media constituents, or culture temperatures that can perturb growth and development are easily accomplished. Thus, fibers grown in culture may provide clues to unraveling the processes of cell elongation and secondary wall deposition.

Ovule Culture As a Model for Fiber Development

The pattern of fiber initiation on ovules cultured 1 day preanthesis is the same as in plant-grown ovules (Lang, 1938; Stewart, 1975; Xie, Trolinder, and Haigler, 1993). Fiber initiation begins at the funicular crest and proceeds toward the micropyle. The rate of fiber elongation per day, along with the duration of the elongation period, is altered in ovule culture. Elongation declines sharply 22 days post-anthesis (dpa) in plant-grown fiber and at 18 dpa in culture-grown fiber (Meinert and Delmer, 1977). A compressed schedule of fiber growth and development in culture-grown fibers was manifest in the transition from primary to secondary wall synthesis, which occurred earlier and extended over a longer period of time in culture-grown fibers (Meinert and Delmer, 1977). Plant-grown fibers exposed to cool night temperatures had growth rings in their secondary walls visible after swelling the fibers with cuprammonium hydroxide. Culture-grown fibers exhibited a growth ring for each time the temperature was cycled from 34°C (celsius) to 22 or 15°C (Haigler et al., 1991). In green fiber cotton, suberin is deposited in concentric cell wall layers alternately with cellulose (Ryser, Meier, and Holloway, 1983). The monomer composition of suberin in culture-grown green fibers was found to be qualitatively similar to the monomer composition in plant-grown fibers (Schmutz, Buchala, and Ryser, 1996). Neutral sugar compositions of plant-grown fiber cell walls and culture-grown cell walls were similar, although the glucose component in plant-grown fiber exceeded the glucose component in culture-grown fiber (Meinert and Delmer, 1977). A comparison of cell wall polymer molecular weight distributions analyzed by gel permeation chromatography revealed that fibers grown in culture for 21 days were similar to 30 dpa plant-grown fibers (Triplett and Timpa, 1995). Some additional comparisons are presented in Table 3.1.

The final fiber length and degree of secondary wall deposition depend on the time of ovule excision. When ovules are excised from bolls around the day of anthesis (1 day before to 2 days after), final fiber length is about one-half the length of plant-grown fiber. Cotton ovules excised 10 or 15 dpa produce fibers that are around two-thirds the length of plant-grown fiber (DeLanghe and Eid, 1971; Baert, DeLanghe, and Waterkeyn, 1975; Waterkeyn, DeLanghe, and Eid, 1975). Baert, DeLanghe, and Waterkeyn (1975) concluded that the presence of the embryo-endosperm complex was linked with the presence of auxin needed for fiber growth. Transferring ovules to fresh medium containing gibberellic acid (GA) and naphthalene acetic acid (NAA) every 7 days yielded fibers of the same length as plant-grown fibers (Kosmidou-Dimitropoulou, 1976). The length of fibers at the time of culture initiation influenced secondary wall depo-

TABLE 3.1. Comparison of Plant-Grown and Culture-Grown Fiber

Attribute	Culture-Grown Fiber Response	References
Fiber perimeter in micropylar region larger than in chalazal region	same	G. Davidonis, unpublished
Fiber length	reduced	Kosmidou-Dimitropoulou, 1976; Meinert and Delmer 1977; and others
Cell wall thickening	reduced	Kosmidou-Dimitropoulou, 1976; Davidonis and Hinojosa, 1994
Fiber bundle tensile strength similar for fibers grown under different temperature-cycling schedules	same	Haigler et al., 1991
Fiber bundle tensile strength	reduced	R. B. Turley (personal communication), 1997

sition. The longer the fibers at the time of ovule excision, the thicker the cell wall (weight per mm [millimeter]) after 60 days in culture (Kosmidou-Dimitropoulou, 1976).

OVULE CULTURE METHODS

Field-grown or greenhouse-grown plants can serve as ovule donors. Ovules can be collected preanthesis or postanthesis. Unfertilized ovules are obtained by collecting ovules around the time of flower opening or by removing the stamens and stigma on the morning of anthesis and allowing the ovules to remain on the plant for 2 days prior to ovule collection (Beasley, 1973; Beasley and Ting, 1974a). Fertilization is usually completed 48 h (hours) postanthesis, and ovules can be harvested at this time (Beasley, 1971; Beasley and Ting, 1973). Because elongating fibers on adjacent ovules adhere to one another and primary walls can easily be ruptured, ovules older than 2 dpa cannot be separated without damage to the ovules. Entire locules can be cultured as a unit. Currently, most ovule culture media used are based on the medium developed by Beasley and Ting (1973). Ovules are cultured in the dark at 32°C to 34°C. Detailed discussions of plant material, boll surface sterilization, ovule excision techniques, and culture conditions are found in three reviews (Beasley, Ting, et al., 1974; Beasley, 1977b, 1984). Additional modification of culture conditions has included culturing ovules in a 5 percent CO_2 atmosphere (Marden and Stewart, 1984).

Ideal growing conditions for donor plants have been emphasized as a way to reduce the amount of variability in ovule cultures (Beasley and Ting, 1973; Beasley, 1977b). Other methods have been used to reduce ovule variability, including removal of branches after bolls designated for ovule culture had been harvested (Beasley, 1977b), limiting the number of bolls on the plant (Meinert and Delmer, 1977), and placing ovules from different bolls in the same culture vessel (Graves and Stewart, 1988).

MEDIA CONSTITUENTS

Extensive fiber development was achieved when cotton ovules harvested 2 dpa were floated on a high-salt liquid medium (Beasley, 1971). The successful culture of 5 and 10 dpa ovules was reported around the same time (Eid, DeLanghe, and Waterkeyn, 1973). Further refinements made it possible to culture preanthesis ovules (Beasley and Ting, 1974b; Graves and Stewart, 1988). The difficulty of discerning the contribution of media constituents to growth is compounded by interactions with hormones. The presence of thiamine promoted GA-induced fiber production and ovule development more than indole-3-acetic acid (IAA) (Beasley, 1977b). Boron was required for the maintenance of fiber growth during a 14-day culture period (Birnbaum, Beasley, and Dugger, 1974). Boron deficiency in the presence of IAA permitted some aerial fiber growth, but no fiber development, on the submerged portion of the ovule (Birnbaum, Beasley, and Dugger, 1974). The percentage of ovules with fiber and the amount of fiber per ovule increased with temperature (28°C to 34°C) (Beasley, 1977a). The addition of NH_4NO_3 (ammonium nitrate) to the culture medium increased the percentage of ovules producing fiber at nonpermissive temperatures, but not fiber production per ovule (Beasley, 1977a).

Complex interactions between genotype and media constituents were apparent when fiber production on *Gossypium hirsutum* ovules was compared to that on *G. barbadense* and *G. arboreum*. Fiber production on *G. barbadense* ovules was the same as *G. hirsutum* only when cultured 4 dpa rather than 1 dpa and when in a medium in which the GA concentration was elevated (Peeters et al., 1985). Ovules of *G. hirsutum* and *G. arboreum* (2 dpa) produced fibers in the absence of exogenous hormones, whereas ovules of *G. barbadense* (2 dpa) produced fibers only if GA was added to the culture medium (Beasley and Egli, 1976).

METHODS OF FIBER MEASUREMENT

A stain-destain method for the estimation of fiber production on culture-grown ovules was developed by Beasley, Birnbaum, and colleagues (1974). Briefly, Toluldine blue 0 was used to selectively stain fibers subsequently destained in an acid-alcohol solution. One total fiber unit was equal to one absorbance unit of stain absorbed to fibers on a set of twenty ovules. The staining procedure can be adjusted to accommodate different numbers of ovules per set. Fiber production was expressed in terms of total fiber units (TFU). These arbitrary units have been correlated with fiber length (through 14 dpa), but not with fiber number or with the extent of lint versus fuzz fibers (see Figure 3.1). It was observed that from about 6 dpa to 14 dpa, the rate of increase in TFU was more rapid than the rate of elongation (Beasley, Birnbaum, et al., 1974). The fiber-staining procedure has been modified by Graves and Stewart (1988). When attached to the ovule, fibers can be straightened in a stream of water and measured (Gipson and Joham, 1969). A variation on this method, for very short fibers, involves cutting an ovule in half and pulling the halves apart to separate fibers (Davidonis, 1993b). Individual fibers can be removed from the ovule and placed on a velvet cloth to straighten them prior to measurement. Relative measurements of wall thicknesses can be done with never-dried fibers using a regular microscope, phase contrast or

FIGURE 3.1. Average Total Fiber Units and Fiber Length Obtained at 2-Day Intervals from Day of Anthesis to 14 Days Postanthesis, from Ovules Produced on Greenhouse Plants

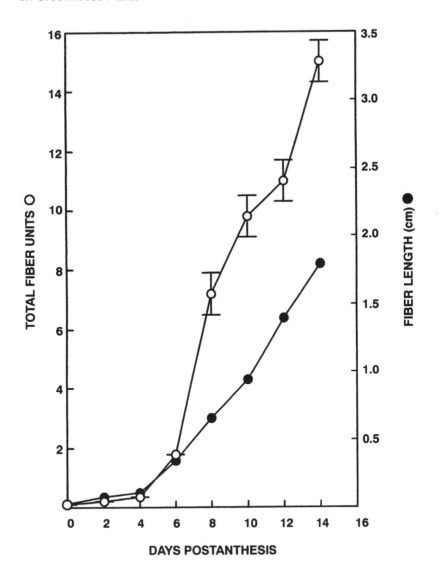

Source: Reproduced with permission from Beasley, Ting, et al. (1974, p. 89).

one equipped with polarization optics (Baert, DeLanghe, and Water-keyn, 1975; Kosmidou-Dimitropoulou, 1976; Davidonis, 1993b); also, fibers can be mounted and sectioned (Haigler et al., 1991). Small fiber samples (150 mg [milligrams]) can be analyzed using the Advanced Fiber Information System (AFIS) or X-ray fluorescence spectroscopy (Chu and Shofner, 1992; Wartelle et al., 1995).

HORMONAL CONTROL OF GROWTH

Fiber Initiation and Elongation

Cotton ovule cultures lend themselves to probing the events associated with fiber differentiation and its manifestation as the initiation of fiber growth. Ovules at 1 and 2 days preanthesis and grown in media containing IAA and GA produced fibers (Beasley and Ting, 1974b; Marden and Stewart, 1984; Gould and Dugger, 1986; Graves and Stewart, 1988). Epidermal cells differentiate into fiber cells around 3 days preanthesis. Ovules that were placed in culture 6 days preanthesis initiated fiber development 4 days later (Graves and Stewart, 1988). The time period during which fiber growth could be initiated was determined to be anytime between 2 days preanthesis and anthesis (Graves and Stewart, 1988). Although fiber cells have the ability to initiate growth 2 days preanthesis, they do not begin growth until the day of anthesis. Therefore, ovules grown in culture maintain an internal clock that signals when anthesis would occur in the donor plant (Graves and Stewart, 1988). Fiber initiation starts in the region of the funicular crest, then proceeds around the lateral circumference of the ovule toward the chalazal cap, and finishes in the micropylar region around 3 to 4 dpa (Lang, 1938; Stewart, 1975). If the fiber elongation rate and duration are the same along the entire length of the ovule, then the delay in fiber initiation in the micropylar region should be seen in fiber length differences. In plant-grown fibers, mean length was shortest in the region within a radius of 2 mm from the micropyle, whereas fibers in the middle of the seed were the longest, although upper-quartile mean lengths were similar in most cultivars for middle and chalazal regions (Vincke et al., 1985). In culture-grown fiber, mean lengths in the middle, chalazal, and micropylar regions were significantly different (Peeters et al., 1985). The shortest fibers were found in the micropylar regions of cultured ovules. In other studies, differences in fiber lengths on seeds were related to locule locations (Iyengar, 1939, 1941, 1947; Krishnan and Iyengar, 1960; Davidonis and Hinojosa, 1994).

Total fiber production was enhanced with GA when fertilized ovules (2 dpa) were grown in culture, whereas unfertilized ovules required IAA and/or GA for fiber production (Beasley and Ting, 1973, 1974a). Short-term exposure of day-of-anthesis ovules to medium with GA (24 h), followed by transfer to medium containing IAA, produced the same amount of fiber (TFU) as ovules grown in both hormones during the 2-week culture period (Beasley, Birnbaum, et al., 1974). Increasing the GA concentration from 0.5 to 5.0 µM (micromolar) during a 24-hour and 48-hour preculture period prior to culture in IAA confirmed that ovules must be exposed to IAA within the first 24 hours of culture for maximum fiber production (Beasley, 1977b). To further unravel events around fiber initiation and the start of elongation, ovules were collected 2 dpa and examined after 3 days in culture. Mean lengths for fibers on ovules grown continuously in GA were longer than for fibers grown for 16 h in GA and then transferred to a medium containing IAA for an additional 56 h (Davidonis, 1993a). Gibberellin induced the initiation of a normal number of fiber initials, but did not promote elongation to the same extent as a medium containing both GA and auxin when ovules were placed in culture on the day of anthesis (Kosmidou-Dimitropoulou, 1976). It has been concluded that fiber elongation depends more on the availability of auxin than GA (Birnbaum, Beasley, and Dugger, 1974; Dhindsa, 1978).

Fluridone, an inhibitor of abscisic acid biosynthesis, was shown to alter endogenous levels of auxins, gibberellins, and abscisic acid in culture (Nayyer et al., 1989). Fiber production (TFU) on ovules grown in a medium containing fluridone was greater than on control ovules (Kaur et al., 1990). Abscisic acid inhibited fiber production only when applied during the first 4 days in culture (Dhindsa, Beasley, and Ting, 1976). Kinetin also inhibited fiber production (TFU) (Beasley and Ting, 1973).

Ethylene caused excessive callus formation on cultured ovules grown in the presence of GA and decreased the percentage of ovules forming fibers in response to IAA (Beasley and Eaks, 1979). Fiber growth did not occur on ovules treated with ethylene supplied as 2-chloroethylphosphonic acid (CEPA) and examined after 14 days in culture (Hsu and Stewart, 1976). Normal differentiation of the ovule integument was disrupted by combinations of GA and CEPA (Stewart and Hsu, 1977). Shorter culture periods (3 days) were used to minimize callus formation (Davidonis, 1993a). Fiber elongation was inhibited on ovules pretreated with CEPA and returned to medium containing GA and on ovules pretreated with GA and subsequently grown on CEPA-containing medium (Davidonis, 1993a).

Secondary Wall Deposition

Secondary wall synthesis begins earlier in culture-grown fibers than in plant-grown fibers, but the percent cellulose in the cell wall after 24 days is similar (Meinert and Delmer, 1977). Fibers located in the chalazal region of cultured ovules were examined after 60 days in culture (Kosmidou-Dimitropoulou, 1976). When ovules were placed in culture 5 dpa, the addition of GA or NAA into the medium did not change mean wall width or dry weight of fiber per mm length. Ovules older than 5 dpa placed in culture showed greater mean wall widths and dry weights of fiber per mm length when NAA was added to the medium (Kosmidou-Dimitropoulou, 1976).

FIBER AND FUZZ MUTANTS

The fiber development characteristics of fiber and fuzz mutants are manifest in ovule culture (Butenko and Azizkhodzhaev, 1987; Triplett, Busch, and Goynes, 1989; Triplett, 1995). Ovule culture of the Ligon-lintless mutant has led to the conclusion that the short fibers on that mutant are analogous to normal lint fibers (Triplett, Busch, and Goynes, 1989). It was suggested that the initiation of fiber growth by 3 dpa indicated that some of the fibers on this mutant are very short lint fibers. The Ligon-lintless mutant has been

introduced into the genetic marker line Texas Marker-1 (Kohel,
Quisenberry, and Benedict, 1974). TFU was 2.2 times greater in
Texas Marker-1 than in Ligon-lintless after 15 days in culture (Triplett, Busch, and Goynes, 1989). In the naked seed mutant, fiber
development is restricted to the chalazal region of the seed, and
there are no fuzz fibers. Ovule culture of the naked seed mutant in
thirty combinations of IAA and GA revealed that no hormone combination permitted fuzz development, although some hormone combinations doubled TFU (Triplett, 1995).

DONOR PLANT STATUS

The responses of ovules in culture are related to the physiological
status of the donor plants. Fiber development occurred when ovules
of SMA-4, a single-gene mutant from *G. arboreum* L., were cultured at 30°C, but was inhibited in the same medium at 32°C
(Beasley and Egli, 1977). The occasional production of limited
fiber on SMA-4 ovules cultured at 32°C was presented as evidence
that environmental conditions imposed on donor plants altered the
responses of ovules grown in culture. The appearance of ovules
after a culture period of 2 weeks can be classified into these categories: enlarged, white with fibers; enlarged, with no fibers (W-F); or
brown and shriveled (B-S) (Beasley, 1977b). The percent of unfertilized ovules grown in a medium without hormones that were W-F
varied not only from locule to locule but with the time of year the
ovules were excised, leading Beasley (1977b) to suggest that the
responses of the ovules were indicative of endogenous shifts in
hormone balances altered by environmental conditions.

Motes are developmentally arrested seeds, and the percentage of
small motes with short fiber has been used as an indicator of fertilization efficiency (Pearson, 1949). Increases in temperature on the
day of anthesis have been related to increased small-mote formation
(Pearson, 1949). When ovules were cultured in a medium containing 5.7 µM GA and 5 µM IAA, all ovules reached lengths of 6 mm
or larger. A reduction in GA concentration (0.6 to 1.2 µM) and
elimination of IAA from the culture medium revealed a wider distribution of ovule lengths. Variability in ovule size might be related

to the physiological status of the donor plant at the time of anthesis. To test this hypothesis, first-position flowers were tagged on the day of anthesis. Thirty hours later, some of the tagged bolls were taken for ovule culture while others remained on the plants. Ovules were cultured in a reduced GA medium for 6 weeks. Control plant-grown bolls were harvested after boll dehiscence, small short-fiber motes were counted, and bolls were categorized according to mote numbers. Ovules in culture were compared to plant-grown bolls from the same flowering dates (see Table 3.2). The larger the average mote number per boll, the larger the number of small ovules. Fiber production (mg dry wt [milligrams dry weight]) was greater in cultures that had more large ovules.

Variability in boll size and fiber properties is related to assimilate partitioning. Both fiber length and secondary wall deposition decrease in first-position bolls as the season progresses (Kerby and Ruppenicker, 1989). Bolls located closest to the main stem are larger than those at second or third positions on a fruiting branch (Kittock, Pinkas, and Fry, 1979). Differences in boll size and fiber properties may be due entirely to postanthesis assimilate partitioning or to a combination of assimilate partitioning and an intrinsic boll location factor. To test this hypothesis, ovules were grown on a reduced GA medium that might perpetuate intrinsic differences between first- and third-position ovules. Fiber dry weight per culture was not significantly different for first- and third-position bolls. Fiber properties of plant- and culture-grown fibers from first- and

TABLE 3.2. Effect of Donor Plant Status on Ovule Size and Fiber Production in Vitro (Greenhouse-Grown *Gossypium hirsutum* var. Deltapine 50)

Flowering Period	Average Small Mote/Boll Plant-Grown	Average No Ovules < 5 mm Per Culture	Average mg Fiber Per Culture
autumn 1994	5	1	135 ± 21
autumn 1994	11	5	58 ± 39
autumn 1995	7	3	100 ± 18
autumn 1995	10	5	39 ± 16

third-position bolls were compared (see Table 3.3). Theta for a perfect circle is 1.00; therefore, the closer the value is to 1.00, the more the fiber cross section approaches a circle. The higher the theta value, the thicker the cell wall. Micronafis is similar to micronaire (Davidonis and Hinojosa, 1994). Fibers from first-position bolls showed a trend toward thicker walls. Differences in wall thickness did not exist in culture-grown fiber from ovules excised from first- and third-position bólls.

In addition to differences in assimilate partitioning related to boll location, a second level of partitioning occurs among seeds within a boll (Porter, 1936). Differences in seed size and fiber properties may be due to postanthesis assimilate partitioning or to an intrinsic boll seed location factor. To test this hypothesis, ovules were grown on a medium containing 5.7 µM GA and 5 µM IAA. Ovules from the apex, middle, and pedicel regions of bolls were cultured separately (Davidonis and Hinojosa, 1994). In plant-grown material, seeds located in the middle of a locule had the longest fibers. Fiber length distributions were similar for ovules taken from all locations in a locule. Secondary wall deposition was the greatest in fibers located on plant-grown seeds or culture-grown ovules closest to the pedicel. Therefore, fiber length appears to be most affected by assimilate partitioning, whereas secondary wall deposition is influenced by assimilate partitioning and an intrinsic seed location factor.

TABLE 3.3. Cotton Fiber Properties Measured by AFIS from First- and Third-Position Bolls (Greenhouse-Grown *Gossypium hirsutum* var. Deltapine 50)

Fiber Property	Plant-Grown Fiber		Culture-Grown Fiber	
	First-Position Bolls	Third-Position Bolls	First-Position Bolls	Third-Position Bolls
length (mm)	24.6 ± 1.9	24.1 ± 0.7	15.0 ± 0.7	15.2 ± 0.5
theta, degree of circularity	0.633 ± 0.010	0.593 ± 0.033	0.334 ± 0.015	0.335 ± 0.013
micronafis	6.42 ± 0.16	5.77 ± 0.51	1.98 ± 0.20	2.00 ± 0.16

PERTURBATIONS IN THE TIME LINE
OF FIBER DEVELOPMENT

During seed and fiber development, events can occur that alter or terminate seed development. Perturbations in fiber development and growth are eventually translated into low-weight seeds and atypical fiber. The early termination of embryo growth produces motes with short fiber; later termination of growth produces motes with long fiber. The developmental time frame can be perturbed by the addition of tunicamycin, an inhibitor of lipid-linked glycosyl transfer, and monensin, an inhibitor of golgi function (Elbein, 1987; Morré et al., 1983). Inhibitor treatments were limited to a single 24-hour treatment period, followed by removal of media containing the inhibitor and replacement with fresh media. Tunicamycin treatments affected both fiber elongation and secondary wall formation when given prior to the period of rapid fiber elongation. Monesin treatments perturbed fiber elongation and secondary wall formation during the time of rapid fiber elongation (Davidonis, 1993b). At 6 to 10 dpa, fiber expresses proteins putatively involved in secondary wall synthesis (Graves and Stewart, 1988). Therefore, it is not unreasonable to suggest that a perturbation of fiber growth during this period could alter secondary wall deposition. The time-line can be delayed by changes in temperature. When ovules were cultured on the day of anthesis or 1 day preanthesis, fiber initiation and early elongation (up to 0.5 mm long) were independently delayed by cycling cool temperatures (34°C/15°C) (Xie, Trolinder, and Haigler, 1993). Later elongation was temperature independent. After 35 days in culture, fiber lengths in cycled ovules (34°C/13°C) and noncycled ovules (34°C) were the same. The onset of secondary wall synthesis occurred earlier in culture-grown fibers grown under constant temperatures than if they were exposed to diurnal cycles of cool temperatures (Haigler et al., 1991).

COTTON CELL SUSPENSIONS AS MODELS
OF FIBER DEVELOPMENT

Cotton cells in suspension cultures derived from ovule callus display characteristics similar to cotton fibers (Trolinder, Berlin, and Goodin, 1987; Triplett, Busch, and Goynes, 1989; Davidonis, 1989,

1990). Cultures were initiated by placing preanthesis ovules on agar-solidified medium, allowing callus to form, and then transferring the callus to a liquid medium (Davidonis, 1989) or by directly placing ovule fragments in liquid media (Trolinder, Berlin, and Goodin, 1987; Triplett, Busch, and Goynes, 1989). Cell suspension culture media are based on the Beasley and Ting medium used for ovule culture. Cell length distributions were prepared for cultures taken from preanthesis and postanthesis ovules (Trolinder, Berlin, and Goodin, 1987). The longest fibers were in cultures derived from ovules 4 days preanthesis to day of anthesis. Cells reached a maximum length of 2.5 to 3.5 mm. Examination of cells by polarization microscopy revealed that birefringence was present as early as the fifth day of culture, indicating secondary wall synthesis. Cell wall thicknesses of 3 μm were observed, but the majority of cell wall thickness values were between 0.3 to 0.9 μm (Trolinder, Berlin and Goodin, 1987).

Cell length distributions for cultures initiated from Texas Marker-1 (TM-1) and Ligon-lintless (Li) showed that 93 percent of the Li cells were shorter than 0.5 mm, while 57 percent of the TM-1 cells were less than 0.5 mm (Triplett, Busch, and Goynes, 1989). Thus, fiber length differences related to genotype are manifest in plant-grown fiber, culture-grown fiber, and cotton cell suspensions. Ovules grown on agar formed fiber and callus. Callus tissue was subcultured for over 25 weeks, and the degree of cell elongation in suspension cultures derived from the callus was monitored (Davidonis, 1989). Some cotton cells in culture elongated without an increase in cell diameter, while in others, cell expansion was more pronounced in the center of the cell (see Figure 3.2). An examination of cells after 28 days in culture by polarized light revealed birefringent secondary walls (see Figure 3.2). An examination of the frequency distributions of cell lengths demonstrated that the number of cells longer than 0.4 mm decreased when the subculture number increased. Cell lengths in suspensions derived from hypocotyl tissue never exceeded lengths of 0.4 mm. Trolinder, Berlin, and Goodin (1987) found that the maximum cell length was observed when NAA and kinetin were added to the suspension culture medium. When ovule callus from auxin-dependent or -independent cell lines was placed in suspension culture medium, cell elongation was greater in auxin-dependent cultures (Davidonis, 1990). The percentages of

FIGURE 3.2. Cotton (*Gossypium hirsutum* L. Texas Marker-1) Suspension Cells Derived from Ovule Callus Tissue

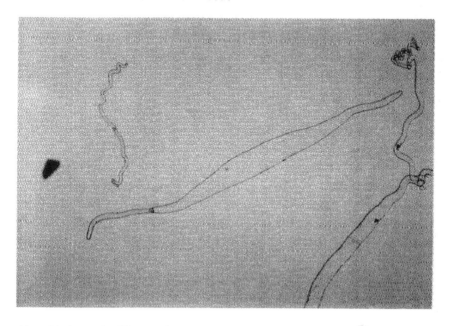

Note: Horizontal cell is 2 mm long. Cells were cultured for 28 days and examined for birefringence.

cells longer than 0.5 mm were the same for cultures in which GA was present throughout the culture period or only during the first 3 days (Davidonis, 1990). Cell elongation was inhibited in the presence of CEPA while lateral expansion was promoted (Davidonis, 1993a). Suspension cultures contained a variety of cell shapes, and the most atypical cells were elliptical or ovate and generally had a diameter greater than 100 μm. In CEPA cultures, 22 percent of the cells had atypical shapes, while only 4 percent of the control cells had atypical shapes.

CONCLUSIONS

Basic features of fiber development have been elucidated through the use of ovule culture in a way that greenhouse or field experiments could not address. Since fiber quality begins at flowering, the physiological status of the donor plant becomes an important component in predicting fiber development. Ovule cultures have made it possible to distinguish between the time that fiber cells have the ability to initiate growth and the actual time that growth begins. Fiber development has been fine-tuned to the extent that many temperature-sensitive processes in fiber development have been identified. Complex interactions among hormones, genotype, and fiber development have been investigated, and a more thorough understanding is necessary. Ovule cultures may also reveal clues about assimilate partitioning.

REFERENCES

Baert, T., E. DeLanghe, and L. Waterkeyn (1975). In vitro culture of cotton ovules III. Influence of hormones upon fiber development. *La Cellule* 71: 55-63.

Beasley, C.A. (1971). In vitro culture of fertilized cotton ovules. *BioScience* 21: 906-907.

Beasley, C.A. (1973). Hormonal regulation of growth in unfertilized cotton ovules. *Science* 179: 10003-10005.

Beasley, C.A. (1977a). Temperature dependent response to indoleacetic acid is altered by NH_4^+ in cultured ovules. *Plant Physiology* 59: 203-206.

Beasley, C.A. (1977b). Ovule culture: Fundamental and pragmatic research for the cotton industry. In *Plant Cell, Tissue and Organ Culture*, Eds. J.R. Reinert and J.P.S. Bajaj. Berlin, Heidelberg, New York: Springer-Verlag, pp. 160-178.

Beasley, C.A. (1984). Culture of cotton ovules. In *Cell Culture and Somatic Cell Genetics of Plants* Volume I, Ed. I.K. Vasil. London, New York: Academic Press, pp. 232-240.

Beasley, C.A., E.H. Birnbaum, W.M. Dugger, and I.P. Ting (1974). A quantitative procedure for estimating cotton fiber growth. *Stain Technology* 49: 85-92.

Beasley, C.A. and I.L. Eaks (1979). Ethylene from alcohol lamps and natural gas burners: Effects on cotton ovules cultured in vitro. In Vitro 15: 263-269.

Beasley, C.A. and E. Egli (1976). Differences in the hormonal regulation of fiber development from ovules of *Gossypium hirsutum, G. barbadense,* and *G. arboreum* cultured in vitro. In *Proceedings of the Beltwide Cotton Production Research Conferences,* Ed. J.M. Brown. Memphis, TN: National Cotton Council of America, pp. 39-44.

Beasley, C.A. and E. Egli (1977). Fiber production in vitro from a conditional fiberless mutant of cotton. *Developmental Biology* 57: 234-237.

Beasley, C.A. and I.P. Ting (1973). The effects of plant growth substances on in vitro fiber development from unfertilized cotton ovules. *American Journal of Botany* 60: 130-189.

Beasley, C.A. and I.P. Ting (1974a) Effects of plant growth substances on *in vitro* fiber development from unfertilized cotton ovules. *American Journal of Botany* 61: 188-194.

Beasley, C.A. and I.P. Ting (1974b). Phytohormone effects on in vitro cotton seed development. In *Plant Growth Substances 1973*, Tokyo: Hirokawa Publishing Company, pp. 917-924.

Beasley, C.A., I.P. Ting, A.E. Linkins, E.H. Birnbaum, and D.P. Delmer (1974). Cotton ovule culture: A review of progress and a preview of potentiality. In *Tissue Culture and Plant Science*, Ed. H.E. Street. London, New York: Academic Press, pp. 169-192.

Birnbaum, E.H., C.A. Beasley, and W.M. Dugger (1974). Boron deficiency in unfertilized cotton (*Gossypium hirsutum*) ovules grown in vitro. *Plant Physiology* 54: 931-935.

Butenko, R.G. and A. Azizkhodzhaev (1987). Fiber growth in the culture of isolated cotton ovules. *Fiziologiya i Biokhimiya Kul'turnykh Rastenii* 19: 235-239.

Chu, Y.T. and F.M. Shofner (1992). Progress report on fineness and maturity distributions by AFIS. In *Proceedings of the Beltwide Cotton Conferences*, Ed. D. Herber. Memphis, TN: National Cotton Council of America. pp. 1017-1019.

Davidonis, G.H. (1989). Fiber development in preanthesis cotton ovules. *Physiologia Plantarum* 75: 290-294.

Davidonis, G.H. (1990). Gibberellic acid-induced cell elongation in cotton suspension cultures. *Journal of Plant Growth Regulation* 9: 243-247.

Davidonis, G.H. (1993a). A comparison of cotton ovule and cotton cell suspension cultures: Response to gibberellic acid and 2-chloroethylphosphonic acid. *Journal of Plant Physiology* 141: 505-507.

Davidonis, G.H. (1993b). Cotton fiber growth and development in vitro: Effects of tunicamycin and monensin. *Plant Science* 88: 229-236.

Davidonis, G.H. and O. Hinojosa (1994). Influence of seed location on cotton fiber development *in planta* and in vitro. *Plant Science* 103: 107-113.

DeLanghe, E. and A.A.H. Eid (1971). Preliminary studies on the in vitro culture of fertilized cotton ovules. *Annales Scientifiques Textiles Belges* 4: 65-73.

Dhindsa, R.S. (1978). Hormonal regulation of cotton ovule and fiber growth: Effects of bromodeoxyuridine, AMO-1618, and p-chlorophenoxyisobutyric acid. *Planta* 141: 269-273.

Dhindsa, R.S., C.A. Beasley, and I.P. Ting (1976). Effect of abscisic acid on in vitro growth of cotton fiber. *Planta* 130: 197-201.

Eid, A.A.H., E. DeLanghe, and L. Waterkeyn (1973). In vitro culture of fertilized cotton ovules. I. The growth of cotton embryos. *La Cellule* 63: 361-371.

Elbein, A.D. (1987). Inhibitors of biosynthesis and processing of N-linked digosaccharide chains. *Annual Review of Biochemistry* 56: 497-537.

Gipson, J.R. and H.E. Joham (1969). Influence of night temperature on growth and development of cotton (*Gossypium hirsutum* L.) III. Fiber elongation. *Crop Science* 9: 127-129.

Gould, J.H. and W.M. Dugger (1986). Events surrounding fiber initiation in *G. hirsutum* var., Acala SJ-2. In *Proceedings of the Beltwide Cotton Production*

Research Conferences, Eds. J.M. Brown and T.C. Nelson. Memphis, TN: National Cotton Council of America, pp. 81-82.

Graves, D.A. and J.M. Stewart (1988). Chronology of the differentiation of cotton (*Gossypium hirsutum* L.) fiber cells. *Planta* 175: 254-258.

Haigler, C.H., N.R. Rao, E.M. Roberts, J-Y. Huang, D.R. Upchurch, and N.L. Trolinder (1991). Cultured ovules as models for cotton fiber development under low temperatures. *Plant Physiology* 95: 88-96.

Hsu, C.L. and J.M. Stewart (1976). Callus induction by (2-chloroethyl) phosphonic acid on cultured cotton ovules. *Physiologia Plantarum* 36: 150-153.

Iyengar, R.L.N. (1939). Variations in the measurable characters of cotton fiber. I. Variation with respect to the length of the fiber. *Indian Journal of Agricultural Science* IX: 305-327.

Iyengar, R.L.N. (1941). Variation in the measurable characters of cotton fibers. II. Variation among seeds within a lock. *Indian Journal of Agricultural Science* XI: 703-735.

Iyengar, R.L.N. (1947). Variation of fiber length in a sample of cotton. *Indian Cotton Growers Review* I: 179-183.

Kaur, K., H. Nayyar, A.S. Basra, and C.P. Malik (1990). Stimulation of enzymes of non-photosynthetic C_4 metabolism in cultured cotton ovules by fluridone. *Acta Physiologiae Plantarum* 12: 3-6.

Kerby, T.A. and G.F. Ruppenicker (1989). Node and fruiting branch position effects on fiber and seed quality characteristics. In *Proceedings of the Beltwide Cotton Production Research Conferences*, Ed. J.M. Brown. Memphis, TN: National Cotton Council of America, pp. 98-100.

Kittock, D.L., L.L.H. Pinkas, and K.E. Fry (1979). Branch node location and competition effects on yield components of Pima cotton bolls. In *Proceedings of the Beltwide Cotton Production Research Conferences*, Ed. J.M. Brown. Memphis, TN: National Cotton Council of America, pp. 53-56.

Kohel, R., J. Quisenberry, and C. Benedict (1974). Fiber elongation and dry weight changes in mutant lines of cotton. *Crop Science* 14: 471-474.

Kosmidou-Dimitropoulou, K. (1976). Physiological identification of cotton fiber technological characteristics: Hormonal action in fiber initiation and development. PhD Dissertation, State University of Ghent, Belgium.

Krishnan, T.V. and R.L.N. Iyengar (1960). Study of the variation of fiber length within and between seeds of the same strain. *Indian Cotton Growers Review* XIV: 395-406.

Lang, A.G. (1938). The origin of lint and fuzz fibers of cotton. *Journal of Agricultural Research* 56: 507-521.

Marden, L.L. and J.M. Stewart (1984). Influence of inhibitors and atmospheric CO_2 on cultured cotton ovules (abstract). In *Proceedings of the Beltwide Cotton Production Research Conferences*, Ed. J.M. Brown. Memphis, TN: National Cotton Council of America, p. 57.

Meinert, M.C. and D.P. Delmer (1977). Changes in biochemical composition of the cell wall of the cotton fiber during development. *Plant Physiology* 59: 1088-1097.

Morré, D.J., W.F. Boss, H. Grimes, and H.H. Mollenhauer (1983). Kinetics of Golgi apparatus membrane flux following monesin treatment of embryogenic carrot cells. *European Journal of Cell Biology* 30: 25-32.

Nayyar, H., K. Kaur, A.S. Basra, and C.P. Malik (1989). Hormonal regulation of cotton fibre elongation in *Gossypium arboreum* L. in vitro and *in vivo*. *Biochemie und Physiologie der Pflanzen* 185: 415-421.

Pearson, N.L. (1949). Mote types in cotton and their occurrence as related to variety, environment, position in lock, lock size, and number of lock per boll. *USDA Technical Bulletin 1000*. Washington, DC: U.S. Government Printing Office, pp. 1-37.

Peeters, M.C., J. Wijsmans, H. Vincke, and E. DeLanghe (1985). The nucleolus: A driving force in fiber development. In *Cotton Fibres: Their Development and Properties*, Manchester, UK: International Institute for Cotton, pp. 8-12.

Porter, D.D. (1936). Positions of seeds and motes in locks and lengths of cotton fibers from bolls borne at different positions on plants at Greenville, TX. *USDA Technical Bulletin 509*. Washington, DC: U.S. Government Printing Office, pp. 1-13.

Ryser, U., H. Meier, and P.J. Holloway (1983). Identification and localization of suberin in the cell walls of green cotton fibres (*Gossypium hirsutum* L., var. green lint). *Protoplasma* 117: 196-205.

Schmutz, A., A.J. Buchala, and U. Ryser (1996). Changing the dimensions of suberin lamellae of green cotton fibers with a specific inhibitor of the endoplasmic reticulum-associated fatty acid elongases. *Plant Physiology* 110: 403-411.

Stewart, J.M. (1975). Fiber initiation on the cotton ovule (*Gossypium hirsutum*). *American Journal of Botany* 62: 723-730.

Stewart, J.M. and C.L. Hsu (1977). Influence of phytohormones on the response of cultured cotton ovules to (2-chloroethyl) phosphonic acid. *Physiologia Plantarum* 39: 79-85.

Triplett, B.A. (1995). Fiber formation in N_1 (naked seed): Influence of phytohormones on fiber production in ovule culture (abstract). In *Proceedings of the Beltwide Cotton Conferences*, Ed. D. Richter. Memphis, TN: National Cotton Council of America, p. 1141.

Triplett, B.A., W.H. Busch, and W.R. Goynes (1989). Ovule and suspension culture of a cotton fiber development mutant. *In Vitro Cellular and Developmental Biology* 25: 197-200.

Triplett, B.A. and J.D. Timpa (1995). Characterization of cell-wall polymers from cotton ovule culture fiber cells by gel permeation chromatography. *In Vitro Cellular and Developmental Biology* 31: 171-175.

Trolinder, N.L., J. Berlin, and J. Goodin (1987). Differentiation of cotton fibers from single cells in suspension culture. *In Vitro Cellular and Developmental Biology* 23: 789-794.

Vincke, H., E. DeLanghe, T. Fransen, and L. Verschraege (1985). Cotton fibres are uniform in length under natural conditions. In *Cotton Fibres: Their Devel-*

opment and Properties, Manchester, UK: International Institute for Cotton, pp. 2-5.

Wartelle, L.H., J.M. Bradow, O. Hinojosa, and A.B. Pepperman (1995). Quantitative cotton fiber maturity measurements by x-ray fluorescence spectroscopy and Advanced Fiber Information System. *Journal of Agricultural and Food Chemistry* 43: 1219-1223.

Waterkeyn, L., E. DeLanghe, and A.A.H. Eid (1975). In vitro culture of fertilized cotton ovules. II. Growth and differentiation of cotton fiber. *La Cellule* 71: 41-51.

Xie, W., N.L. Trolinder, and C.H. Haigler (1993). Cool temperature effects on cotton fiber initiation and elongation clarified using in vitro cultures. *Crop Science* 33: 1258-1264.

Chapter 4

Cellulose Biosynthesis in Developing Cotton Fibers

Deborah P. Delmer

It has often occurred to this writer that the fibers of cotton and the erythrocytes of animals share strikingly similar destinies. Both the erythrocyte and cotton fiber are terminally differentiating cells programmed to carry out one specific process at massive rates. Thus, the erythrocyte synthesizes massive amounts of a single protein, hemoglobin, while the cotton fiber devotes much of its later life to the synthesis of secondary wall cellulose. Because cotton fibers are unique cells that differentiate synchronously and synthesize so much cellulose at one stage of development, they have become one of the model systems of choice in plants for the study of the mechanism and regulation of cellulose synthesis. Furthermore, the study of cellulose synthesis in the cotton fiber has obvious practical implications because the timing, extent, and pattern of deposition of this polymer has a marked influence on fiber quality. This chapter will begin with a brief summary of what is known about the processes relating to cellulose synthesis as they occur in vivo during fiber development. Thereafter, the bulk of this chapter will concentrate on details of the

The author wishes to acknowledge the contributions of all her former colleagues who have been involved with her in studies on fiber development and glucan synthesis in cotton. In particular, she wishes to acknowledge the recent collaborations with Dr. David Stalker and his colleagues at Calgene, Inc., of Davis, CA, that have proved so fruitful in identifying genes that may be involved in cotton fiber development. Our recent research in this area has been supported by contract DE-AC02-76ERA-1338 from the U.S. Department of Energy and grants IS-2282-93 and IS-2428-94 from the United States-Israel Binational Agricultural Research and Development Fund.

85

biochemistry and molecular biology of the process of cellulose synthesis itself. In addition to other chapters in this book, the reader may
wish to consult other sources on the process of cellulose biosynthesis
and/or fiber development (Basra and Malik, 1984; Buchala and Meier,
1985; Ryser, 1985; Ross, Mayer, and Benziman, 1991; Delmer and
Amor, 1995; Volman, Ohana, and Benziman, 1995; Brown, Saxena,
and Kudlicka, 1996; Kawagoe and Delmer, 1997). Kawagoe and
Delmer (1997) cite many older reviews as well, for those interested
in studying some of the earlier literature on the topic.

PATTERNS OF CELLULOSE SYNTHESIS
IN VIVO DURING FIBER DEVELOPMENT

Figure 4.1 outlines the general events known to occur during fiber
development that particularly relate to the deposition of cellulose.
During the phase from anthesis until about 15 days postanthesis
(dpa), the fibers elongate and deposit a thin primary wall. This wall
resembles, in composition, a typical dicot primary wall that contains
uronic acid-rich polymers, xyloglucan, some protein, and cellulose
(Meinert and Delmer, 1977; Huwyler, Franz, and Meier, 1978, 1979;
Hayashi et al., 1987; Hayashi and Delmer, 1988). The level of cellulose in these very thin walls is relatively constant during elongation,
representing about 20 to 25 percent of the dry weight of the wall.
Microfibril diameters (about 6 to 8 nm [nanometers]) do not appear
to differ considerably in size throughout fiber development and often
do appear to exist as bundles of many fibers (Willison and Brown,
1976). The pattern in which this cellulose is laid down, however,
does differ, depending upon the stage of development. During the
time of primary wall deposition, microfibril deposition parallels the
roughly transverse orientation of the cortical microtubule (MT) network (Seagull, 1990, 1992a,b), which is believed to play a role in
orienting deposition. Attempts in using the freeze-fracture techniques
that were used successfully to demonstrate the rosettes and linear
terminal complexes believed to represent cellulose synthase complexes in plants and algae (see Brown, Saxena, and Kudlicka, 1996)
have met with limited success with cotton fibers. The only possible
putative complexes thus far observed may be the large globules that
fractured either to the E-face of the plasma membrane or to the inner

FIGURE 4.1. The Three Major Stages of Cotton Fiber Development

Elongation, Primary Wall Synthesis   1-15 dpa

PW→

a. Cortical MTs aligned largely transverse to direction of cell elongation; pattern of cellulose deposition parallels MT alignment.
b. Fibers apparently grow by overall elongation.
c. Intusseseption of a primary wall similar in composition to that of other dicots.
d. Cellulose constitutes about 20 to 25 percent of the dry weight of the wall; rates of synthesis are low.

Transition Phase   16-21 dpa

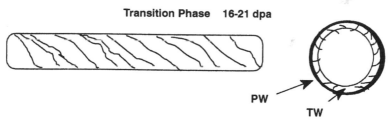

PW

TW

a. Signal transduction events; degradation of uronic acid-containing wall polymers; optimal expression of small GTPase Rac; H_2O_2 production begins; rise in intracellular Ca^{2+}; transient synthesis of callose.
b. Reorientation of MTs to helical arrays; pattern of cellulose deposition similarly reorients.
c. Secondary wall synthesis begins; rates of cellulose synthesis rise > 100-fold.
d. *CelA-1* and *CelA-2* genes start to be expressed at high levels.
e. Elongation continues, with little change in rate until end of transition.

Phase of Secondary Wall Synthesis   22-32 dpa

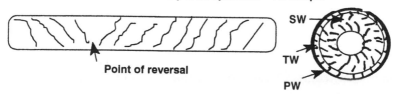

SW

TW

Point of reversal

PW

a. Elongation, primary wall synthesis cease. Cellulose synthesis at high rates is the predominant process.
b. Occasional realignments of MTs lead to reversals to layered wall patterns.
c. *CelA-1* and *CelA-2* genes continue to be expressed at high rates.
d. At the end of this phase, is programmed cell death initiated? Is synthesis and activity of nucleases and proteases enhanced?

wall face of cotton fibers (Willison and Brown, 1976). Whether this lack of information relates more to difficulties in preserving fibers for freeze fracture or to a possibility that the complexes of cotton may not resemble the usual proposed rosettes of other plants is not clear.

Perhaps the most interesting phase of fiber development is that characterized by the transition from primary to secondary wall synthesis. This phase initiates in most varieties of *Gossypium hirsutum* at about 16 dpa and is characterized by an abrupt change in the rate of cellulose synthesis, which rises over 100-fold at this stage of development (Meinert and Delmer, 1977). One of the great peculiarities of the transition period is that elongation of the fiber continues unabated for at least 5 days in the short-staple *G. hirsutum* varieties. This elongation occurs even though secondary wall cellulose begins to be deposited at very high rates and might be expected to hinder elongation. Furthermore, the direction of cellulose microfibril deposition, in line with the MT network, undergoes a shift from transverse to helical arrays (Seagull, 1990, 1992a,b) that should also be expected to limit elongation capacity. This might well have practical importance, as the length of this overlap period seems to be a major determinant of fiber length, that length being considerably longer in the long-staple *G. barbadense* cottons compared to the short-staple *G. hirsutum* cottons (Benedict, Smith, and Kohel, 1973; Schubert et al., 1973, 1976). If, as proposed, the fiber elongates by deposition of wall material throughout its length (Ryser, 1992; Tiwari and Wilkins, 1995), then one would expect an abrupt cessation, or at least a substantial reduction, in the rate of elongation at this time (see Basra and Saha, Chapter 2, this volume, for more details on fiber elongation). One possibility is that the fibers preferentially grow at the tip at this time; alternatively, it may be that the rates of cellulose synthesis differ among varieties at this stage and are still low enough not to limit elongation completely. Thus, understanding how elongation can continue during the transition period becomes quite important in understanding how fiber quality is determined.

In addition to reorganization of the cortical MT network, other studies indicate that there is a substantial elevation in levels of mRNA derived from tubulin genes and an accompanying rise in the level of tubulin protein (Seagull, 1992a; Dixon, Seagull, and Triplett, 1994; J. Pear, D. Stalker, and D. P. Delmer, unpublished). Little is known in plants of how the realignment of the MT network is controlled nor how MTs may direct patterns of cellulose deposition (Giddings and Staehelin, 1991; Cyr, 1994). However, evidence from both cotton fibers and tracheary elements suggests that realignment of the actin networks precedes MT alignment and is necessary; thus, actin patterns may set the patterns of MTs, which, in turn, set the pattern of cellulose deposition (Kobayashi, Fukuda, and Shibaoka, 1987; Seagull, 1990, 1992b; Fukuda, 1991). We reported, in collaboration with David Stalker and Julie Pear from Calgene, Inc., Davis, California, the highly enhanced expression, just at the time of transition, of genes encoding two small GTPases of the Rho subfamily (Delmer et al., 1995), called *Rac9* and *Rac13*. In animals, the Rac protein is involved in signal transduction pathways leading to actin reorganization (Hall, 1994; Symons, 1996). In addition to showing high expression during the time of cytoskeletal reorganization, a potential role for cotton Rac in actin organization is suggested by our studies, together with colleagues Doug Johnson and Cheryl Collins at the University of Vermont, wherein we have recently observed that overexpression of the cotton *Rac13* gene in yeast leads to an abnormal phenotype that mimics mutations that affect actin organization (unpublished).

The rise in the rate of cellulose synthesis that occurs during the transition appears to involve induction of expression of genes encoding components of the cellulose synthase complex. New evidence to support this conclusion has now come from another collaboration of ours with colleagues at Calgene, Inc., in which we identified two cotton genes, termed *CelA-1* and *CelA-2*, that are highly expressed in the fiber beginning at the time of transition (Pear et al., 1996). These genes have been shown to be homologs of the bacterial genes that encode the UDP (uredine diphosphate)-Glc binding catalytic subunit of the bacterial cellulose synthases; they will be discussed in more detail later in this chapter. In addition to changes in the rate of synthesis, the chain length (DP [degree of polymerization]) of newly deposited cellulose also dramatically rises during this period from a DP of < 5,000 to about 13,000

(Marx-Figini, 1966). Unfortunately, nothing is known about how chain length is determined.

Another peculiarity of the transition to secondary wall formation in cotton fibers is the transient deposition of callose (β-1,3-glucan) that continues almost throughout the transition period (Meinert and Delmer, 1977; Huwyler, Franz, and Meier, 1978, 1979; Maltby et al., 1979). Our characterization of the cotton fiber callose synthase (Hayashi et al., 1987) indicates that this enzyme is activated by Ca^{2+}, thus suggesting that a transient rise in Ca^{2+} is also associated with this transition period. Other studies (Roberts and Haigler, 1989, 1990; Fukuda, 1991, 1996) also suggest that elevation of Ca^{2+} is associated with this transition in the tracheary element in *Zinnia* and that a gene for calmodulin is also induced at this time. Other events that might be involved in signaling the transition to secondary wall synthesis will be discussed later in this chapter.

At one time, Meier and colleagues argued very strongly that the callose produced serves as a precursor to cellulose (Meier et al., 1981; Jaquet, Buchala, and Meier, 1982). The arguments were based upon the observation that rates of synthesis of callose from radioactive substrates supplied in vivo were higher than those predicted by the levels of callose that accumulated and that this suggested turnover of the glucan. Other work did show an apparent turnover of radioactivity in callose that looked consistent with its conversion to cellulose, a reaction suggested to occur by transglycosylation via a β-1,3-glucanase (Meier et al., 1981). This idea was further supported by the finding of β-1,3-glucanase activity in the walls of fibers (Buchala and Meier, 1985). However, Maltby and colleagues (1979) could not demonstrate callose turnover using cultured ovules/fibers, and the Swiss group later also could not show an absolute correlation between callose degradation and cellulose synthesis under all conditions (Meier, 1983; Buchala and Meier, 1985). To date, no new evidence has surfaced to support the notion that callose is a mandatory precursor to cellulose in the fiber, and role of callose deposition in fiber development remains unclear.

Once fibers enter the phase of active cellulose synthesis, fiber elongation does eventually cease, and the fiber at this stage is essentially a factory devoted to cellulose synthesis. MT rearrangements apparently continue to occur, leading to the frequently observed "reversals," as well as to the characteristic layering of the secondary cell wall. Synthesis of callose and rates of *Rac* gene expression decline considerably, whereas those for *CelA-1* and *CelA-2* genes continue to remain high during this period.

By analogy with tracheary elements (Fukuda, 1996), one might speculate that the final phase of fiber development consists of a "programmed cell death" wherein specific nucleases and proteases are induced, intracellular contents are degraded, and cells eventually die. Unfortunately, little attention has been paid to this phase of fiber development.

TRACING THE PATH OF CARBON INTO CELLULOSE IN VIVO

For many years, it was assumed by many that the fiber might be a cell that was symplastically isolated from its neighbors. Although it was generally accepted that sucrose from phloem tissue was the major source of carbon supplied to fibers, it was thought that this sucrose was not transported symplastically via plasmodesmata directly into fibers, but rather unloaded from the phloem and taken up again by fibers, either intact by use of a sucrose symporter or by hexose transporters following cleavage by a cell wall invertase. However, in 1992, Ryser published photographs that clearly showed plasmodesmatal connections between the foot of the cotton fiber and its neighbor cells within the ovule, thus supporting the notion that carbon as sucrose enters the fiber via a symplastic route. Nevertheless, this issue is not entirely resolved, and it is not clear if alternate routes might be available and used also by the fiber. Fiber cells do possess a cell wall-localized acid invertase; however, measurements of its level during development suggest that it is present only during the stage of elongation and primary wall depositon (Waefler and Meier, 1994). Cultured cotton ovules with associated fibers do not grow well on supplied sucrose and prefer glucose (Beasley, 1992); this may be due to the sealing of funiculus in culture and prevention of entry of materials via the normal phloem pathway. Although glucose serves well as a carbon source for ovule and fiber development up to the stage of early secondary wall deposition, cultured fibers cannot be successfully maintained to full maturity for unknown reasons. Whether this might be due to loss of a hexose transport system in older fibers has never been investigated. Furthermore, when glucose is supplied to cultured ovules and fibers, a great deal of it is subsequently converted back to sucrose in the fiber (Carpita and Delmer, 1981), presumably via the enzymes sucrose-P synthase and phosphatase that have been shown to exist in fairly high levels in the fibers (C. Haigler and S. Holaday, personal communication).

UDP-Glc is known to be the substrate for the cellulose synthase of the well-characterized bacterium *Acetobacter xylinum* (Ross, Mayer, and Benziman, 1991; Volman, Ohana, and Benziman, 1995). However, as will be discussed later in this chapter, it has been exceedingly difficult to demonstrate direct synthesis of cellulose in vitro from UDP-Glc. Because of these difficulties with in vitro synthesis of cellulose, Carpita and Delmer (1981) attempted to trace the path of carbon in vivo using cultured cotton ovules/fibers supplied with ^{14}C-glucose. In pulse and pulse-chase experiments, the concentrations and extent of radioactivity accumulated in various phosphorylated sugars, free sugars, and sugar nucleotides were measured; computer modeling was applied to the results, and the conclusion was drawn that the rate flow of carbon through UDP-Glc was sufficient to account for the rates of cellulose and callose synthesis observed, and therefore, UDP-Glc is most likely a precursor to both of these glucans. In addition, these and other studies (Franz, 1969) indicate that pools of UDP-Glc are quite high in fibers. No other nucleotide sugar was found in sufficient levels and/or had a sufficient turnover rate to be a likely candidate as precursor for glucan synthesis.

UDP-Glc can potentially be synthesized by either of two reversible reactions—that catalyzed by UDP-Glc pyrophosphorylase (for a review of this enzyme in plants, see Kleczkowski, 1994) or by sucrose synthase (SuSy). Cotton fibers contain high levels of both enzymes (Waefler and Meier, 1994; Amor et al., 1995; Nolte et al., 1995). Recently, a PhD student in our laboratory, Yehudit Amor, made the fascinating discovery that more than half of the total SuSy of cotton fibers is tightly associated with the plasma membrane fraction (Amor et al., 1995). (Since that work was published, we have found that a great deal of the membrane-associated form was discarded by us in the low-speed pellet, and this was particularly the case during harvests from fibers engaged in secondary wall synthesis. Therefore, the percent SuSy in the membrane fraction is even higher than estimated by Amor and colleagues [1995] during secondary wall synthesis.) Since SuSy had previously been studied only as a soluble enzyme, the finding of the membrane-associated form was surprising. (Hereafter, we shall refer to the membrane-associated and soluble forms as M-SuSy and S-SuSy, respectively.) Furthermore, immunolocalization studies done in collaboration with Haigler's group on fibers taken at the time of active secondary wall synthesis showed that M-SuSy is found in helical arrays that parallel the patterns of cellulose deposition (Amor et al., 1995). These findings suggested to us a model in which M-SuSy is associated with the cellulose synthase complex, whereby it serves to channel carbon from sucrose via UDP-Glc to the catalytic subunit of the synthase. We note that pathways involving sucrolysis via SuSy have been proposed previously (Huber and Akazawa, 1986; Xu et al., 1989) and that a coupled reaction between SuSy and glucan synthases was also proposed years ago (Rollit and Maclachlan, 1974). However, none of those researchers recognized that M-SuSy, in addition to S-SuSy, might exist in plant cells.

More recently, we have been thinking of a more general model in which the ratio of M- to S-SuSy in nonphotosynthetic plant cells may be a major factor in determining the pattern of partitioning of carbon between cellulose to the wall or to processes in other compartments such as noncellulosic polysaccharide synthesis in the Golgi, respiration in mitochondria, and starch synthesis in amyloplasts. In this regard, cotton fibers engaged in secondary wall deposition would represent an extreme example of partitioning to cellulose. Figure 4.2 presents a model of a proposed pathway of carbon metabolism in fibers actively engaged in cellulose synthesis.

Some Considerations Regarding This Model

Early work (Mutsaers, 1976) has suggested that up to 80 percent of total carbon entering the fiber at this most active stage of wall synthesis is partitioned to cellulose. This, in combination with our finding of a high percentage of M-SuSy at this stage, has led us to estimate in the model that sucrose entering the cell is partitioned in a ratio of 3:1 between M- and S-SuSy.

Degradation of sucrose by M-SuSy leads to production of UDP-Glc that can be directly channeled to the catalytic subunit (presumably that encoded by the *CelA* genes; see later discussion). Upon transfer of the glc to the growing β-1,4-glucan chain, UDP is regenerated for immediate reuse by M-SuSy. This cycle results in net synthesis of cellulose from sucrose, with no additional energy input; it also ensures that the UDP produced does not accumulate, an important point, since UDP is a potent inhibitor of glucan synthases (Morrow and Lucas, 1986; Ross, Mayer, and Benziman, 1991). Finally, the UDP-Glc produced is in a sequestered pool that does not equilibrate with its soluble counterpart.

The model also attempts to account for the relatively high levels of UDP-Glc pyrophosphorylase (Waefler and Meier, 1994) and sucrose-P synthase (C. Haigler and S. Holaday, personal communication; D. Delmer, unpublished) that are found in fibers. Others have noted cases in which nonphotosynthetic cells possess both SuSy and sucrose-P synthase in the same cell and have suggested that this leads to a futile cycle in which sucrose is continually synthesized by sucrose-P synthase and sucrose phosphatase and degraded again by SuSy, with the cost of 1 UTP (uridine triphosphate) per cycle (Wendler et al., 1990; Geigenberger and Stitt, 1991, 1993; Huber and Huber, 1996). This strikes us as being so wasteful as to require further explanation. It is obvious that heavy partitioning of sucrose to M-SuSy would lead to rapid accumulation of fructose that could cause feedback inhibition if not removed (Doehlert, 1987). If the goal of such cells is to channel as much carbon as possible to cellulose, then it would be most logical to resynthesize sucrose from this fructose. Thus, the model proposes roles for fructokinase, sucrose-P synthase and phosphatase, phosphoglucomutase, phosphoglucose isomerase, and UDP-Glc pyrophosphorylase in this "not so futile" cycle. Sucrose, thus synthesized, would enter a common pool where it once again would be repartitioned in a 3:1 ratio to M- and S-SuSy.

Note that the UDP-Glc pyrophosphorylase is functioning in the direction of UDP-Glc synthesis, something not normally proposed in schemes for nonphotosynthetic tissues. This results in production of PPi (pyrophosphate); thus, we have proposed that the major route of conversion of fructose-6-P to fructose 1,6-di-P is via the PPi-dependent phosphofructokinase; this fits with data that show cotton fibers have higher levels of this enzyme than of the ATP-dependent phosphofructokinase (Waefler and Meier, 1994).

The amount of carbon partitioned to respiration is more than sufficient to regenerate the ATP utilized by phosphorylation of fructose, with presumably sufficient ATP left for other cellular processes that are largely involved in maintenance at this stage of development. Discounting the energy costs for the phosphorylation of fructose, six glc residues are polymerized to cellulose with additional input of only 1 UTP. Other possibilities exist for cycling of uridylates and adenylates than those shown; for example, the fructokinase of many plants can also use UTP in place of ATP (see Xu et al., 1989). Furthermore, it should be noted that the attempts to balance all of the equations shown in Figure 4.2 are obviously only estimates.

By contrast, one can consider another extreme case wherein a high percentage of the carbon is partitioned to starch—representative of cells engaged in synthesis of storage material in tubers, seeds, etc. In this situation, there would be no need to invoke a role for M-SuSy for cells that have ceased expansion and, similarly, no need for cycling via sucrose-P synthase. A third situation might characterize what occurs in cells engaged in cell division or expansion and synthesis of the primary cell wall. In this scenario, because cellulose synthesis occurs, but at a much lower rate than during secondary wall synthesis, we would predict low levels of M-SuSy. Carbon must also be partitioned to the Golgi for use in synthesis of noncellulosic polysaccharides, presumably via the use of S-SuSy, and a considerable amount of carbon must also be allocated for respiration and other processes associated with growth.

The previous concepts are also supported by the recent results of Carlson and Chourey (1996) in developing maize endosperm: these researchers also detected M-SuSy and found that the percent of total SuSy in the membrane is about 14 percent during the phase of expansion (primary walls, so rates of cellulose synthesis are not expected to be too high), dropping to only 3 percent at the time of starch synthesis in the endosperm. We also note that the studies of carbon flow in fibers in vivo using cultured ovules (Carpita and Delmer, 1981) can now be reinterpreted in a new light. In the model

FIGURE 4.2. A Model for Carbon Metabolism in the Developing Cotton Fiber at the Stage of Active Secondary Wall Cellulose Synthesis

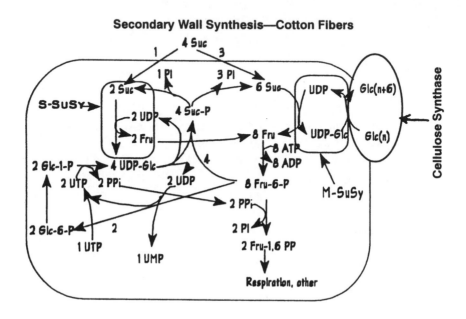

of Figure 4.2, one predicts that the UDP-Glc produced by M-SuSy should exist as a very small pool with rapid turnover that might not have been detected in those studies. However, the fact that ^{14}C-glucose was used as carbon source in those experiments means that essentially all of the radioactivity must have passed through the soluble UDP-Glc pools on its way to M-SuSy if the model shown is correct. Therefore, the finding that the rate of UDP-Glc turnover matched well the combined rates of synthesis of cellulose and callose should still be valid.

Finally, in discussing the path of carbon into cellulose, it should be mentioned that recent studies on cellulose synthesis with the bacterium *Agrobacterium tumefaciens* implicate a role, beyond the level of UDP-Glc, for lipid intermediates, as well as a cellulase acting as a transglycosylase (Matthysse, Thomas, and White, 1995; Matthysse, White, and Lightfoot, 1995). A role for lipid intermediates in plants was once widely discussed, but to date, there is still no strong evidence to support such a pathway. Nevertheless, this still remains a possibility that cannot yet be totally excluded (reviewed in detail by Kawagoe and Delmer, 1997).

STUDIES ON CELLULOSE SYNTHESIS IN VITRO

As early as 1969, Franz and Meier supplied UDP-^{14}C-Glc to detached cotton fibers and claimed synthesis of cellulose. Other early work (Delmer, Heiniger, and Kulow, 1977; Heiniger and Delmer, 1977) showed, by contrast, that in detached fibers the major product from supplied UDP-Glc is callose. Several other early reports suggested that synthesis of both glucans can occur (see Buchala and Meier, 1985). These and other later studies (Carpita and Delmer, 1980) also indicated that intact fibers are apparently impermeable to UDP-Glc, and utilization for glucan synthesis (largely callose) improves gradually as fibers are progressively more wounded by cutting or detergent permeabilization. Callose synthase in cotton fibers is an extremely active enzyme in the presence of two activators, Ca^{2+} (in micromolar levels) and a β-glucoside (cellobiose is normally supplied, but the native activator may be β-furfuryl-β-glucoside [Ohana, Benziman, and Delmer, 1993]). Localized in the plasma membrane and normally latent, it becomes the major activity detected when broken fibers or isolated plasma membranes are supplied with UDP-Glc. Other more recent work has suggested that two types of calcium-binding proteins, annexins (Andrawis, Solomon, and Delmer, 1993; Delmer and Potikha, 1997) and a 65 kDa (kilodalton) membrane polypeptide, later shown to be calnexin (Delmer, Ohana, et al., 1993; D. P. Delmer, unpublished), interact with cotton fiber callose synthase in a cation-dependent manner. The significance of these findings is not yet clear; however, recent results have shown that tomato annexins can associate with actin (Calvert, Gant, and Bowles, 1996). Thus, it may be possible that annexins could serve as a bridge between glucan synthases and the cytoskeleton. Our laboratory has recently cloned genes for these calcium-binding proteins from cotton (Potikha and Delmer, 1997), and we hope these will prove useful for studying their potential role in glucan synthesis in the fiber.

Controversy still surrounds the issue of exactly how much supplied UDP-Glc can also be converted to cellulose, in addition to callose, in these in vitro systems (Delmer, Volokita, et al., 1993). However, the laboratory of Brown has now accumulated substantial evidence indicating that at least some synthesis of cellulose, in addition to large amounts of callose, can be reproducibly demonstrated with isolated cotton fiber membrane preparations (see Kudlicka et al., 1995; Kudlicka, Lee, and Brown, 1996, and references therein), and this represents a significant finding. Also of importance is that the authors can rather convincingly distinguish fibrils of cellulose from those of callose by their differences in morphology. Interestingly, this group claims that the ratio of cellulose to callose synthesized is greatly enhanced when a combination of MOPS [3-(N-morpholino) propanesulphonic acid] buffer and sucrose are used in the membrane isolation and assay procedures. Although this group has rather convincingly shown some limited incorporation of radioactivity from supplied UDP-Glc into cellulosic fibrils, most of their work has relied upon distinguishing the two products by microscopic examination coupled with labeling of the nonradioactive cellulose produced by cellobiohydrolase-gold labeling. Thus, it has occurred to us that at least some of the cellulose produced might have come from the sucrose present via the combined action of M-SuSy and cellulose synthase rather than from the supplied radioactive UDP-Glc. This possibility is supported by our finding that supplying radioactive sucrose to detached and detergent-permeabilized fibers resulted in substantial synthesis of cellulose in addition to callose, whereas use of UDP-Glc as substrate resulted in mostly callose (Amor et al., 1995). When the two activators of callose synthase, Ca^{2+} and cellobiose, were present, synthesis of callose predominated, whereas the reverse was true when Ca^{2+} was removed by chelation with EDTA (ethylene diamine tetraacetic acid). (The finding that such a coupled reaction can also result in callose synthesis suggests that M-SuSy might also associate with the callose synthase and/or that both glucan synthases are part of the same complex. However, it is clear that this synthase can also readily synthesize callose directly from supplied UDP-Glc.) It is necessary to point out that the rates obtained in these coupled reactions are highly variable, at times approaching those observed in vivo and at other times being quite low. Thus, if the coupled reaction between M-SuSy and cellulose synthase occurs, factors that control its rate in vitro still need to be clarified.

Also relevant to these studies is an earlier work by Pillonel, Buchala, and Meier (1980) who supplied the radioactive substrates sucrose, glucose, or UDP-Glc to "intact" locules of *G. arboreum* and measured rates of incorporation into glucan. Although rates of glucan synthesis were similar for glucose or sucrose during the phase of primary wall synthesis, there was a 140-fold preference for the use of sucrose during the phase of secondary wall synthesis. There was little competition between sucrose and UDP-Glc (the latter serving primarily as a substrate for callose synthesis), suggesting that these channel carbon to glucan via distinct pathways. In general, these results support the model in which M-SuSy preferentially channels carbon from sucrose to cellulose, whereas supplied UDP-Glc is used preferentially for callose synthesis. In interpreting the results of Pillonel, Buchala, and Meier (1980), it becomes important to ask if the fibers were truly intact in these experiments. The fact that the fibers were first wetted by immersion in solutions containing 0.01 percent Triton X-100, which is known to permeabilize membranes, and that UDP-Glc could be utilized indicates strongly to us that these authors were working with permeabilized fibers. The authors suggested otherwise, since incorporation from sucrose, but not UDP-Glc, was inhibited by the respiratory inhibitors DNP (2,4 dinitrophenol) and KCN (potassium cyanide). However, examining the model shown in Figure 4.2, it becomes clear that substantial synthesis of cellulose from sucrose could only continue, even in permeabilized fibers, if a continuous energy source is available to recycle the accumulated fructose.

In sum, although some synthesis of cellulose is possible from supplied UDP-Glc in vitro, results are now accumulating that suggest synthesis occurs much more effectively if sucrose, rather than UDP-Glc, is the supplied substrate. Clarifying and optimizing these in vitro reactions still remains a high priority for future research on cellulose synthesis in cotton fibers.

Another open question is whether plant cellulose synthases might be activated by cyclic diguanylic acid—a unique molecule that activates the cellulose synthase of *A. xylinum* (see Ross, Mayer, and

Benziman, 1991; Volman, Ohana, and Benziman, 1995). Polypeptides that bind this activator have been reported for cotton fibers (Amor et al., 1991), and antibodies against the bacterial enzyme that synthesizes the activator have also been shown to cross-react with plant polypeptides (Mayer et al., 1991). However, to date, no one has shown substantial activation of cellulose synthesis in vitro by supplied cyclic diguanylic acid. Another area deserving of further study concerns the, by now, almost-forgotten suggestions that a membrane potential may be required, or at least important, for cellulose synthesis (Carpita and Delmer, 1980; Bacic and Delmer, 1981; Delmer, Padan, and Benziman, 1982). Finally, since data suggest that both the DP of cellulose (Timpa, 1992) and the crystallinity (Benedict, Kohel, and Jividen, 1994) can be related to bundle strength in fibers, mechanisms for control of these parameters deserve study, although it is not yet clear how these topics can be approached experimentally.

THE CelA *GENES OF COTTON*

Studies on cellulose synthesis entered a new era with the recent characterization of the *CelA-1* and *CelA-2* genes of cotton (Pear et al., 1996). Both of these genes, although clearly distinct from each other, are nevertheless highly homologous and also show high homology to a number of cDNA clones identified in the EST (expressed sequence tag) databanks of rice and Arabidopsis, indicating that homologs of these genes can now be easily identified in many plants. The cotton *CelA-1* and *CelA-2* genes show only low expression in nonfiber tissues at the level of Northern analysis of total RNA isolated from roots, leaves, and flower tissues. This, however, does not discount the possibility that they might be expressed selectively in a subset of these tissues synthesizing secondary walls, such as in developing xylem or trichomes of leaves. Induction of expression of both *CelA-1* and *CelA-2* in the fibers begins at the onset of secondary wall synthesis and continues at high levels throughout the stage of secondary wall formation.

What is the evidence that these genes really encode the catalytic subunit of cellulose synthase? Besides their pattern of expression, which fits this role well, homology studies (Pear et al., 1996) show three regions within the deduced CelA proteins that show good homology with the *CelA* genes of bacteria (for reviews of bacterial cellulose synthesis, see Ross, Mayer, and Benziman, 1991; Volman, Ohana, and Benziman, 1995). Figure 4.3 shows a diagram that aligns these regions of homology, called H-1, H-2, and H-3, comparing the cotton CelA-1 protein with that encoded by the *bsA/acsA* gene of *Acetobacter xylinum* and the *CelA* gene of *Agrobacterium tumefaciens*. Within these three regions are four even more highly conserved motifs (areas designated as U-1, U-2, U-3, and U-4 in Figure 4.3) that were previously identified as possibly being important for binding of the substrate UDP-Glc and for catalysis (Delmer and Amor, 1995; Saxena et al., 1994, 1995). Of some interest is the fact that the encoded cotton and rice CelA proteins contain two regions not found in the bacterial genes. The first, called P-CR, is conserved in all the plant CelA proteins identified, whereas the second region, called HVR, is a hypervariable region that differs considerably in the different plant genes.

FIGURE 4.3. Alignment of Regions of Homology Among the Proteins Encoded by Bacterial and Cotton *CelA* Genes

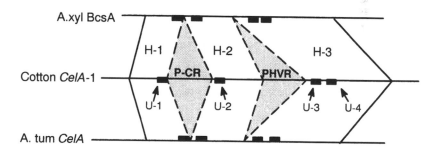

Note: The alignment is derived from data presented by Pear et al. (1996). The regions H-1, H-2, and H-3 show the strongest sequence homology between the deduced cotton CelA protein and those encoded by the bacterial homologs. U-1 through U-4 indicate regions of especially high homology that are believed to be involved in binding of UDP-Glc and/or catalysis. The shaded regions P-CR and HVR represent regions that are plant specific and are not found in the bacterial proteins.

Dr. Yasushi Kawagoe, in our laboratory, succeeded to subclone the region spanning the U-1 through U-4 motifs that is predicted to be located on the cytoplasmic face of the membrane between a series of transmembrane helices in the N- and C-terminal portions of the protein. The overexpressed and purified recombinant fragment was clearly shown to bind UDP-Glc; however, when a region encoding the first conserved U-1 motif believed to be involved in UDP-Glc binding was deleted, the resulting recombinant fragment did not bind the substrate (Pear et al., 1996). Taken together, these results strongly suggest that the *CelA* genes are true functional homologs; however, further work will clearly be necessary to prove that they function in vivo as proposed. We also note that, should involvement of lipid intermediates ever be discovered for plants, by analogy with the proposed function of the CelA homolog of *A. tumefaciens*, the cotton CelA proteins might catalyze synthesis of a lipid-oligosaccharide rather than direct polymerization of cellulose from UDP-Glc.

We also note that Cutler and Somerville (1997) are now studying a different class of *Arabidopsis* genes, identified among the ESTs as showing some homology with the bacterial *CelA* genes and with the cotton *CelA* genes. In comparing sequences, our groups have tentatively concluded that, although related, this class of genes is nevertheless distinct from the *CelA* class identified in cotton and rice. To date, it has been more difficult to assign a possible role for these genes based upon expression patterns or antisense experiments, and their relationship to the process of cellulose synthesis remains unclear (Cutler and Somerville, 1997). The expression patterns of the cotton *CelA* genes suggest that they function primarily in secondary wall synthesis. Still to be determined is whether distinct genes (e.g., those studied by Cutler and Somerville) might encode other CelA proteins for primary wall cellulose synthesis or for synthesis in other specific cell types. Genomic Southern analyses in cotton suggest that the *CelA*s comprise at least a small gene family (Pear et al., 1996).

More recently, we have identified two potential Zn^{2+}-binding motifs in the N-terminal region of the deduced cotton CelA proteins that are not present in the bacterial CelA proteins, and preliminary experiments indicate that a recombinant N-terminal fragment of

CelA-1 does indeed bind Zn^{2+} (Kawagoe and Delmer, 1997). These two motifs together resemble the LIM domains identified in mammalian proteins that seem to be important for protein-protein interactions (Sanchez-Garcia and Rabbitts, 1994). Thus, it will be interesting to see if this region of the CelA protein will be involved in interaction with either other CelA subunits or other potentially interacting proteins such as M-SuSy, regulatory subunits, pore structures involved in glucan secretion, or components of the cytoskeleton. Other regions of the CelA proteins that are likely candidates for interaction with other proteins are the P-CR and HVR regions that are unique to plants (see Figure 4.3). One of the major unanswered questions in cellulose synthesis concerns the relationship between the MT network and the synthase complexes. If the complexes move in the membrane by the force of polymerization, while the direction of movement is controlled by MTs, then we need to understand whether this involves a direct interaction of the synthase complex with the cytoskeleton or an indirect restriction in path of movement imposed by the MTs. If there is a direct interaction with a cytoskeletal component, then one of the plant-specific regions might be expected to be involved in this interaction because the bacteria do not have cytoskeletons.

Having these genes available should provide many new approaches for the study of cellulose synthesis in cotton fiber. Transgenic experiments can be envisioned, wherein rates of cellulose synthesis might be altered. In addition, antibodies to the CelA proteins can be used to probe synthase organization and interaction with M-SuSy. The list is long, and we can predict that there should now be a substantial increase in the use of molecular approaches to the study of cellulose synthesis in cotton and other plants. Genetic approaches are also becoming more feasible as there are now a number of reports of mutants partially impaired in cellulose synthesis in plants. Thus, mutants have now been reported that display reduced cellulose in *Arabidopsis* roots (Baskin et al., 1992), and it is exciting that the mutation (rsw1) has been identified as being in a gene that is very homologous to the cotton *CelA* genes (Arioli et al., 1998). Another Arabdopsis mutant shows reduced cellulose in trichomes (Potikha and Delmer, 1995), culms of barley (Kokubo et al., 1991), and even in cotton fibers (Kohel, Benedict, and Jividen,

1993). This last mutant appears to lack the capacity to synthesize secondary wall cellulose, and it should prove extremely interesting to see if it lacks the ability to express the *CelA* genes.

REGULATION OF THE ONSET
OF SECONDARY WALL FORMATION

As a final topic, possible signal transduction events that might regulate the onset of secondary wall formation in cotton fiber are considered. As mentioned previously, understanding the regulation of this transition may well have practical implications. Thus, knowledge of how to manipulate the timing of the transition might lead to strategies for modification that could affect ultimate fiber properties. There are many clues as to signals that may trigger this transition, and new data have recently been obtained (Potikha et al., 1999) suggesting that the signal transduction pathway may mimic, at least in some ways, that involved in an oxidative burst (for a general review of the oxidative burst in plants, see Low and Merida, 1996). The oxidative burst involved in signaling related to defense responses that occur during pathogenesis in plants first can be triggered by release of pathogen-derived or plant cell wall-derived elicitors such as oligogalacturonans, fragments of glucan, or glycoproteins. Such elicitors further signal the rapid production of H_2O_2, which is most likely derived from an NADPH (nicotinamide adenine dinucteotide phosphate [reduced form]) oxidase at the plasma membrane, which, upon activation, generates superoxide radicals that are either converted spontaneously or through the action of superoxide dismutase to H_2O_2. This is followed by a continuing signal cascade that involves protein phosphorylation/dephosphorylation events, a rise in intracellular Ca^{2+}, synthesis of callose, and, finally, activation of defense-related genes. Thus, such defense responses do resemble, in some ways, events that occur during induction of secondary walls because they involve changes in patterns of wall synthesis that often include deposition of callose and lignification. Whereas tracheary element differentiation does involve cessation of normal primary wall synthesis, coupled with induction of synthesis of xylan, secondary wall cellulose, and lignin, cotton fibers are unique in containing only cellulose as a major secondary wall component, with additional transient synthesis of callose during the transition.

The oxidative burst of plants involved in defense responses resembles, in many ways, that which also occurs in leucocytes (for a review, see Bagglioni and Wymann, 1990). In leucocytes, external signals associated with pathogen invasion lead to activation of the NADPH oxidase. One of the key elements of this signal transduction is the activation of the NADPH oxidase by the small GTPase Rac (Diekman et al., 1994). Thus, of interest is our discovery that cotton fibers also express the two genes *Rac9* and *Rac13*, which are homologs of animal *Rac*s and are highly expressed in the fiber just at the onset of secondary wall formation (Delmer et al., 1995) (see earlier discussion on the potential role of these Rac proteins in regulating actin organization). Reasoning that one of the cotton Racs might be involved in an analogous activation of an NADPH oxidase, we recently measured H_2O_2 production in fibers and found that it is initiated exactly coincident with onset of secondary wall formation, and levels of H_2O_2 stay high throughout the transition and well into the phase of true secondary wall synthesis (Potikha et al., 1999). Furthermore, addition of the NADPH oxidase inhibitor diphenyleneiodonium to cultured ovules/fibers inhibits the production of H_2O_2 and, more important, also inhibits development of secondary wall formation, suggesting that H_2O_2 is indeed serving as a signal. This is further substantiated by our finding that addition of H_2O_2 to young fibers causes premature initiation of secondary wall synthesis. Finally, in thinking of clues as to what could be the initial signaling event, we recalled old data of ours that indicated a net loss of uronic acid from the fiber wall at this stage of development (Meinert and Delmer, 1977). Thus, at present, we speculate that this might generate oligo-galacturonide elicitors that could be one of the earliest initiation events in signaling the transition. Testing of this idea, as well as whether Rac can directly activate the oxidase, is currently in progress. Further downstream from the H_2O_2 signal, one can speculate that elevation of cytoplasmic Ca^{2+} occurs since callose synthesis is induced in the fiber, and evidence exists that elevation of Ca^{2+} and induction of calmodulin occurs at this time in tracheary elements (Fukuda, 1996). Finally, as for the oxidative burst relating to pathogenesis, induction of gene expression (e.g., genes for *Rac*, tubulin, and *CelA-1* and *CelA-2* in fibers) certainly is a major event that characterizes the transition to secondary wall formation. Clearly, this model needs further testing and refinement; however, if true in general, it represents a new way of thinking about regulation of secondary wall formation in plants.

CONCLUSIONS

After many years of frustration, exciting progress is being made in studies on cellulose synthesis in the developing cotton fiber. The ability to use in situ immunolocalization techniques has further clarified the dynamic nature of the cytoskeleton and its relationship to determining patterns of cellulose deposition. However, many questions remain unanswered in this area: What controls cytoskeletal organization? What are the respective roles of actin and microtubules in determining patterns of cellulose deposition? And, most fascinating and least understood of all, what is the mechanism by which the cytoskeleton directs the movement of cellulose synthase complexes? Progress has been made in attempts to synthesize cellulose in vitro. When UDP-Glc is used as substrate in such reactions, callose is still the major product produced, but some cellulose can clearly be synthesized as well, and the morphology of the two types of products can be distinguished by electron microscopy. New evidence suggests that sucrose may be a preferred substrate in such reactions, and this relates to the finding of a membrane-associated form of sucrose synthase that is proposed to be part of the synthase complex where it serves to channel carbon from sucrose, via UDP-Glc, to the glucan synthase catalytic subunit. Perhaps most exciting of all is the discovery of two genes in cotton that are homologs of the genes that encode the catalytic subunit of bacterial cellulose synthases. These genes are highly expressed in the fiber at the time of secondary wall synthesis, and the encoded proteins have been shown to bind the substrate UDP-Glc, making it most likely that they do indeed function in cellulose synthesis. Having these genes, as well as others for sucrose synthase and those related to cytoskeletal organization and callose synthesis (e.g., tubulins, actin, Rac, annexins), now open many new possible approaches for the study of this process.

REFERENCES

Amor, Y., C.H. Haigler, S. Johnson, M. Wainscott, and D.P. Delmer (1995). A membrane-associated form of sucrose synthase and its potential role in synthesis of cellulose and callose in plants. *Proceedings of the National Academy of Sciences USA* 92: 9353-9357.

Amor, Y., R. Mayer, M. Benziman, and D.P. Delmer (1991). Evidence for a cyclic diguanylic acid-dependent cellulose synthase in plants. *Plant Cell* 3: 989-995.

Andrawis, A., M. Solomon, and D.P. Delmer (1993). Cotton fiber annexins: A potential role in the regulation of callose synthase. *Plant Journal* 3: 763-772.

Arioli, T., L.C. Peng, A.S. Betzner, J. Burn, W. Wittke, W. Herth, C. Camilleri, H. Hofte, J. Plazinske, R. Birch, et al. (1998). Molecular analysis of cellulose biosynthesis in *Arabidopsis*. *Science* 279: 717-720.

Bacic, A. and D.P Delmer (1981). Stimulation of membrane-associated polysaccharide synthetases by a membrane potential in developing cotton fibers. *Planta* 152: 346-351.

Bagglioni, B. and M.P. Wymann (1990). Turning on the respiratory burst. *Trends in Biochemical Science* 15: 69-72.

Baskin, T.I., A.A. Betzner, R. Hoggart, A. Cork, and R.E. Williamson (1992). Root morphology mutants in *Arabidopsis Thaliana*. *Australian Journal of Plant Physiology* 19: 427-437.

Basra, A.S. and C.P. Malik (1984). Development of the cotton fiber. *International Review of Cytology* 89: 65-113.

Beasley, C.A. (1992). *In vitro* cotton ovule culture: A review. In *Proceedings from Cotton Fiber Cellulose: Structure, Function and Utilization Conference*, Memphis, TN: National Cotton Council, pp. 65-90.

Benedict, C.R., R.J. Kohel, and G.M. Jividen (1994). Crystalline cellulose and cotton fiber strength. *Crop Science* 34: 147-151.

Benedict, C.R., R.H. Smith, and R.J. Kohel (1973). Incorporation of ^{14}C-photosynthate into developing cotton bolls, *Gossypium hirsutum* L. *Crop Science* 13: 88-91.

Brown, R.M. Jr., I.M. Saxena, and K. Kudlicka (1996). Cellulose biosynthesis in higher plants. *Trends in Plant Science* 1: 149-156.

Buchala, A.J. and H. Meier (1985). Biosynthesis of β-glucans in growing cotton (*Gossypium arboreum* L. and *Gossypium hirsutum* L.) fibres. In *Biochemistry of Plant Cell Walls*, Eds. C. Brett and J. Hillman. Cambridge, UK: University Press Cambridge, pp. 221-241.

Calvert, C.M., S.J. Gant, and D.J. Bowles (1996). Tomato annexins p34 and p35 bind to F-actin and display nucelotide phosphodiesterase activity inhibited by phospholipid binding. *Plant Cell* 8: 333-342.

Carlson, S.J. and P.S. Chourey (1996). Evidence for plasma membrane-associated forms of sucrose synthase in maize. *Molecular and General Genetics* 252: 303-310.

Carpita, N.C. and D.P. Delmer (1980). Protection of cellulose synthesis in detached cotton fibers by polyethylene glycol. *Plant Physiology* 66: 911-916.

Carpita, N.C. and D.P. Delmer (1981). Concentration and metabolic turnover of UDP-glucose in developing cotton fibers. *Journal of Biological Chemistry* 256: 308-315.

Cutler, S. and C. Somerville (1997). Cellulose synthesis: Cloning *in silico. Current Biology* 7: R108-R111.

Cyr, R.J. (1994). Microtubules in plant morphogenesis: Role of the cortical array. *Annual Review of Cell Biology* 10: 153-180.

Delmer, D.P. and Y. Amor (1995). Cellulose biosynthesis. *Plant Cell* 7: 987-1000.

Delmer, D.P., U. Heiniger, and C. Kulow (1977). UDP-glucose: Glucan synthetase from developing cotton fibers. I. Kinetic and physiological properties. *Plant Physiology* 59: 713-718.

Delmer, D.P., P. Ohana, L. Gonen, and M. Benziman (1993). *In vitro* synthesis of cellulose in plants: Still a long way to go. *Plant Physiology* 103: 307-308.

Delmer, D.P., E. Padan, and M. Benziman (1982). Requirement for a membrane potential for cellulose synthesis in intact cells of *Acetobacter xylinum. Proceedings of the National Academy of Sciences USA* 79: 5282-5286.

Delmer, D.P., J. Pear, A. Andrawis, and D. Stalker (1995). Genes for small GTP-binding proteins analogous to mammalian Rac are highly expressed in developing cotton fibers. *Molecular and General Genetics* 248: 43-51.

Delmer, D.P. and T. Potikha. (1997). Function of annexins in plants. In *Annexins*, Ed. J. Mollenhauer. Cellular and Molecular Life Sciences 53: 546-553.

Delmer, D.P., M. Volokita, M. Solomon, U. Fritz, W. Delphendahl, and W. Herth (1993). A monoclonal antibody recognizes a 65 kDa higher plant membrane polypeptide which undergoes cation-dependent association with callose synthase *in vitro* and co-localizes with sites of high callose deposition *in vivo*. *Protoplasma* 176: 33-42.

Diekman, D., A. Abo, C. Johnston, Segal, and A.W. Hall (1994*)*. Interaction of Rac with p67phox and regulation of the phagocytic NADPH oxidase activity. *Science* 265: 531-533.

Dixon, D.C., R.W. Seagull, and B.A. Triplett (1994). Changes in accumulation of α- and β-tubulin isotypes during cotton fiber development. *Plant Physiology* 105: 1347-1353.

Doehlert, D.C. (1987). Substrate inhibition of maize endosperm sucrose synthase by fructose and its interaction with glucose inhibition. *Plant Science* 52: 153-157.

Franz, G. (1969). Soluble nucleotides in developing cotton hair. *Phytochemistry* 8: 737-741.

Franz, G. and H. Meier (1969). Biosynthesis of cellulose in growing cotton hairs. *Phytochemistry* 8: 579-583.

Fukuda, H. (1991). Tracheary element formation as a model system of cell differentiation. *International Review of Cytology* 136: 289-332.

Fukuda, H. (1996). Xylogenesis: Initiation, progression, and cell death. *Annual Review of Plant Physiology and Plant Molecular Biology* 47: 299-325.

Geigenberger, P. and M. Stitt (1991). A "futile" cycle of sucrose synthesis and degradation is involved in regulating partitioning between sucrose, starch, and respiration in cotyledons of germinating *Ricinus communis* L. seedlings when phloem transport is inhibited. *Planta* 185: 81-90.

Geigenberger, P. and M. Stitt (1993). Sucrose synthase catalyses a readily reversible reaction *in vivo* in developing potato tubers and other plant tissues. *Planta* 189: 329-339.

Giddings, T.H. and L.A. Staehelin (1990). Microtubule-mediated control of microfibril deposition: A re-examination of the hypothesis. In *The Cytoskeletal Basis of Plant Growth and Development*, Ed. C.W. Lloyd. London: Academic Press, pp. 85-99.

Hall, A. (1994). Small GTP-binding proteins and the regulation of the actin cytoskeleton. *Annual Review of Cell Biology* 10: 31-54.

Hayashi, T. and D.P. Delmer (1988). Structural analysis of the xyloglucan of cotton fiber cell walls. *Carbohydrate Chemistry Research* 181: 273-277.

Hayashi, T., S.M. Read, J. Bussell, M. Thelen, F.C. Lin, R.M. Brown Jr., and D.P. Delmer (1987). UDP-glucose: (1,3)-β-glucan synthases from mung bean and cotton: Differential effects of Ca^{2+} and Mg^{2+} on enzyme properties and on macromolecular structure of the glucan product. *Plant Physiology* 83: 1054-1062.

Heiniger, U. and D.P. Delmer (1977). UDP-glucose: glucan synthetase in developing cotton fibers. II. Structure of the reaction product. *Plant Physiology* 59: 719-723.

Huber, S.C. and T. Akazawa (1986). A novel sucrose synthase pathway for sucrose degradation in cultured sycamore cells. *Plant Physiology* 81: 1008-1013.

Huber, S.C. and J.L. Huber (1996). Role and regulation of sucrose-P synthase in higher plants. *Annual Review of Plant Physiology and Plant Molecular Biology* 47: 431-444.

Huwyler, H.R., G. Franz, and H. Meier (1978). β-1,3-glucans in the cell walls of cotton fibres (*Gossypium arboreum* L.). *Plant Science Letters* 12: 55-62.

Huwyler, H.R., G. Franz, and H. Meier (1979). Changes in the composition of cotton fibre cell walls during development. *Planta* 46: 635-642.

Jaquet, J.P., A.J. Buchala, and H. Meier (1982). Changes in the non-structural carbohydrate content of cotton (*Gossypium* spp.) fibres at different stages of development. *Planta* 156: 481-486.

Kawagoe, Y. and D.P. Delmer (1997). Pathways and genes involved in cellulose biosynthesis. In *Genetic Engineering,* Volume 19., Ed. J.K. Setlow. New York: Plenum Press, pp. 63-87.

Kleczkowski, L.A. (1994). Glucose activation and metabolism through UDP-glucose pyrophosphorylase in plants. *Phytochemistry* 37: 1507-1515.

Kobayashi, H., H. Fukuda, and H. Shibaoka (1987). Reorganization of actin filaments associated with the differentiation of tracheary elements in *Zinnia* mesophyll cells. *Protoplasma* 138: 69-71.

Kohel, R.J., C.R. Benedict, and G.M. Jividen (1993). Incorporation of [^{14}C] glucose into crystalline cellulose in aberrant fibers of a cotton mutant. *Crop Science* 33: 1036-1040.

Kokubo, A., N. Sakurai, S. Kuraishi, and K. Takeda (1991). Culm brittleness of barley *(Hordeum vulgare* L.) mutants is caused by smaller number of cellulose molecules in cell wall. *Plant Physiology* 97: 509-514.

Kudlicka, K., R.M. Brown Jr., L. Li, J.H. Lee, H. Shen, and S. Kuga (1995). β-glucan synthesis in the cotton fiber. IV. *In vitro* assembly of the cellulose I allomorph. *Plant Physiology* 107: 111-123.

Kudlicka, K., J.H. Lee, and R.M. Brown Jr. (1996). A comparative analysis of *in vitro* cellulose synthesis from cell-free extracts of mung bean *(Vigna radiata,* Fabaceae) and cotton *(Gossypium hirsutum,* Malvaceae). *American Journal of Botany* 83: 274-284.

Low, P.S. and J.R. Merida (1996). The oxidative burst in plant pathogenesis: Function and signal transduction. *Physiologia Plantarum* 96: 535-542.

Maltby, D., N.C. Carpita, D. Montezinos, C. Kulow, and D.P. Delmer (1979). β-1,3-glucan in developing cotton fibers: Structure, localization, and relationship of synthesis to that of secondary wall cellulose. *Plant Physiology* 63: 1158-1164.

Marx-Figini, M. (1966). Comparison of the biosynthesis of cellulose *in vitro* and *in vivo* in cotton bolls. *Nature* 210: 754-755.

Matthysse, A.G., D.O.L. Thomas, and A.R. White (1995). Mechanism of cellulose synthesis in *Agrobacterium tumefaciens. Journal of Bacteriology* 177: 1076-1081.

Matthysse, A.G., S. White, and R. Lightfoot (1995). Genes required for cellulose synthesis in *Agrobacterium tumefaciens. Journal of Bacteriology* 177: 1069-1075.

Mayer, R., P. Ross, H. Weinhouse, D. Amikam, G. Volman, P. Ohana, R.D. Calhoon, H.C. Wong, A.W. Emerick, and M. Benziman (1991). Polypeptide composition of bacterial cyclic diguanylic acid-dependent cellulose synthase and the occurrence of immunologically cross-reacting proteins in higher plants. *Proceedings of the National Academy of Sciences USA*: 88: 5472-5476.

Meier, H. (1983). Biosynthesis of (1→3)-β and (1→4)-β-D-glucans in cotton fibers *(Gossypium arboreum* and *Gossypium hirsutum). Journal of Applied Polymer Science: Applied Polymer Symposium* 37: 123-130.

Meier, H., L. Buchs, A.J. Buchala, and T. Homewood (1981). (1→3)-β-D-glucan (callose) is a probable intermediate in biosynthesis of cellulose of cotton fibers. *Nature* 289: 821-822.

Meinert, M. and D.P. Delmer (1977). Changes in the biochemical composition of the cell wall of the cotton fiber during development. *Plant Physiology* 59: 1088-1097.

Morrow, D.L. and W.J. Lucas (1986). (1,3)-β-glucan synthase from sugar beet. I. Isolation and solubilization. *Plant Physiology* 81: 171-176.

Mutsaers, H.J. (1976). Growth and assimilate conversion of cotton bolls *(Gossypium hirsutum* L.). I. Growth of fruits and substrate demand. *Annals of Botany* 40: 300-315.

Nolte, K.D., D.L. Hendrix, J.W. Radin, and K.E. Koch (1995). Sucrose synthase localization during initiation of seed development and trichome differentiation in cotton ovules. *Plant Physiology* 109: 1285-1293.

Ohana, P., M. Benziman, and D.P. Delmer (1993). Stimulation of callose synthesis *in vivo* correlates with changes in intracellular distribution of the callose synthase activator β-furfuryl-β-glucoside. *Plant Physiology* 101: 187-191.

Pear, J., Y. Kawagoe, W. Schreckengost, D.P. Delmer, and D. Stalker (1996). Higher plants contain homologs of the CelA genes that encode the catalytic subunit of the bacterial cellulose synthases. *Proceedings of the National Academy of Sciences USA* 93: 12637-12642.

Pillonel, C., A.J. Buchala, and H. Meier (1980). Glucan synthesis by intact cotton fibres fed with different precursors at the stages of primary and secondary wall formation. *Planta* 149: 306-312.

Potikha, T., C. Collins, D.I. Johnson, D.P. Delmer, and A. Levine (1999). The involvement of hydrogen peroxide in the differentiation of secondary walls in cotton fibers. *Plant Physiology* 119: 849-858.

Potikha, T. and D.P. Delmer (1995). A mutant of *Arabidopsis thaliana* displays altered patterns of cellulose deposition. *Plant Journal* 7: 453-460.

Potikha, T. and D.P. Delmer (1997). cDNA clones for annexin AnnGh-1 and AnnGh-2 from *Gossypium hirsutum* (cotton) (PGR 97-003). *Plant Physiology* 113: 305.

Roberts, A.W., and C.H. Haigler (1989). Rise in chlorotetracycline fluorescence accompanies tracheary element differentiation in suspension cultures of *Zinnia*. *Protoplasma* 152: 37-45.

Roberts, A.W. and C.H. Haigler (1990). Tracheary-element differentiation in suspension-cultured cells of *Zinnia* requires uptake of extracellular Ca^{2+}. *Planta* 180: 502-509.

Rollit, J. and G.A. Maclachlan (1974). Synthesis of wall glucan from sucrose by enzyme preparations from *Pisum sativum*. *Phytochemistry* 13: 367-374.

Ross, P., R. Mayer, and M. Benziman (1991). Cellulose biosynthesis and function in bacteria. *Microbiological Review* 55: 35-58.

Ryser, U. (1985). Cell wall biosynthesis in differentiating cotton fibres. *European Journal of Cell Biology* 39: 236-265.

Ryser, U. (1992). Ultrastructure of the epidermis of developing cotton (*Gossypium*) seeds: Suberin, pits, plasmodesmata, and their implication for assimilate transport into cotton fibers. *American Journal of Botany* 79: 14-22.

Sanchez-Garcia, I. and T.H. Rabbitts (1994). The LIM domain: A new structural motif found in zinc-finger-like proteins. *Trends in Genetics* 10: 315-320.

Saxena, I.M., K. Kudlicka, K. Okuda, and R.M. Brown Jr. (1994). Characterization of genes in the cellulose-synthesizing operon(acs Operon) of *Acetobacter xylinum*: Implication for cellulose crystallization. *Journal of Bacteriology* 176: 5735-5752.

Saxena, I.M., R.M. Brown Jr., M. Fevre, R.A. Geremia, and B. Henrissat (1995). Multidomain architecture of β-glycosyl transferases: Implications for mechanism of action. *Journal of Bacteriology* 177: 1419-1424.

Schubert, A.M., C.R. Benedict, J.D. Berlin, and R.J. Kohel (1973). Cotton fiber development—Kinetics of cell elongation and secondary wall thickening. *Crop Science* 13: 704-709.

Schubert, A.M., C.R. Benedict, C.E. Gates, and R.J. Kohel (1976). Growth and development of the lint fibers of Pima S-4 cotton. *Crop Sciences* 16: 539-543.

Seagull, R. (1990). The effects of microtubule and microfilament disrupting agents on cytoskeletal arrays and wall deposition in developing cotton fibers. *Protoplasma* 159: 44-59.

Seagull, R. (1992a). A quantitative electron microscopic study of changes in microtubule arrays and wall microfibril orientation during *in vitro* cotton fiber development. *Journal of Cell Science* 101: 561-577.

Seagull, R. (1992b). Cytoskeletal involvement in microfibril organization during cotton fiber development. In *Proceedings from Cotton Fiber Cellulose: Structure, Function and Utilization Conference*. Memphis, TN: National Cotton Council, pp. 171-192.

Symons, M. (1996). Rho family GTPases: The cytoskeleton and beyond. *Trends in Biochemical Science* 21: 178-181.

Timpa, J.D. (1992). Molecular chain length distributions of cotton fiber: Developmental, varietal, and environmental influences. In *Proceedings from Cotton Fiber Cellulose: Structure, Function and Utilization Conference*. Memphis, TN: National Cotton Council, pp. 199-209.

Tiwari, S.C. and T.A. Wilkins (1995). Cotton (*Gossypium hirsutum*) seed trichomes expand via a diffuse growth mechanism. *Canadian Journal of Botany* 73: 746-757.

Volman, G., P. Ohana, and M. Benziman (1995). Biochemistry and molecular biology of cellulose biosynthesis. *Carbohydrates in Europe* 12: 20-27.

Waefler U. and H. Meier (1994). Enzyme activities in developing cotton fibres. *Plant Physiology and Biochemistry* 32: 697-702.

Wendler, R., R. Veith, J. Dancer, M. Stitt, and E. Komor (1990). Sucrose storage in cell suspension cultures of *Saccharum* sp. (sugarcane) is regulated by a cycle of synthesis and degradation. *Planta* 183: 31-39.

Willison, J.H.M and R.M. Brown Jr. (1976). An examination of the developing cotton fiber: Wall and plasmalemma. *Protoplasma* 92: 21-41.

Xu, D.P., S.J. Sung, T. Loboda, P.P. Kormanik, and C.C. Black (1989). Characterization of sucrolysis via the uridine diphosphate and pyrophosphate-dependent sucrose synthase pathway. *Plant Physiology* 90: 635-642.

Chapter 5

Noncellulosic Carbohydrates in Cotton Fibers

Antony J. Buchala

INTRODUCTION

The advantages of using the cotton fiber as a model system to study the biosynthesis of cellulose have been discussed in other chapters of this book (Delmer, 1999; Ryser, 1999) and have been presented before (inter alia, Buchala and Meier, 1985; Ryser, 1985). Emphasis has been placed on cotton fiber's suitability due to its high cellulose content. However, during the phase of elongation, when primary cell wall is laid down, and during the transition phase from primary to secondary wall synthesis, polysaccharides other than cellulose are mainly synthesized (Meinert and Delmer, 1977; Huwyler, Franz, and Meier, 1979). Interpretation of the results concerning the synthesis of cellulose in cotton fibers is complicated by the fact that the cell walls contain significant amounts of endogenous (1→3)-β-glucan, callose (Meinert and Delmer, 1977; Huwyler, Franz, and Meier, 1978; Buchala and Meier, 1985). Often, the biosynthesis of cellulose has been studied in vitro by feeding precursors to fiber homogenates, detached fibers, or cotton ovules cultured on liquid medium. In the last of these systems, callus tissue often forms on the

The author wishes to thank his colleagues who stimulated his interest in the cotton fiber cell wall; in particular, Hans Meier, Ueli Ryser, Yvan Francey, Christian Pillonel, Jean-Pierre Jaquet, Peter Bucheli, Urs Waefler, and Thierry Genoud. Our work on the cotton fiber was mainly funded by the Swiss National Science Foundation.

damaged seed coat and may contaminate the fibers studied, thereby complicating interpretation of the results. Noncellulosic polysaccharides from the cell wall of suspension-cultured cotton cells, derived from meristem tissue or seed coat callus, have also been studied and are supposed to be similar to those in the primary cell walls of cotton fibers. The first section of this chapter gives an account of the noncellulosic polysaccharides that have been identified in cotton, and some speculation is made on their roles. The second part describes the metabolism of low molecular weight carbohydrates in cotton fiber and attempts to relate their presence to the synthesis of polysaccharides found in the cell wall. An account of the role of sucrose synthase is given by Delmer elsewhere in this book (see Chapter 4). In a third short section, the roles of some other low molecular weight compounds are discussed.

NONCELLULOSIC POLYSACCHARIDES

The presence of noncellulosic polysaccharides in cotton fibers can be deduced from the presence of monosaccharides other than glucose in hydrolysates of cell wall preparations (Meinert and Delmer, 1977; Huwyler, Franz, and Meier, 1979) and methylation analysis of soluble polysaccharide fractions derived from the same type of material (A. Buchala, unpublished observations). More direct evidence comes from studies in which such polysaccharides have been isolated and subjected to structural analysis. Thus, $(1{\rightarrow}3)$-β-glucan (Huwyler, Franz, and Meier, 1978), type II arabinogalactan (Buchala and Meier, 1981), xyloglucan (Hayashi and Delmer, 1988; Buchala et al., 1993), and xylan (see Buchala and Meier, 1985) have been identified. It has been proposed (Darvill et al., 1980) and widely accepted that the polysaccharides found in the cell walls of plant cells cultured in vitro and those which are secreted into the culture medium are similar to those in the primary cell wall of the corresponding plant, but only in one case (Marti, 1991) did the cultured cells have an appearance similar to that of cotton fibers (Trolinder, Berlin, and Goodin, 1987). However, Thompson and Fry (1997) have shown that xyloglucan, initially laid down in the cell wall of suspension-cultured rose cells, was partially degraded and then released into the culture medium. The soluble material was of lower molecular weight, but no structural comparisons were made. When all types of cotton cells cultured in vitro are considered, then more details become available on the fine structure of xyloglucan (El Rassi et al., 1991) and rhamnogalacturonan I (Komalavilas and Mort, 1989).

Xyloglucans

Xyloglucans have been attributed diverse roles apart from just being structural components of primary cell walls (Fry, 1989). The structures of xyloglucans from different dicotyledonous plants are similar concerning the main structural features, but differences are evident when one considers molecular weight, the distribution of various side chains, and the stage of plant tissue development. It was suspected that xyloglucans or xyloglucan metabolism could play a role in the control of cotton fiber elongation, and it has been shown that deposition of xyloglucan only takes place during the elongation phase of cotton fiber development (Delmer et al., 1985).

The structures of xyloglucans from cell walls of cotton fibers have been reported (Buchala and Meier, 1985; Hayashi and Delmer, 1988), and the presence of xyloglucan in suspension culture medium has been established (Buchala, Genoud, and Meier, 1987). These results were compared to those obtained for xyloglucans purified from cotton fibers (differentiated, elongating cells), cotton leaf (differentiated cells), and suspension-cultured cells (undifferentiated and essentially spherical), as well as from the culture medium of the latter (Buchala et al., 1993). Methylation analysis and ^{13}C-NMR (nuclear magnetic resonance) spectroscopy confirmed the presence of a main chain of β-1,4-linked glucopyranosyl residues substituted at C(O)6 by α-xylopyranosyl, 2-O-β-galactopyranosyl-α-xylopyranosyl-, and 2-O-α-fucopyranosyl-2-O-β-galactopyranosyl-α-xylopyranosyl side chains. There were differences in the fine structure, for example, the frequency of single α-xylopyranosyl side chains, but no evident relationship between xyloglucan structure and cell type could be deduced. Hydrolysis of xyloglucan-containing material from the medium of cultured cells suggested that $(Glc)_4Xyl_3$ and $(Glc)_4Xyl_3Gal_1Fuc_1$ [the latter partly acetylated, mainly at C(O)6] represented the main structural repeating units (El Rassi et al., 1991). The predominance of the single xylosyl side chains for the xyloglucan from the culture medium was also observed by Buchala and colleagues (1993).

The $(Glc)_4Xyl_3Gal_1Fuc_1$ nonasaccharide has been found, in the free form, in the medium of suspension-cultured spinach (Fry, 1988) and cotton cells (Buchala et al., 1993) and shows an antiauxin effect (York, Darvill, and Albersheim, 1984). The inactive, unfucosylated, octasaccharide (McDougal and Fry, 1989) was apparently not detected, but since a chromatographic standard was unavailable for $(Glc)_4Xyl_3Gal_1$, no clear conclusion could be drawn as to the eventual role of the nonasaccharide in the culture medium. Warneck, Haug, and Seitz (1996) were able to show that treatment of pea epicotyls with the nonasaccharide simultaneously activated peroxidase and inhibited growth, leading to cell wall tightening.

Pectic Polysaccharides

The sugars arabinose, galactose, and uronic acid, found in the hydrolysates of cell wall preparations from elongating cotton fibers, suggest the presence of significant amounts of pectin. An acidic arabinogalactan (Aspinall type II) has been isolated from the fibers of *Gossypium arboreum* and has been shown to have a main chain of $(1{\rightarrow}3)$-linked galactopyranosyl residues to which side chains are attached at C(O)6. The latter consist of $(1{\rightarrow}6)$-linked galactopyranosyl residues substituted at C(O)3 by $(1{\rightarrow}5)$-linked arabinofuranosyl chains. Terminal glucopyranuronosyl residues are also present (Buchala and Meier, 1981). The presence of similar polysaccharides in the culture medium of cotton cells has been established by methylation analysis and ^{13}C-NMR (Buchala, Genoud, and Meier, 1987) and in the culture medium of cotton ovules cultured in vitro (Buchala, Roulin, and Meier, 1989). The similarity of the results suggests that the polysaccharides secreted into the culture medium of the ovules are actually derived from callus tissue forming at the ovule epidermis. Kreuger and van Holst (1996) state that arabinogalactans of the previous type are not covalently linked in the cell wall and do not have a structural function. They are supposed to be involved in signal mechanisms in developmental processes, and possibly some evidence for this has been found in cotton (Marti, 1991).

Solvolysis with HF (hydrofluoric acid) of cell walls from cotton cells cultured in vitro gave rise to a mixture of acetylated and nonacetylated disaccharides of galacturonic acid and rhamnose. (Komalavilas and Mort, 1989). The authors concluded that the only known polysaccharide that would produce such disaccharides is rhamnogalactoronan I (Darvill et al., 1980) and that this pectic polysaccharide is acetylated on C(O)3 of about one in three of the galacturonic residues in its backbone.

Xylans

An acidic arabinoxylan has been isolated from the cell wall of fibers from *Gossypium arboreum* (A. Buchala, unpublished results) and is similar in structure to the known arabinoxylans (Darvill et al., 1980). The presence of substituted xylan in the culture medium of both cotton cells and cotton ovules can also be deduced from ^{13}C-NMR data and methylation analysis (Buchala, Genoud, and Meier, 1987; Buchala, Roulin, and Meier, 1989).

Glucomannans

Small amounts of mannose have been detected in hydrolysates of cell wall from cotton fibers (Meinert and Delmer, 1977; Huwyler, Franz, and Meier, 1979) and in the culture medium of cotton cells (Buchala, Genoud, and Meier, 1987; Buchala, Roulin, and Meier, 1989). This may suggest the presence of small amounts of (galacto)glucomannan, too low, however, to be detected by ^{13}C-NMR or by the methylation techniques used. Indirect evidence for the presence of glucomannan comes from incorporation experiments using labeled UDP (uridine diphosphate)-Glc (glucose) or GDP (guanine diphosphate)-Glc as precursors and detached cotton fibers. It could be shown that the incorporation of GDP-Glc diminished markedly with fiber development and was almost negligible at the stage of secondary wall formation, whereas incorporation from UDP-Glc into alkali-insoluble material increased (Delmer, Beasley, and Ordin, 1974). These results were later confirmed, and it could be shown that incorporation of radioactivity from GDP-Man correlated with that from GDP-Glc (A. Allenbach and A. Buchala, unpublished observations). After enzymic hydrolysis of the polysaccharides solubilized from the fiber cell walls, the radioactivity from

GDP-Man was found in oligosaccharides chromatographically identical to glucosylmannose and mannosylglucose. Mannose may also be present in the side chains of a yet unidentified glycoprotein similar to that studied by Forsee and Elbein (1975).

β-glucans

The β-glucan (callose) isolated from *Gossypium arboreum* was found to have $(1{\rightarrow}3)$-β-glucopyranosyl linkages, and a small amount of branching at $C(O)6$ was deduced from the methylation analysis (Huwyler, Franz, and Meier, 1978). Later work (methylation analysis and enzymic hydrolysis) carried out in this laboratory invariably showed the presence of 5 to 10 percent of $(1{\rightarrow}6)$-β-glucopyranosyl linkages in the β-glucan (Meier et al., 1981; Pillonel and Meier, 1985). However, branching has not always been found in the products formed in vitro (Maltby et al., 1979).

Localization of Noncellulosic Polysaccharides in Fibers

Substantial amounts of β-glucan can be found in cotton fibers at the transition phase between primary and secondary wall formation and at the beginning of secondary wall formation (Meinert and Delmer, 1977; Huwyler, Franz, and Meier, 1979; Jaquet, Buchala, and Meier, 1982). Fluorescence microscopy has shown that β-glucan is almost exclusively located in the innermost wall layer bordering the plasma membrane (Waterkeyn, 1981), and the strongest fluorescence was obtained at the beginning of secondary wall deposition. Rowland and Howley (1988) attempted to obtain similar information by studying the relative ease of extraction of β-glucan by chemical means. They concluded that the decrease in solubility found with increasing fiber maturity was consistent with high exposure and accessibility in the relatively disordered primary wall, decreasing progressively within the highly ordered secondary wall, with increasing deposition of cellulose due to entrapment and association with the cellulose microfibril. Their results were consistent with the location of fractions of the β-glucan in the lumen and the primary cell wall and of a substantial fraction ("presumably between layers") in the secondary wall of near mature fibers.

Fibers at the stage of primary wall formation are rich in pectic material and xyloglucan. Using specific antibodies, S. C. Tiwari and T. Wilkins (unpublished observations) have been able to show that prior to, and at, the onset of fiber initiation, pectin was dispersed throughout the cell wall to become organized later into fibrils. As elongation proceeded, these fibrils were first orientated tranversally and then parallel to the fiber axis. Xyloglucans were found to be evenly dispersed throughout the cell wall. Both polysaccharides were localized at the Golgi apparatus at fiber initiation and onset of elongation.

Analysis of Noncellulosic Polysaccharides During Fiber Development

Whereas it is clear from experiments in which $^{14}CO_2$ was fed to intact cotton plants that all the noncellulosic polysaccharides other than the β-glucan are synthesized uniquely during the deposition of primary cell wall (Meier et al., 1981), β-glucan is synthesized throughout fiber development (Buchala and Meier, 1985; Ryser, 1985; Delmer, 1987). The work of Meinert and Delmer (1977) and of Huwyler, Franz, and Meier (1979) showed that the absolute amounts of most of the monosaccharides, with the exception of xylose, in hydrolysates of noncellulosic polysaccharides (whether calculated on a cell length or, even better, on a constant cell number basis) were maximum at the end of primary wall formation or at the beginning of secondary wall formation. These results indicate that a turnover of the noncellulosic polysaccharides occurs, particularly in arabinose-, galactose-, and glucose-containing polysaccharides. The fate of most of these polysaccharides is unknown, but given the availability of the antibodies, it appears a good idea to try to localize the polysaccharides at later stages of fiber development.

Cell wall-bound hydrolytic activities (measured using synthetic substrates) that could eventually be involved in the turnover have been the subject of a preliminary study (Huwyler, 1978), but the values obtained appeared so low that further work was not pursued. Nevertheless, it is worth noting that the main activity found was that of β-galactosidase and that it, too, exhibited a maximum at the end of elongation of the fibers and could thus be involved in turnover of pectic polysaccharides. When ovules are cultured in vitro, the amount of polysaccharide material secreted into the culture medium increases with culture age, and the amounts of each monosaccharide in the polysaccharides increase more or less in parallel (Buchala, Roulin, and Meier, 1989). Cells of callus tissue, isolated from cotton ovules cultured in vitro in the presence of gibberellic acid and ethylene, elongate to about 1 mm (millimeter) and resemble short cotton fibers (Marti, 1991). Attempts to stimulate the synthesis of cellulose and to induce the formation of a secondary wall were unsuccessful. During cell elongation, polysaccharides typical of the primary wall were synthesized; in particular, the proportion of an arabinose-containing polysaccharide, probably a type-II arabinogalactan, was found to increase with elongation. During elongation of fibers, the amount of putative arabinogalactan increases to diminish toward the stage of secondary wall formation (Huwyler, Franz, and Meier, 1979). These observations constitute part of the circumstantial evidence for the role of arabinogalactans (or arabinogalactan proteins) in developmental processes (Kreuger and van Holst, 1996).

Analysis of cell wall polymers has been carried out during fiber development by solubilizing cell wall components, without prior extraction or derivatization (Timpa and Triplett, 1993), with the aim of minimizing any degradation. Gel permeation chromatography showed that cell wall polymers (of unknown identity) from fibers at primary cell wall stages of growth had lower molecular weights than cellulose from fibers at the secondary wall stages. Some high molecular weight material was detected as early as 8 days postanthesis (dpa), and the amount of this material decreased during the later stages of cell elongation, indicating that hydrolysis had occurred. In a subsequent study, the same authors (Triplett and Timpa, 1995) compared their results with those obtained for fibers grown on ovules cultured in vitro. The chromatographic pattern of the polymers from the latter was quite similar to those grown in vivo, but some subtle differences were observed that the authors claim may explain why ovule culture fibers rarely reach their full genetic potential in length.

The metabolism of the noncellulosic β-glucan has been the subject of numerous publications, particularly since most attempts to synthesize cellulose in vitro have led to the synthesis of $(1\rightarrow3)$-β-glucan

(Delmer, 1987; 1999). The β-glucan appears to undergo turnover in fibers grown both in vivo (Huwyler, Franz, and Meier, 1979; Meinert and Delmer, 1977; Jaquet, Buchala, and Meier, 1982) and in vitro (Francey et al., 1989), although Maltby and colleagues (1979) could find no evidence for turnover in the latter. The results of Jaquet and colleagues (1982) could be confirmed by Rowland, Howley, and Anthony (1984) who used a derivatization procedure to estimate (1→3)-β-glucan.

Cell wall-bound β-glucanase activity in cotton fiber has been examined and both exo- and endo-(1→3)-β-glucanase activity, the former predominating, determined during fiber development (Bucheli et al., 1985). The endo activity remained more or less constant during fiber development and may, thus, be associated uniquely with primary walls, whereas the exo activity showed a maximum during the deposition of secondary wall. Little or no (1→4)-β-glucanase activity was detected. The β-glucanase activity could be solubilized in buffers containing 3 M (moles) LiCl (lithium chloride) and was found to exhibit transglucosylase activity by introducing a (1→6)-linkage onto cellobiose. When fiber wall fragments were incubated in vitro, free glucose was released (Bucheli, Buchala, and Meier, 1987). Such autolysis was low when the endogenous (1→3)-β-glucan content was low, increased markedly at the beginning of secondary wall formation, and remained more or less constant until fiber maturity. Clearly, these glucanases could be involved in the observed turnover of the (1→3)-β-glucan, and it was proposed that this polysaccharide could be an intermediate in the synthesis of cellulose (Meier et al., 1981). Although the (1→3)- β-glucanase preparation does show transglucosylation activity (Bucheli, Buchala, and Meier, 1986), this is considered to be thermodynamically unfavorable for the synthesis of cellulose (Stone and Clarke, 1992). An alternative mechanism would involve hydrolytic breakdown of (1→3)-β-glucan to glucose, which could, in part, be used for the synthesis of cellulose. Again, this corresponds to a situation in which energy is required. The turnover does occur, and it has been calculated that the turnover rate of the callose pool should lie between 2 to 8 times (depending on the stage of development) within 24 h (hours) to deliver the glucose necessary to synthesize the cellulose laid down during the same period (Buchala and Meier, 1985), but it has not been shown that it is directly related to the synthesis of cellulose.

Biosynthesis of Noncellulosic Polysaccharides

In general, the biosynthesis of hemicelluloses and pectic polysaccharides (cf. Gibeaut and Carpita, 1994) has attracted much less attention, and the cotton fiber is no exception. In the relatively few studies carried out, the products were often poorly characterized. This is due to the low incorporation of radioactivity from the precursor and the fact that hemicelluloses do not have well-defined, unique structures. Additional problems arise from the fact that most of the polysaccharides are composed of several different sugars, and their synthesis requires the cooperation of several synthases. Given the heterogeneity of hemicellulose structures, it is not really known whether such synthases are under strict control in vivo. However, it is generally accepted that the substrates are the corresponding sugar nucleotides and, apparently, that the rate of synthesis may depend on the sugar nucleotide pool or the activity of the transferases that are developmentally regulated (Northcote, 1985). More recently, Robertson, Beech, and Bolwell (1995) showed that UDP-Glc dehydrogenase showed a pronounced maximum during the elongation phase of French bean hypocotyls and that less significant changes were obtained for other enzymes involved in sugar nucleotide interconversion. It is well documented that, whereas cellulose and $(1\rightarrow3)$-β-glucan are synthesized at the plasma membrane (Delmer, 1987), hemicelluloses and pectic polysaccharides are synthesized by endomembrane systems, particularly the Golgi apparatus. However, modifications may occur after the initial polymerization process, for example, methyl esterification of uronic acid residues, acetylation, or cross-linkage to other cell wall components. Some of these processes probably occur in the cell wall itself.

When $^{14}CO_2$ was fed to intact cotton plants, noncellulosic polysaccharides other than $(1\rightarrow3)$-β-glucan were labeled in fibers during the formation of primary cell wall (L. Buchs, unpublished observation). This work was not followed up. Feeding of ^{14}C-labeled monosaccharides (arabinose, galactose, and glucose) to the culture medium of cotton ovules cultured in vitro led to labeling of polysaccharides in both the culture medium and the fibers (Buchala,

Roulin, and Meier, 1989). With glucose, only β-glucans (cellulose and callose) in the fibers were labeled (cf. Francey et al., 1989), whereas all the sugars, but mainly glucose, were labeled in the xyloglucan from the culture medium; no labeled $(1\rightarrow3)$-β-glucan was found in the culture medium. With galactose or arabinose, labeling was found in an acidic arabinogalactan of both the fibers and the culture medium and also in xylan of the culture medium. After transport to the cytoplasm, the monosaccharides are probably phosphorylated and converted to the corresponding uridine sugar nucleotides. Thus, only in the case of the fibers, when glucose was fed, could synthesis of polysaccharides typical of secondary walls be detected.

Mention has previously been made of the synthesis of glucomannan from GDP-Man and GDP-Glc in detached fibers in the primary walls of fibers (cf. Delmer, Beasley, and Ordin, 1974), but it should be noted that the authors characterized their product as cellulose, which is now thought to be unlikely. Although UDP-Glc is the recognized precursor for the synthesis of xyloglucan (Ray, 1980), radioactivity from UDP-Glc fed to detached fibers at the primary wall stage was poorly incorporated (Delmer, Beasley, and Ordin, 1974; A. Allenbach and A. Buchala, unpublished) and was not detected in fractions known to contain xyloglucan. However, no UDP-Xyl was added to the incubation medium.

The eventual role of callose in the synthesis of cellulose has been discussed previously and by Delmer (see Chapter 4, this volume). The precursor for the synthesis of $(1\rightarrow3)$-β-glucan is UDP-Glc; the synthase activity requires micromolar amounts of Ca^{2+} and is stimulated by β-glucosides (Delmer, 1987; Ohana, Benziman, and Delmer, 1993). It is tempting to suggest that the latter may act as primers for polymerization, but no incorporation from labeled cellobiose (a commonly used β-glucoside in the synthase assays) was observed A. Buchala, (unpublished observation). The endogenous β-glucoside has been identified as β-furfuryl-β-glucoside (Ohana, Benziman, and Delmer, 1993), and calcium-binding proteins may be involved (Andrawis, Solomon, and Delmer, 1993). The idea that callose and cellulose are synthesized by the same enzymic system, depending on the conditions prevailing in the plant cell (Delmer et al., 1985), is attractive. Cellulose (and callose) synthase are considered to be constituted by a complex of many different polypeptides that are supposed to be responsible for, inter alia, orientation of the complex in the plasma membrane, binding of the substrate, and transfer of glucose to the growing polymer chain. Disruption of the complex (stressed plasmalemma) in many ways, for example, disturbed ion homeostasis, wounding, and so forth, would lead to synthesis of callose.

$(1{\rightarrow}3)$-β-glucan synthase remains active after solubilization in detergents such as digitonin or CHAPS (3-[(3-cholamidopropyl)-dimethylammonio]-propane sulfate), and this property has been exploited in attempts to purify the synthase complex. A 52-kDa (kilodalton) peptide that shows Ca^{2+}-dependent binding of UDP-Glc has been recognized in the membranes of cotton fibers (Delmer, Solomon, and Read, 1991), and Li and colleagues (1993) also identified a 52 kDa polypeptide as the most likely candidate for the catalytic subunit of $(1{\rightarrow}3)$-β-glucan synthase, as well as a 37 kDa UDP-Glc-binding polypeptide supposed to be involved in the synthesis of $(1{\rightarrow}4)$-β-glucan in vitro (Okuda et al., 1993; but see Delmer et al., 1993).

Both membrane-bound and detergent-solubilized preparations from suspension-cultured cotton cells incorporated radioactivity from UDP-Glc into a product that was found to be a slightly branched $(1{\rightarrow}3)$-β-glucan, with traces of $(1{\rightarrow}4)$-β-glucan (T. Genoud and P. Buchala, unpublished observations). The product existed in the form of fibrils about 1 nm (nanomater) in section that tended to associate into units of about 10 nm. The synthase was purified (25-fold) and examined by SDS-PAGE (sodium dodecyl sulphate-polyacrylamide gel electrophoresis) which showed that the preparation was highly enriched in polypeptides of 84, 58, 45, 34, 31, and 13 kDa. Photoaffinity labeling of the microsomal membrane fraction with ^{125}I-5-[3-(4-azidosalicylamide)]allyl-UDP-Glc, under conditions leading to the synthesis of β-glucans, labeled two peptides of about 30 kDa and an additional peptide of 27 kDa that are believed to be involved in the synthesis of the β-glucans. Three of the peptides (84, 58, and 34 kDa) are similar in molecular weight to those observed in cotton fibers (Delmer, Solomon, and Read, 1991), whereas the 34 kDa peptide was later identified as being of the annexin type (Andrawis, Solomon, and Delmer, 1993). Ramsden and Meier

(1990) detected a 45 kDa UDP-Glc-binding peptide in plasma membrane fractions that were phosphorylated. Other groups have identified analogous peptides with a molecular weight of 52 to 57 kDa in other plant cells (Lawson et al., 1989; Frost et al., 1990; Mason et al., 1990), whereas Meikle and colleagues (1991) identified a 31 kDa-peptide in *Lolium multiflorum*. Thus, it is likely that important callose synthase components belong to the two size classes of 30 to 31 kDa and 52 to 57 kDa.

UDP-glucose:protein transglucosylase has been found in the culture medium of cotton cells cultured in vitro (T. Genoud and A. Buchala, unpublished observations). Two 38 kDa peptides mainly present in the cytosol were found to be glucosylated by UDP-Glc. Digestion experiments with protease suggested that the glucose was covalently bound to the protein. Transfer of glucose from UDP-$[^{14}C]$Glc was greatly stimulated by the addition of ADP-Glc and Mn^{2+}, and ADP-Glc had a synergistic effect on the activation. However, ADP-Glc was neither the substrate for protein glucosylation nor for addition of a second glucose moiety to the glucosylated protein. In addition, the glucose incorporated could be chased by adding an excess of UDP-Glc or UDP-Xyl to the reaction medium, but not by UDP-Gal nor GDP-Glc. UDP-Xyl is thus thought to be another possible substrate for the transglucosylase. The glucosylation takes place on a large complex (> 1000 kDa) that was partially purified. Dhugga and colleagues (1991) postulated a role for a similar polypeptide glucosylation activity in the synthesis of xyloglucan. The essentially cytosolic localization of the labeled peptides in cotton does not support this idea since the synthesis of xyloglucan is known to be membrane associated (Ray, 1980). No direct evidence was obtained for the involvement of the transglucosylase in the synthesis of cell wall polysaccharides in cotton cells, and it is conceivable that starch is the relevant product.

LOW MOLECULAR WEIGHT CARBOHYDRATES IN COTTON FIBERS

The presence of sugars or honeydew in cotton lint may cause considerable trouble in processing the lint and may lower the quality of the yarn. Fibers with more than 0.3 percent of free sugars on a dry weight basis are termed sticky, and this characteristic depends on the rate of drying of the fibers once the boll has opened (Elsener, Hani, and Lubenevskaya, 1980). In part, the stickiness is due to honeydew formed after insect infestation, typically by *Bemesia* or *Aphis*, and the fibers are found to contain, in addition to the endogenous sugars (glucose, fructose, sucrose, glycerol, erythritol, and inositol), other sugars, such as trehalose, melizitose, and maltose, derived from the insects (Bourély, Gutknecht, and Fournier, 1984).

The changes in the nonstructural carbohydrate content of fibers of several species of cotton, at different stages of development, were examined by Jaquet, Buchala, and Meier (1982). The monosaccharides glucose and fructose showed maxima at the end of elongation or at the beginning of secondary wall formation, whereas the values for fructose-6-P and glucose-6-P continued to increase for several more days. Glucose and fructose were found in the ratio of about 1:1, but with slightly more glucose before the maxima and slightly more fructose after the maxima. The maxima for glucose and fructose were found to occur at the same time as those for $(1\rightarrow3)$-β-glucan, whereas the sucrose content continued to increase gradually until fiber maturity. These results were interpreted to mean that during primary wall formation, sucrose transported to the fiber is inverted and glucose and fructose are stored to be used later for the formation of secondary wall. Maximum ^{14}C-assimilate partitioning in the ovule and lint fiber also occurred at about the same time as the maxima for glucose and fructose (Benedict and Kohel, 1975). Invertase activity was found in the fibers and believed to be involved.

The mechanism of transport of assimilates (sucrose) from the leaf to the cotton fiber has been a subject of discussion. Benedict, Kohel, and Schubert (1976) showed that the rate of incorporation of ^{14}C-labeled assimilates into fibers was similar to the rate of dry weight increase of those fibers. Assimilate was also incorporated into the seed coat, sugars, oil, and protein of the ovule, showing that the funiculus was intact during this period. Absorption through the carpel wall into the locule cavity by the ovules was probably not important since transport to the carpel is complete by about the beginning of secondary wall formation. The same argument can be applied to the fibers. In addition, Ryser (1992) was able to show that many plamodesmata occur in the base of the cotton fiber, thereby providing a large capacity for symplastic transport processes. Acid invertase in fibers of *Gossypium arboreum* was found to be present in both soluble and cell wall-bound forms (Buchala, 1987). The total activity in fibers was found to be maximum at the onset of massive secondary cell wall formation, corresponding to the maxima found for glucose and fructose. Essentially similar results were later reported by Basra and colleagues (1990). Buffer-soluble invertase in *Gossypium hirsutum* showed an earlier maximum (Waefler and Meier, 1994). Radioactivity from both glucose and sucrose fed to the apoplast of intact seed clusters is incorporated into cell wall β-glucans (cf. Pillonel and Meier, 1985), but no direct relationship between invertase activity and β-glucan synthesis could be established. On the other hand, by feeding asymetrically labeled sucrose or labeled 1'-fluorosucrose, it could be shown that inversion of sucrose is not a prerequisite for its uptake from the cotton fiber apoplast.

Two pathways are feasible for the breakdown of sucrose: the more energetically favorable, reversible reaction mediated by sucrose synthase (cf. Basra et al., 1990) that yields UDP-Glc, the precursor for β-glucan synthesis and a starting point for the synthesis of other sugar nucleotides in the cytoplasm (for a full account see Delmer, Chapter 4, this volume), and, of course, hydrolysis by invertase to glucose and fructose. The latter can then be activated by hexokinases and be used for the glycolysis or the pentose phosphate pathways. Basra and Malik (1984) have proposed that it is likely that the rate of sucrose import and growth of fibers may be controlled by an interplay of invertase and sucrose synthase. Sucrose 6-phosphate synthase activity has also been detected in cotton fibers (unpublished observations of the author and C. Haigler), but this reaction is essentially irreversible and leads to the synthesis of sucrose.

The respiratory changes in the cotton fiber have been reviewed in detail by Basra and Malik (1984). Both the glycolysis and pentose phosphate pathways have been studied in cotton fibers, and the activities of key enzymes were found to increase during the periods of rapid elongation. Buffer-soluble glucose 6-phosphate dehydrogenase, the first committed enzyme of the pentose phosphate pathway, from *Gossypium hirsutum,* showed two maxima—one corresponding to the elongation period and a second during secondary wall formation (Waefler and Meier, 1994), that is, when energy requirements are high. Glutathione reductase activity drops progressively throughout fiber development. Waefler and Meier (1994) proposed that in elongating fibers, glutathione could act as hydrogen donor for peroxidase and that the level of glutathione is held more or less constant by the glutathione reductase, with consumption of NADPH (nicotinamide adenine dinucleotide phosphate [reduced form]) provided by the oxidation of glucose 6-phosphate. With decreasing glutathione reductase activity, less carbon would be directed through the pentose phosphate pathway, leaving more available for the synthesis of cellulose. However, no measurement of glutathione levels were carried out.

The main sugar nucleotide in cotton fibers is UDP-Glc (Franz, 1969). Carpita and Delmer (1981) investigated the carbon flow from labeled glucose in fibers grown on ovules cultured in vitro. The size of pools of various sugars (it should be noted that fructose was ignored) was examined, and computer predictions were made concerning their possible roles. They concluded that UDP-Glc was most likely the precursor of both cellulose and $(1\rightarrow3)$-β-glucan and that the turnover of UDP-Glc was more than sufficient to account for the continued rates of accumulation of the β-glucans, sucrose, and glucosylated lipids. Their work was taken a step further by Francey and colleagues (1989) who could show in pulse-chase experiments with ovule-grown fibers that turnover of $(1\rightarrow3)$-β-glucan, UDP-sugars (mainly UDP-Glc), sugar phosphates, and neutral sugars (glucose, fructose, and sucrose) all took place. The presence of the 2,6-dichlorobenzonitrile, known to inhibit the synthesis of cellulose, did not inhibit synthesis of $(1\rightarrow3)$-β-glucan, but led to increases in the radioactivity of low molecular weight sugar pools.

Waefler and Meier (1994) reported difficulties in measuring buffer-soluble sucrose synthase activity in *Gossypium hirsutum*, probably due to the very high invertase activity in their extracts. On the other hand, the activity of UDP-glucose pyrophosphorylase, another possible source of UDP-Glc in the cytoplasm, was found to increase with fiber development, and it was suggested that this enzyme could be the source of the precursor for β-glucans. More recently, Amor and colleagues (1995) (see also Delmer, Chapter 4, this volume) discovered that high amounts of sucrose synthase were associated with membrane fractions (we have to admit that our membrane fractions were often found to be contaminated with sucrose synthase activity and that this finding was ignored). Thus, it has been proposed that sucrose (capable of delivering UDP-Glc via membrane-bound sucrose synthase) is a better precursor for the synthesis of cellulose than UDP-Glc itself. High amounts of free UDP-Glc favor the synthesis of $(1\rightarrow3)$-β-glucan. Readers are referred to Figure 2 in D. P. Delmer's excellent account (Chapter 4) of the state of the art, for her explanations of the carbon metabolism leading to the synthesis of cellulose, since more or less the same considerations apply to the synthesis of $(1\rightarrow3)$-β-glucan. However, the finding that detached cotton fibers can still incorporate radioactivity into cellulose from sucrose does not correspond well with our own unpublished results that show poor incorporation occurred with detached fibers and carelessly handled seed clusters (again we have to admit that characterization of the product was not regularly carried out under such circumstances since the experiment was deemed to have "failed"). Sucrose synthase has also been localized in rapidly elongating fibers (Nolte et al., 1995). The authors state that the timing of onset for the cell-specific localization in fiber initials indicates a close association between this enzyme and sucrose import at a cellular level, as well as a potentially integral role in cell wall biosynthesis.

Glycolipids (glycosyl-acylsterols and glycosyl-polyprenols) are also labeled when particulate enzyme fractions (Forsee and Elbein, 1973; Forsee, Laine, and Elbein, 1974) or detached fibers (Delmer, Beasley, and Ordin, 1974) are incubated with labeled UDP-Glc or GDP-Glc. Similar lipids are also synthesized when sucrose is fed to the apoplast of intact seed clusters (A. Buchala, unpublished observations). There is no convincing evidence for the implication of such lipids as intermediates in the synthesis of β-glucans; several inhibitors of lipid glycosylation are without effect on the synthesis of β-glucans. However, one study (Palmer and Dugger, 1992) found labeled glucose from sterylglucosides in β-glucans of cotton fibers.

OTHER LOW MOLECULAR WEIGHT COMPOUNDS OF INTEREST

Developing cotton fibers possess an active system (PEP-carboxylase) for assimilating CO_2 (Dhindsa, Beasley, and Ting, 1975; Basra and Malik, 1983) that may complicate the interpretation of data from labeled photoassimilate (cf. Meier et al., 1981). The same authors also show that fiber growth depends on the turgor pressure in the fiber and that K^+ and malate are the main osmotically active solutes. The correlation between turgor pressure and fiber growth has already been reviewed (Basra and Malik, 1984) and will not be discussed here. Of particular interest, however, are the enzymes responsible for the metabolism of malate, for example, malate dehydrogenase. Our results involving an increase with fiber maturity (Waefler and Meier, 1994) for this activity are the opposite of those observed by Basra and Malik (1983). Since the malate content of the fibers rises with increasing age (9 to 46) and shows a turnover (Dhindsa, Beasley, and Ting, 1975; Jaquet, 1986), a stabilization, or even an increase, might be expected. In pulse-chase experiments using $^{14}CO_2$ and ovule cultures, it could be shown that fibers at the primary wall stage of growth incorporated significantly more radioactivity into high molecular weight material than fibers at the secondary wall stage. The absolute incorporation was poor compared to that from sugars fed to the culture medium, and therefore, this approach was not followed up.

CONCLUSIONS

The recent work in the laboratories of D. P. Delmer (membrane-bound sucrose synthase, characterization of *CelA* genes) and of R. M. Brown Jr. [synthesis in vitro of $(1{\rightarrow}4)$-β-glucan] should certainly stimulate progress in the field of β-glucan biosynthesis. It is to be hoped that some of the enthusiasm may be transferred to the biosynthesis of matrix polysaccharides. Just what prompts the cotton fiber cell to pass from the elongation (primary wall formation) stage to the onset of secondary wall formation still remains largely a mystery. Screening of cDNA libraries of cotton fibers showed that no major changes occur in the mRNA population during primary and secondary wall formation (John and Crow, 1992). No subset of genes was exclusively expressed during a given developmental stage, indicating that most of the fiber genes are active throughout development or that gene transcription ceases early in fiber development, but differential mRNA utilization occurs during growth. Little has been done other than the work on the regulation of transferases in Northcote's research group (cf. Northcote, 1985). UDP-

Glc appears to occupy a central point of control. The switch-over may involve the synthesis of enzymes that are more efficient in metabolizing UDP-Glc, directing synthesis toward cellulose. Thaker and colleagues (1986), Waefler and Meier (1994), and Ryser (1985) have shown that cell wall-bound peroxidase activity increases with fiber maturity and is maximum at about the end of cell elongation. To what extent, if any, is this activity involved in introducing cross-linkage of polysaccharides (Fry, 1988) to restrict fiber elongation? The assembly of the primary wall poses questions completely different from those concerning the secondary wall with its relatively highly ordered structure, for example, the transport of polysaccharide material from the endomembrane system to the cell wall (see Gibeaut and Carpita, 1994).

REFERENCES

Amor, Y., C.H. Haigler, S. Johnson, M. Wainscott, and D.P. Delmer (1995). A membrane-associated form of sucrose synthase and its potential role in synthesis of cellulose and callose in plants. *Proceedings of the National Academy of Sciences USA* 92: 9353-9357.

Andrawis, A., M. Solomon, and D.P. Delmer (1993). Cotton fiber annexins: A potential role in the regulation of callose synthesis. *Plant Journal* 3: 763-772.

Basra, A.S. and C.P. Malik (1983). Dark metabolism of CO_2 during fibre elongation of two cottons differing in fibre length. *Journal of Experimental Botany* 34: 1-9.

Basra, A.S. and C.P. Malik (1984). Development of the cotton fiber. *International Review of Cytology* 87: 65-113.

Basra, A.S., R.S. Sarlach, H. Nayyar, and C.P. Malik (1990). Sucrose hydrolysis in relation to development of cotton (*Gossypium* spp.) fibres. *Indian Journal of Experimental Biology* 28: 985-988.

Benedict, C.R. and R.J. Kohel (1975). Export of ^{14}C-assimilate in cotton leaves. *Crop Science* 15: 367-372.

Benedict, C.R., R.J. Kohel, and A.M. Schubert (1976). Transport of ^{14}C-assimilates to cotton seed: Integrity of funiculus during the seed filling stage. *Crop Science* 16: 23-27.

Bourély, J., J. Gutknecht, and J. Fournier (1984). Etude chimique du collage des fibres de coton. *Coton et Fibres Tropicales* 39: 47-51.

Buchala, A.J. (1987). Acid β-fructofuranoside fructohydrolase (invertase) in developing cotton (*Gossypium arboreum* L.) fibres and its relationship to β-glucan synthesis from sucrose fed to the fibre apoplast. *Journal of Plant Physiology* 127: 219-230.

Buchala, A.J., T. Genoud, and H. Meier (1987). Polysaccharides in the culture medium of cotton cells cultured *in vitro*. *Food Hydrocolloids* 1: 359-363.

Buchala, A.J., T. Genoud, S. Roulin, and K. Summermatter (1993). Xyloglucans in different types of cotton (*Gossypium* sp.) cells. *Acta Botanica Neerlandica* 42: 213-219.

Buchala, A.J. and H. Meier (1981). An arabinogalactan from the fibres of cotton (*Gossypium arboreum* L.). *Carbohydrate Research* 89: 137-143.

Buchala, A.J. and H. Meier (1985). Biosynthesis of β-glucans in growing cotton (*Gossypium arboreum* L. and *Gossypium hirsutum* L.) fibres. In *Biochemistry of Plant Cell Walls*, Eds. C.T. Brett and J.R. Hillman. Cambridge, UK: Cambridge University Press, Society of Experimental Biology Seminar Series, 28, pp. 221-241.

Buchala, A.J., S. Roulin, and H. Meier (1989). Polysaccharides in the culture medium of cotton (*Gossypium hirsutum* L.) ovules cultured in vitro. *Plant Cell Reports* 8: 25-28.

Bucheli, P., A.J. Buchala, and H. Meier (1986). β-glucanases in cotton fibres can function as transglucosylases. In *Cell Walls '86. Proceedings of the Fourth Cell Wall Meeting*, Eds. B. Vian, D. Reis, and R. Goldberg. Paris: Université Pierre et MarieCurie-Ecole Normale Supérieure, pp. 368-369.

Bucheli, P., A.J. Buchala, and H. Meier (1987). Autolysis in vitro of cotton (*Gossypium hirsutum* L.) fibre cell walls. *Physiologia Plantarum* 70: 633-638.

Bucheli, P., M. Dürr, A.J. Buchala, and H. Meier (1985). β-glucanases in developing cotton (*Gossypium hirsutum* L.) fibres. *Planta* 166: 530-536.

Carpita, N.C. and D.P. Delmer (1981). Concentration and metabolic turnover of UDP-glucose in developing cotton fibers. *Journal of Biological Chemistry* 256: 308-315.

Darvill, A., M. McNeil, P. Albersheim, and D.P. Delmer (1980). The primary cell walls of flowering plants. In *The Biochemistry of Plants*, Volume 1, Ed. N.E. Tolbert. New York: Academic Press, pp. 91-162.

Delmer, D.P. (1987). Cellulose biosynthesis. *Annual Review of Plant Physiology* 38: 259-290.

Delmer, D.P. (1999). Cellulose biosynthesis in developing cotton fibers. In *Cotton Fibers: Developmental Biology, Quality Improvement, and Textile Processing*, Ed. A.S. Basra. Binghamton, NY: Food Products Press, pp. 85-112.

Delmer, D.P., C.A. Beasley, and L. Ordin (1974). Utilization of nucleoside diphosphate glucoses in developing cotton fibers. *Plant Physiology* 53: 149-153.

Delmer, D.P., G. Cooper, D. Alexander, J. Cooper, and T. Hayashi (1985). New approaches to the study of cellulose biosynthesis. *Journal of Cell Science* (suppl.) 2: 33-50.

Delmer, D.P., P. Ohana, L. Gonen, and M. Benziman (1993). In vitro synthesis of cellulose in plants: Still a long way to go. *Plant Physiology* 103: 307-308.

Delmer, D.P., M. Solomon, and S.M. Read (1991). Direct photolabeling with [^{32}P]UDP-glucose for identification of a subunit of cotton fiber callose synthase. *Plant Physiology* 95: 556-563.

Dhindsa, R.S., C.A. Beasley, and I.P. Ting (1975). Osmoregulation in the cotton fiber. Accumulation of potassium and malate during growth. *Plant Physiology* 56: 394-398.

Dhugga, K.S., P. Ulvskov, S.R. Gallagher, and P.M. Ray (1991). Plant polypeptides reversibly glycosylated by UDP-glucose. *Journal of Biological Chemistry* 266: 21977-21984.

El Rassi, Z., D. Tedford, J. An, and A. Mort (1991). High-performance reversed-phase chromatographic mapping of 2-pyridylamino derivatives of xyloglucan oligosaccharides. *Carbohydrate Research* 215: 25-38.

Elsener, O., J. Hani, and E. Lubenevskaya (1980). The sugar content in cotton lint of growing bolls. *Coton et Fibres Tropicales* 37: 223-225.

Forsee, W.T. and A.D. Elbein (1973). Biosynthesis of mannosyl- and glucosyl-phosphoryl-polyprenols in cotton fibers. *Journal of Biological Chemistry* 248: 2858-2867.

Forsee, W.T. and A.D. Elbein (1975). Glycoprotein biosynthesis in plants. Demonstration of lipid-linked oligosaccharides of mannose and N-acetylglucosamine. *Journal of Biological Chemistry* 250: 9283-9293.

Forsee, W.T., R.A. Laine, and A.D. Elbein (1974). Solubilization of a particulate UDP-glucose: sterol β-glucosyl-transferase in developing cotton fibres and seeds and characterization of steryl 6-acyl-D-glucosides. *Archives of Biochemistry and Biophysics* 161: 248-259.

Francey, Y., J.P. Jaquet, S. Cairoli, A.J. Buchala, and H. Meier (1989). The biosynthesis of β-glucans in cotton (*Gossypium hirsutum* L.) fibres of ovules cultured in vitro. *Journal of Plant Physiology* 134: 485-491.

Franz, G. (1969). Soluble nucleotides in developing cotton hair. *Phytochemistry* 8: 737-741.

Frost, D.J., S.M. Read, R.R. Drake, B.E. Haley, and B.P. Wasserman (1990). Identification of the UDP-glucose-binding polypeptide of callose synthase from *Beta vulgaris* L. by photoaffinity labeling with 5-azido-UDP-glucose. *Journal of Biological Chemistry* 265: 2162-2167.

Fry, S.C. (1988). *The Growing Plant Cell Wall: Chemical and Metabolic Analyses.* Harlow, UK: Longman.

Fry, S.C. (1989). The structure and function of xyloglucans. *Journal of Experimental Botany* 40: 1-11.

Gibeaut, D.M. and N.C. Carpita (1994). Biosynthesis of plant cell wall polysaccharides. *The Federation of American Societies for Experimental Biology Journal* 8: 904-915.

Hayashi, T. and D.P. Delmer (1988). Structural analysis of the xyloglucan of cotton fiber cell walls. *Carbohydrate Research* 181: 273-277.

Huwyler, H.R. (1978). Die Zellwandkomponenten und Glykosidasenaktivitäten in Baumwollhaaren (*Gossypium arboreum* L.) verschiedener Entwicklungsstadien. PhD Thesis, Faculty of Science, University of Fribourg, Switzerland.

Huwyler, H.R., G. Franz, and H. Meier (1978). β-glucans in the cell walls of cotton fibers (*Gossypium arboreum* L.). *Plant Science Letters* 12: 55-62.

Huwyler, H.R., G. Franz, and H. Meier (1979). Changes in the composition of cotton fibre cell walls during development. *Planta* 146: 635-642.

Jaquet, J.P. (1986). Le rôle du fructose, du glucose, du saccharose et du L-malate dans le métabolisme des hydrates de carbone et dans la synthèse de la cellulose et de la callose dans les fibres provenant d'ovules de cotton (*Gossypium hirsutum* L.) cultivées *in vitro*. PhD Thesis, Faculty of Science, University of Fribourg, Switzerland.

Jaquet, J.P., A.J. Buchala, and H. Meier (1982). Changes in the non-structural carbohydrate content of cotton (*Gossypium* spp.) fibres at different stages of development. *Planta* 156: 481-486.

John, M.E. and L.J. Crow (1992). Gene expression in cotton (*Gossypium hirsutum* L.) fiber: Cloning of the mRNAs. *Proceedings of the National Academy of Sciences USA* 89: 5769-5773.

Komalavilas, P. and A.J. Mort (1989). The acetylation at O-3 of galacturonic acid in the rhamnose-rich portions of pectins. *Carbohydrate Research* 189: 261-272.

Kreuger, M. and G.J. van Holst (1996). Arabinogalactan proteins and plant differentiation. *Plant Molecular Biology* 30: 1077-1086.

Lawson, S.G., T.L. Mason, R.D. Sabin, M.E. Sloan, R.R. Drake, B.E. Haley, and B.P. Wasserman (1989). UDP-glucose:(1,3)-β-glucan synthase from *Daucus carota* L. Characterization, photoaffinity labeling and solubilization. *Plant Physiology* 90: 101-108.

Li, L., R.R. Drake Jr., S. Clement, and R.M. Brown Jr. (1993). β-glucan synthesis in the cotton fiber. III. Identification of UDP-glucose-binding subunits of β-glucan synthases by photoaffinity labeling with [β-^{32}P]5′-N$_3$-UDP-glucose. *Plant Physiology* 101: 1149-1156.

Maltby, D., N.C. Carpita, D. Montezinos, C. Kulow, and D.P. Delmer (1979). β-1,3-glucan in developing cotton fibers. *Plant Physiology* 59: 1088-1097.

Marti, R. (1991). In-vitro-Kulturen von Samenanlagen der Baumwolle (*Gossypium hirsutum* L.): Kallusbildung, Zellwandentwicklung von suspendierten Kalluszellen, Einfluss von Ethylen auf die Kallusbildung. PhD Thesis, Faculty of Science, University of Fribourg, Switzerland.

Mason, T.L., S.M. Read, D.J. Frost, and B.P. Wasserman (1990). Inhibition and labeling of red beet uridine 5′-diphosphoglucose:(1,3)-β-glucan (callose) synthase by chemical modification with formaldehyde and uridine 5′-diphosphopyridoxal. *Physiologia Plantarum* 79: 439-447.

McDougal, G.J. and S.C. Fry (1989). Structure-activity relationships for xyloglucan oligosaccharides with anti-auxin activity. *Plant Physiology* 89: 883-887.

Meier, H., L. Buchs, A.J. Buchala, and T. Homewood (1981). (1→3)-β-D-glucan (callose) is a probable intermediate in the biosynthesis of cellulose of cotton fibres. *Nature* 289: 821-822.

Meikle, P.J., K.F. Ng, E. Johnson, N.J. Hoogenraad, and B.A. Stone (1991). The β-glucan synthase from *Lolium multiflorum*: Detergent solubilization, purification using antibodies, and photoaffinity labeling with a novel photoreactive pyrimidine analogue of uridine-5′-diphosphoglucose. *Journal of Biological Chemistry* 266: 22569-22581.

Meinert, M. and D.P. Delmer (1977). Changes in the biochemical composition of the cell wall of the cotton fiber during development. *Plant Physiology* 59: 1088-1097.

Nolte, K.D., D.L. Hendrix, J.W. Radin, and K.E. Koch (1995). Sucrose synthase localization during initiation of seed development and trichome differentiation in cotton ovules. *Plant Physiology* 109: 1285-1293.

Northcote, D.H. (1985). Cell organelles and their function in the biosynthesis of cell wall components: Control of wall assembly during differentiation. In *Biosynthesis and Degradation of Wood Components*, Ed. T. Higuchi. Orlando, FL: Academic Press, pp. 87-108.

Ohana, P., M. Benziman, and D.P. Delmer (1993). Stimulation of callose synthesis *in vivo* correlates with changes in intracellular distribution of the callose activator β-furfuryl-β-glucoside. *Plant Physiology* 101: 187-191.

Okuda, K., L. Li, K. Kudlicka, S. Kuga, and R.M. Brown Jr. (1993). β-glucan synthesis in the cotton fibre. I. Identification of β-1,4- and β-1,3-glucans synthesized *in vitro*. *Plant Physiology* 110: 1131-1142.

Palmer, R.L. and W.M. Dugger (1992). Recovery of glucose from glucolipids in cell wall glucans in cotton fibres. *Phytochemistry* 31: 2631-2633.

Pillonel, C. and H. Meier (1985). Influence of external factors on callose and cellulose synthesis during incubation *in vitro* of intact cotton fibres with [^{14}C]sucrose. *Planta* 165: 76-84.

Ramsden, L. and H. Meier (1990). Phosphorylation of proteins in the plasma membrane fractions from developing cotton fibres. *Journal of Plant Physiology* 136: 313-317.

Ray, P.M. (1980). Cooperative action of β-glucan synthase and UDP-xylosyl transferase of Golgi membranes in the synthesis of xyloglucan-like polysaccharide. *Biochemica et Biophysica Acta* 629: 431-444.

Robertson, D., I. Beech, and G.P. Bolwell (1995). Regulation of the enzymes of UDP-sugar metabolism during differentiation of French bean. *Phytochemistry* 39: 21-28.

Rowland, S.P. and P.S. Howley (1988). Extractability of the (1→3)-β-glucan from developing cotton fiber. *Carbohydrate Research* 148: 162-167.

Rowland, S.P., P.S. Howley, and W.S. Anthony (1984). Specific and direct measurement of the β-1,3-glucan in developing cotton fiber. *Planta* 161: 281-287.

Ryser, U. (1985). Cell wall biosynthesis in differentiating cotton fibres. *European Journal of Cell Biology* 39: 236-265.

Ryser, U. (1992). Ultrastructure of the epidermis of developing cotton (*Gossypium*) seeds: Suberin, pits, plasmodesmata, and their implication for assimilate transport into cotton fibers. *American Journal of Botany* 79: 14-22.

Ryser, U. (1999). Cotton fiber initiation and histodifferentiation. In *Cotton Fibers: Developmental Biology, Quality Improvement, and Textile Processing*, Ed. A.S. Basra. Binghamton, NY: Food Products Press, pp. 1-45.

Stone, B.A. and A.E. Clarke (1992). β-glucans. In *Chemistry and Biology of (1→3)-β-glucans*. Bundoora, Australia: La Trobe University Press.

Thaker, V.S., S. Saroop, P.P. Vaisnav, and Y.D. Singh (1986). Role of peroxidase and esterase activities during cotton fibre development. *Journal of Plant Growth Regulation* 5: 17-27.

Thompson, J.E. and S.C. Fry (1997). Trimming and solubilization of xyloglucan after deposition in the wall of cultured rose cells. *Journal of Experimental Botany* 48: 297-305.

Timpa, J.D. and B.A. Triplett (1993). Analysis of cell-wall polymers during cotton fibre development. *Planta* 189: 101-108.

Triplett, B.A. and J.D. Timpa (1995). Characterization of cell-wall polymers from cotton ovule culture fiber cells by gel permeation chromatography. *In Vitro Cellular and Developmental Biology* 31: 171-175.

Trolinder, N.L., J.D. Berlin, and J.R. Goodin (1987). Differentiation of cotton fibers from single cells in suspension culture. *In Vitro Cellular and Developmental Biology* 23: 789-794.

Waefler, U. and H. Meier (1994). Enzyme activities in developing cotton fibres. *Plant Physiology and Biochemistry* 32: 697-702.

Warneck, H.M., T. Haug, and H.U. Seitz (1996). Activation of cell wall-associated peroxidase isoenzymes in pea epicotyls by a xyloglucan-derived nonasacchariode. *Journal of Experimental Botany* 47: 1897-1904.

Waterkeyn, L. (1981). Cytochemical localization and function of the 3-linked glucan callose in the developing cotton fibre cell wall. *Protoplasma* 106: 49-67.

York, W.S., A.G. Darvill, and P. Albersheim (1984). Inhibition of 2,4-dichlorophenoxyacetic acid stimulated elogation of pea stem segments by a xyloglucan oligosaccharide. *Plant Physiology* 75: 295-297.

Chapter 6

Structural Development of Cotton Fibers and Linkages to Fiber Quality

You-Lo Hsieh

Cotton fibers have economic significance in the global market because of their large share (over 50 percent) among fibers for apparel and textile goods (Harig, 1992). This share is being challenged by the recent commercialization of the new microdenier (polyesters and nylons), elastomeric (spandex), and lyocell fibers, among others. The market value of cotton fibers and the quality of cotton products are directly related to fiber quality. The interfiber competitiveness of cotton fibers and the share of cotton fibers in the global apparel and textile market depend on significant improvement in fiber quality, as well as on process innovation and product differentiation.

The most essential cotton fiber qualities related to dry processing (yarn spinning, weaving, and knitting) are length, strength, and fineness. For wet processing, such as scouring, dyeing, and finishing, the fiber structure related to maturity, or the level of development, plays a major role. Knowledge of cotton fibers has been derived mainly from studies of dried and matured fibers of different varieties. There is, however, much less understanding regarding the relationship between cell wall development and fiber structure and properties. To develop effective strategies for fiber quality improvement and for innovative processing, the developmental linkages to fiber qualities need to be identified.

The author would like to extend special appreciation to Xiao-Ping Hu, Trucmy Dong, Ewa Honic, Viet Nguyen, Phuong-Anh Nguyen, and M. Michelle Hartzell for their assistance in the laboratory. Support for this research from Cotton Incorporated (#93-069) is also appreciated.

This chapter reviews the structural development of cotton fibers and discusses the most recent findings regarding the developmental changes in fiber quality. Structural development and properties of cotton fibers during primary wall formation (elongation) and secondary wall thickening (cellulose synthesis), as well as during desiccation (transition from mobile to highly hydrogen-bonded structure), are reported.

CHEMICAL COMPOSITION

Cotton fibers are composed of mostly α-cellulose (88.0 to 96.5 percent) (Goldwaith and Guthrie, 1954). The rest is noncellulosics that are located on the outer layers and inside the fiber lumen. The specific chemical composition of cotton fibers varies by their varieties and growth conditions. The noncellulosics include proteins (1.0 to 1.9 percent), waxes (0.4 to 1.2 percent), pectins (0.4 to 1.2 percent), inorganics (0.7 to 1.6 percent), and other substances (0.5 to 8.0 percent). These levels are higher in the immature fibers.

The primary walls of cotton fibers contain less than 30 percent cellulose, noncellulosic polymers, neutral sugars, uronic acid, and various proteins (Huwyler, Franz, and Meier, 1979; Meinert and Delmer, 1977). The amounts of noncellulosic components change during fiber elongation and the transition from primary to secondary wall, but discrepancies remain in the exact quantities of these changes (see also Buchala, Chapter 5, this volume). Some of the protein constituents (enzymatic, structural, or regulatory) are unique to cotton fibers and have been found to be developmentally regulated (Meinert and Delmer, 1977). Among the inorganic substances, the presence of phosphorus in the form of organic and inorganic compounds is of importance to the scouring process used to prepare fibers for dyeing. These phosphorus compounds are soluble in hot water but become insoluble in the presence of alkali earth metals. The use of hard water, therefore, can precipitate alkali earth metal phosphates on the fibers instead of eliminating them (Hornuff and Richter, 1964).

MACROSTRUCTURE

The macrostructure of cotton fibers can be viewed along the fiber axis and across the fiber section. Current understanding of cotton fiber macrostructure has emerged mainly from research on the matured fibers in their dried state. Although the biochemical nature of fiber growth or development has been extensively studied, the macrostructure during growth is not as well understood. Prior to boll dehiscence and fiber desiccation, matured cotton fibers have been shown to exhibit high intrinsic mobility and porosity in their structure. The accessibility of water in fiber structure at this stage is higher than after desiccation.

Drying of the fibers involves the removal of fluids from the lumen and of intermolecular water in the cellulose. The fluid loss from the lumen causes the cylindrical fibers to collapse. The loss of intermolecular water allows the cellulose chains to come closer together and form intermolecular hydrogen bonds. The collapse of cell walls and hydrogen bond formation cause irreversible morphological changes, including structural heterogeneity, decreasing porosity, and sorption capacity in the fibers (Stone and Scallan, 1965; Ingram et al., 1974). These changes increase molecular strains and reduce chain mobility and may have influence over properties essential to strength and dyeing/finishing processes. As these irreversible changes determine the utility of fibers, an understanding of the structural changes caused by desiccation is essential to fiber quality research.

The matured fibers dry into flat, twisted ribbon forms. The twist or convolution directions reverse frequently along the fibers. The number of twists in cotton fibers varies between 3.9 and 6.5 per mm (millimeter) (Warwicker et al., 1966), and the spiral reversal changes 1 to 3 times per mm length (Rebenfeld, 1977). The convolution angle has been shown to be variety dependent (Peterlin and Ingram, 1970). Differences in reversal frequency have been observed among different species and varieties of cotton, between lint and fuzz on the same ovule, and along a single fiber (Balls, 1928).

The reversals in cotton fibers are related to the orientation of the secondary wall microfibrils, whose organization is critically important to fiber strength. The orientation angles and shifts of microfibrils along the fiber axis change with cell development stages and have been related to the cellular organization of the cortical microtubules during cotton fiber development (Yatsu and Jacks, 1981; Seagull, 1986, 1992). At the beginning of fiber development, that is, 1 dpa (days postanthesis), cortical microtubules have a random orientation, but a shallow pitched helical orientation of 75 to 80 degrees develops during the transition between initiation and elongation, that is, 2 to 3 dpa. Such angles, which are nearly perpendicular to the fiber axis, are maintained throughout primary wall synthesis. An abrupt shift in orientation to a steeply pitched helical pattern occurs between primary and secondary wall synthesis. As secondary cell walls thicken, the angles reduce further. In early secondary wall synthesis, a fourfold increase occurs in the number of microtubules. The fibrillar orientation reverses along the fiber periodically. The mechanism that regulates the synchronized shifts in microtubule orientation is not yet understood.

The concomitant shifts in orientation of microtubules and microfibrils indicate a strong relationship between the two. However, differing degrees of variability between the two suggest that other factors may modify the order imparted by the microtubules. During secondary wall synthesis, microfibrils exhibit variability in orientation, or undulations. Interfibril hydrogen bonding and differential rigidity of microtubules and microfibrils have been suggested as possible factors influencing the final microfibril organization.

The spiral fibrillar structure can be observed on the surface of mature fibers underneath the primary wall. Parallel ridges and grooves are seen at 20- to 30-degree angles to the fiber axis. Scouring exposes the fibrils of the primary and secondary walls. Neither soaking in water nor slack mercerization removes surface roughness (Tripp, Moore, and Rollins, 1957; Muller and Rollins, 1972; deGruy, Carra, and Goynes, 1973). However, stretching a swollen fiber can smoothen the surface and make residual ridges more parallel to the fiber axis secondary walls. Immature fibers have smooth surfaces.

The molecular packing densities along the fiber, particularly near the spiral reversals, are believed to vary. The packing of fibrils at the reversals is denser (Patel et al., 1990). The adjacent fibrillar structures are less densely packed and often have different dimensions (fineness). Therefore, these adjacent regions are believed to be the weak points on the fiber rather than the reversals themselves. It has also been suggested that the reversals may be growth points in the fibers (Raes, Fransen, and Verschraege, 1968). However, this has not been confirmed by others.

The dried cotton fibers have a bean-shaped cross section. The bilateral structure is thought to originate from the asymmetry of mechanical forces in the fibers during drying. Heterogeneity of molecular packing in cotton fibers has been demonstrated by light and electron microscopy and by the accessibility to reagents (Nelson and Mares, 1965; Ingram et al., 1974; Patel et al., 1990; Peeters and DeLanghe, 1986; Tsuji et al., 1990). The two highly curved ends of the bean-shaped cross section have the highest density of molecular packing and the least accessibility to reagents. The structure of the convex part is less dense and more accessible. The concave section of the cross section is the most accessible and the most reactive portion of the fibers. The higher density and parallel membranes in the curved extremes and convex parts are thought to result from radial compressive forces, whereas the concave portion of the cross section is subject to tangential compressive forces. The sections between the curved ends and concave parts are denoted as neutral zones that are, by far, the most accessible. These differential structures in the cotton cross section have been confirmed by enzymatic attacks (Kassenback, 1970).

FIBER QUALITY

Cotton fibers include the long lint and short fuzz fibers, both of which are epidermal hairs on the cotton ovules or seeds. Although both types of fibers have extensive secondary cell wall development, fuzz fibers initiate between 4 to 10 dpa and never reach final lengths nor degree of maturity (secondary wall thickness) (Beasley, 1977). Fuzz fibers are sometimes pigmented and are less twisted than lint fibers (Joshi, Wadhwani, and Johri, 1967). The fuzz fibers have characteristic alternating ringlike patterns in their cross-sectional structure that are absent in the lint fibers (Berlin, 1986).

The ginning process removes the lint fibers, leaving the seeds covered with short fuzz fibers or "linters." The quality of lint fibers is judged mainly by their spinnability, or how well the fibers can be spun into yarns. Fiber length, fiber fineness and length distribution, and fiber strength are the most important fiber quality factors for textile processing. The ranked importance of these fiber qualities varies with the type of yarn-spinning method, such as ring, rotor, and air jet. These fiber qualities also determine the yarn strength, yarn regularity, and the handle and luster of fabrics. Fiber quality traits, that is, length, fineness, and strength, are determined by both the genetic and environmental variables.

Fiber length can be affected by ginning. The ginning process detaches the long lint fibers from the seeds at the base of the fiber. Distinctive elbow, shank, and foot regions at the basal end of fibers exist near the seed coat (Fryxell, 1963). In the elbow region of fibers, the thickened primary and secondary cell walls expand laterally during development and desiccation, exerting pressure on the neighboring lint cells (Berlin, 1986). The compression by neighboring thick-walled cells makes the elbow the weak point of lint fiber attachment at the seed coat surface (Vigil et al., 1996). Ideally, ginning should remove fibers by breaking the lint fibers along a transact just above the elbow region, preserving the longest fiber lengths. In the spinning process, fibers shorter than 5 mm are also removed because fibers shorter than the average length contribute to poorly spun yarns with excessive hairiness, low uniformity, and low strength.

It has been shown that fibers taper toward thinner tip ends. Using light microscopy, we have observed that about 15 percent of fiber length from the tip has smaller dimensions on fully developed SJ-2 fibers. As much as one-third of fiber length has been reported on Delta 61 fibers (Boylston, Thibodeaux, and Evans, 1993). Fibers shrink in proportion to the amount of cellulose present in the cell wall. The perimeter of less mature fibers (thinner cell wall) is larger than that of a more mature (thicker cell wall) fiber. Thinner cell wall is found near the fiber tip than in the rest of the fiber. Fiber fineness directly determines yarn fineness. Although as few as thirty fibers can be spun into yarns in ring spinning, approximately 100 fibers in the yarn cross section are usually the lower limit. Therefore, reduction in fiber fineness is the only way to achieve fine yarns within the limits of spinning processes.

Fiber maturity is the proportion of cotton cell wall thickness compared to the maximum wall thickness when growth is completed. Therefore, maturity is a growth characteristic and represents the development of the secondary cell wall. The maturity level can be complicated by both developmental and environmental factors. Mature cotton fibers have tenacities in the range of 15 to 40 cN/tex, exceeding the 6 cN/tex minimum tenacity for fiber processing. The strength of yarns, developed through twisting, utilizes only about 30 to 70 percent of total fiber strength. The strength relationship between single fibers and yarns is complex. For example, fibers shorter than 12.5 mm make little contribution to yarn strength. Fiber elastic behavior (elongation) and interfiber frictional characteristics may also contribute to yarn strength, but their association with strength has not yet been systematically studied.

Most work on cotton fiber strength has been reported using fiber bundle measurement. In 1969, the High Volume Instrument (HVI) system was developed by the United States Department of Agriculture (USDA) for classifying cotton fibers (Muller, 1991). The HVI has become an efficient and reliable system for quality assessment and productivity improvement in the textile industry. This fast and economical testing system made possible the "every bale testing" of cotton produced in the United States in 1992.

The HVI system measures several properties of fiber bundles, such as fiber fineness, color, trash content, fiber length, and tensile properties, in a short time. Fiber length measurements can be performed on the "texLAB" system, which consists of a fiber alignment unit (P710), a fiber length measurement unit (AL101), and a data processing unit (P810). The principle of measurement is based on the dielectric properties of a fiber bundle that forms a wedge shape as a result of the distribution of short fibers in the tuft. Such a measurement generates a diagram of fiber frequency against fiber length and information on the fiber lengths at the various percentiles of fibers. The micronaire, or fiber fineness, measure is based on the flow characteristics of gases through a plug of fibers. The flow resistance increases with coarser fibers at a constant fiber mass and flow volume. A higher micronaire value indicates either coarser fibers or thinner fibers with thick cell walls. Fiber bundle strength, measured by the Stelometer or HVI, has to be translated to single fiber strength using established correlations. None of the bundle strength methods, however, can measure differences in varietal traits or developmental effects and, thus, are not as useful to researchers (molecular biologists, breeders, and fiber scientists) as they are to textile processors (Green and Culp, 1990).

STRENGTH

The mechanical properties of cotton fibers have been studied extensively and reviewed (Rebenfeld, 1977; Zeronian, 1991). In general, the strength of cotton fibers is attributed to the rigidity of the cellulosic chains, the highly fibrillar and crystalline structure, and the extensive intermolecular and intramolecular hydrogen bonding. The genotype link to fiber strength has been well documented. The majority of strength data has been based on bundle strength, such as that generated in recent years by the Stelometer or the HVI.

In general, cotton fiber strength has also been shown to be associated with the molecular weight of the cellulose, the crystallinity in the fibers, and the reversal and convolution structure of the fibers. Because of the polydisperse nature of polymer sizes, molecular weights of polymers are commonly described by the number average or weight average molecular weights, depending upon the methods of determination. Positive relationships have been reported on the HVI bundle strengths of cotton fibers from several cultivars and on the number average molecular weight of crystalline cellulose liberated through acid hydrolysis (60 minutes in boiling acetic acid/nitric acid) (Benedict, Kohel, and Jividen, 1994). The HVI bundle strengths of several calibration cottons have also been related to their weight average molecular weights (Timpa and Ramey, 1994).

Positive relationships between bundle strength by the Stelometer and crystallinity (53 percent to 69 percent) have been reported in mature fibers from eight Egyptian cottons (Hindeleh, 1980). However, the relationships between strength and crystallinity may not be easily compared among studies. One of the reasons is that the extent of crystallinity of matured cottons ranges from 50 percent to close to 100 percent, depending upon the measurement techniques. The differences are further complicated by the inevitable variations among cotton fibers due to a combination of varietal and environmental factors.

Much less is known about the strength of developing cotton fibers. Kulshreshtha and colleagues (1973a,b) have shown that the Stelometer bundle strength increases gradually with fiber growth between 30 and 70 dpa. The youngest age, in that case, is about 2 weeks into, or about halfway, through secondary cell wall development. As bundle strength is not sensitive to strength variability, measurement of single fiber strength is necessary. For better understanding of the strength-development relationship, fibers from a wider range of cell developmental stages need to be studied.

We have investigated the single fiber tensile strengths and the crystalline structure of fibers from two Acala cotton varieties, SJ-2 and Maxxa, at varying stages of development and at maturity. Flowers were tagged on the day of flowering (anthesis), and green bolls representing developmental stages ranging from 14 dpa to 50 dpa and bolls that were matured and opened on plants were used. Fibers from randomly sampled first-position (closest to the main stem) bolls between the fourth and the twelfth fruiting branches were used. Positional effects on single fiber strength have been found to be negligible with this boll-sampling method (Hsieh, Honic, and Hartzell, 1995).

Tensile measurements of both hydrated and dried fibers were made. Sources of fiber development variations were minimized by using the fibers from the medial section of the ovules and the most developed ovules from each boll. This sampling approach targets the fibers that most closely represent each of the developmental stages. Variations along single fibers were further minimized by measuring only the middle section of each fiber. Single fiber tensile measurements were performed using either an Instron tensile tester (1122 TM) equipped with standard pneumatic and rubber-faced grips or a Mantis single fiber tester. The much higher rate of measurement on the Mantis instrument enabled collections of large single fiber strength data. Furthermore, the tensile measurements of hydrated fibers (at 20 dpa and older) can also be performed on the Mantis instrument.

The single fiber tensile properties measured by the Mantis instrument show that the forces required to break both hydrated and dried single Maxxa fibers increase with fiber development (see Figure 6.1a). Hydrated fibers have higher breaking forces than dried fibers, but their breaking elongation values are lower than those for dried fibers (see Figure 6.1b). The breaking forces of the dried fibers appear to increase at a higher rate between 20 and 30 dpa than in the later stages. The work to break values (microjoule) are similar between the hydrated and dried fibers (see Figure 6.1c). The decreasing forces and increasing strain at break from the hydrated state to the dry state may be explained by the increased convolution angles resulting from cell collapsing and dehydration.

The average seed fiber weights increase linearly with fiber development (see Figure 6.2a) in every series of developing fibers we have studied (Hsieh, Honic, and Hurtzell, 1995; Hsieh, Hu, and Nguyen, 1997). It has been confirmed that seed fiber weight values are reliable indicators for the stages of fiber development. With this series of developing fibers, the strength and seed fiber weight relationship is similar to that of the strength-dpa relationship (see Figure 6.2b). Therefore, it is quite conceivable that fiber strength may be projected from seed fiber weight at early stages of secondary cell wall development by using such established strength-seed fiber weight correlations.

The breaking forces measured by the Mantis instrument appear to be slightly higher than those measured by the Instron (see Figure 6.3a), whereas the opposite is observed with the breaking elongation values (see Figure 6.3b). These observations are likely due to the differences in the ways fibers are handled for these two measurement techniques. On the Mantis, fibers are positioned manually. The instrument automatically straightens, clamps down, and exerts a preload on individual fibers. Single fiber measurements conducted on the Instron tensile instrument require extensive handling to prepare each fiber in a paper holder (Hsieh, Honic, and Hartzell, 1995). The extra handling involved in preparing fibers for measurements on the Instron can disturb the integrity of the fibers, causing strength reduction. The absence of preload when preparing fibers for measurement using the Instron explains the higher breaking elongation values.

The linear densities of developing Maxxa fibers increase with fiber development (see Figure 6.2c). The breaking tenacities generated from the Mantis and Instron instruments are similar. The most significant observation is that Maxxa fibers from the two 21 dpa bolls have lower tenacities than those which are 24 dpa or older.

FIGURE 6.1. Single Fiber Tensile Properties of Developing Maxxa Fibers Measured by the Mantis Instrument: (a) Breaking Force, (b) Breaking Elongation, and (c) Work to Break

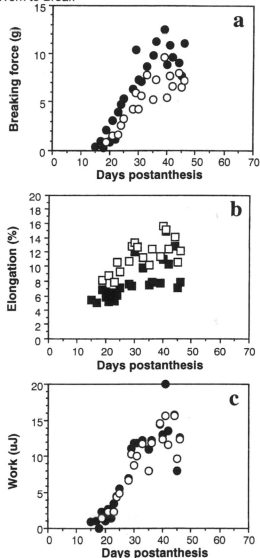

Note: Solid and open symbols indicate data from the hydrated and dried fibers, respectively. Single fiber tensile measurements were performed using a Mantis single fiber tester. A 3.2 mm gauge length and a 60 mm/minute strain rate were employed. All measurements were performed at a constant temperature of 21°C and a 65 percent relative humidity. For each sample, the number of single fiber measurements was 100.

FIGURE 6.2. Relationships Between (a) Seed Fiber Weight and Age, (b) Breaking Force and Seed Fiber Dry Weight, and (c) Linear Density and Age of Developing Maxxa Fibers

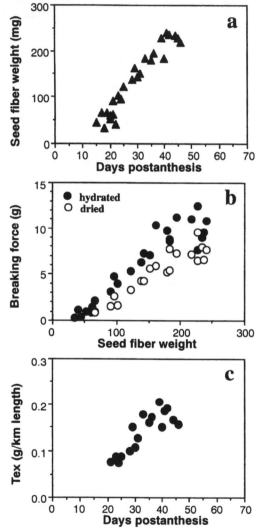

Note: The linear density or tex (grams per kilometer length) of dried fibers was derived from the weights of 100 1 cm sections of fibers. Fibers were sampled from the medial sections of the same ovules as those for the tensile measurements. The middle 1-cm sections of fibers were cut from an array of combed and aligned fibers. One-hundred 1 cm fibers were weighed to 0.1 μg (microgram), and five such measurements were made from ovules selected from each boll. Seed fiber weight (6.2a) and linear density (6.2c) can only be obtained on dried samples.

FIGURE 6.3. Single Fiber Tensile Properties of Developing Maxxa Fibers Measured by the Mantis and the Instron Instruments: (a) Breaking Force, (b) Breaking Elongation, and (c) Tenacity at Break

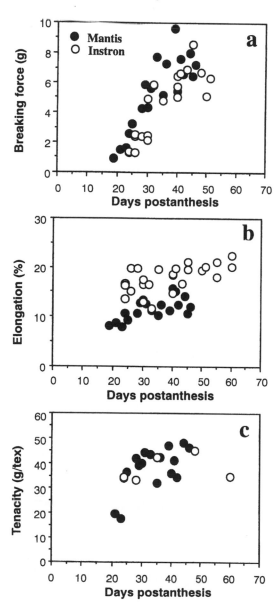

Although we have previously observed that single fiber breaking tenacities of the SJ-2 variety are developed by the beginning of the third week (Hsieh, Honic, and Hartzell, 1995), the data on the Maxxa fibers suggest possible varietal differences in how single fibers gain strength during the early phase of secondary cell development.

In a recent study (Hsieh, Hu, and Nguyen, 1997), we have reported that the forces required to break a single fiber are similar between SJ-2 and Maxxa varieties during the first 4 weeks of fiber development. Between 30 and 50 dpa, both the breaking forces and the breaking tenacities of the Maxxa fibers are higher than those of the SJ-2 fibers. Although the single fiber breaking tenacities of the SJ-2 cotton fibers do not appear to vary beyond 21 dpa, the breaking tenacities of the Maxxa cotton fibers appear to increase with the secondary cell wall thickening process. Because of the small numbers of samples collected so far, the relationship between tenacity and fiber development is not clear.

The single fiber breaking force depends on both fiber development and drying. Our data on Acala varieties have shown that fibers at about 21 to 24 dpa exhibit significant strength when fiber elongation is nearly completed and secondary wall formation is in the beginning stage. Using estimated linear densities for SJ-2 fibers at 14 to 16 dpa, the breaking tenacities of dried fibers were calculated to be 21.5 and 31.3 g/tex, respectively (Hsieh, Honic, and Harzell, 1995). The primary cell wall appears to contribute toward two-thirds or more of the fiber strength. These fiber strength estimations are consistent with the report on 14 dpa *Gossypium hirsutum* Delta Pine 61 fibers whose breaking tenacity was slightly more than half of fibers that were 49 dpa old (Hebert, 1993). Both the breaking forces and fiber mass or linear densities continue to increase throughout the first 36 to 40 days of fiber development. Therefore, the secondary wall thickening continues to contribute to the single fiber breaking force. However, the breaking tenacities of single fibers reach the maximal level when the fibers are about 21 to 24 dpa.

TWIST OR CONVOLUTION

How twists are formed with fiber development is important since twists are characteristic of mature and dried cotton fibers. Dimensions of hydrated and dried SJ-2 fibers have been measured using a microscope equipped with a calibrated ocular micrometer having a resolution of 0.2 μm (micrometer) (Hsieh, Honic, and Harzell, 1995).

On the SJ-2 variety, the 21 dpa fibers tend to roll and fold onto themselves. Typical twists could be observed on fibers about 28 dpa old. The twist thickness increases with age, from approximately 6.5 μm at 21 dpa to 10.5 μm at 40 dpa and maturity. The twist frequency, or the lengths between twists, was found to be irregular along a single fiber, as well as among fibers, and was categorized into either "short" twist lengths for those closely spaced or "long" twist lengths for those separated farther apart. At 21 dpa, the average lengths between long and short twists were 240 μm and 110 μm, respectively. As fibers developed, long twist lengths reduced by nearly one-half to 130 μm (see Figure 6.4a), whereas short twist lengths show only a slight decrease with fiber development to about 90 μm.

Upon drying, fiber widths between the long twists are reduced slightly, but not significantly more than those of the hydrated fibers (see Figure 6.4b). Fiber widths between the long twists are slightly higher than those between the short twists, but these differences diminish with fiber development. There appears to be no difference in the twist thickness between the long and short twists, but it increases slightly between 20 and 36 dpa, reaching a plateau after 36 dpa.

Using the average W (fiber width; see 6.4b) and L (length between twist; see 6.4a) values, the convolution angle (θ) of the twist is calculated as θ $\theta = \tan^{-1}(2W/L)$. A distinct increasing trend with fiber development to about 40 dpa has been observed on the convolution angles of the long twists, whereas only a slight increase is observed on the short twists during the same period (see Figure 6.4c). This trend is expected from the fiber width and twist length data. The insignificant changes in convolution angles after 36 dpa coincide with little change in the linear densities.

Although the spiral reversal changes only 1 to 3 times per mm length (Rebenfeld, 1977), the number of twists in cotton fibers is higher, or between 3.9 and 6.5 per mm (Warwicker et al., 1966). It is obvious that the lateral alignment of the fibrillar reversals in the concentric cellulose layers and the ultimate twists in the dried fibers are related. Twists are formed from cell collapse and the number of

FIGURE 6.4. Fiber Dimensions of Developing SJ-2 Fibers: (a) Lengths Between Twists, (b) Fiber Width, and (c) Convolution Angle

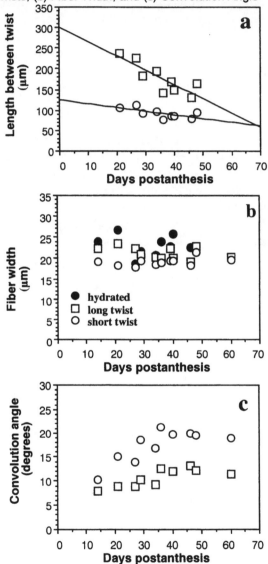

Source: Figures 6.4a and 6.4c are from Hsieh, Honic, and Hartzell (1995).

Note: A total of twelve fibers from each boll was measured. Measurements of the hydrated fiber diameter and the flat ribbon or fiber width (W), twist thickness, and distance between adjacent twists (L) of the dried fibers were made along the middle section of each fiber. Twenty measurements of these dimensions were made on each fiber.

twists increases with fiber maturity or with increasing secondary cell wall thickness. Upon drying, lateral dimensions of the fibers reduce. The spiral reversals of the cellulose fibrils are where buckling most likely occurs in each layer. As the secondary cell wall thickens, the probability for reversals in the cellulose layers to coincide laterally in the fiber increases. This implies that when a threshold level of overlapped reversals is reached, stress from buckling leads to twist formation, meaning that the twist is the region where substantial reversals overlap among concentric cellulose layers. This would also explain the higher frequency of twists over spiral reversals. Hence, the reversals exist, but at a much lower extent, in the cellulose layers in the spans between twists. How reversals in the concentric layers relate to the varying packing density or accessible regions in the fiber cross section is also an important question. Nevertheless, twist characteristics are excellent indicators of fiber maturity. For SJ-2 fibers, the single fiber breaking forces increase with decreasing twist lengths or an increasing number of twists (see Figure 6.5).

FIGURE 6.5. Relationships Between Single Fiber Breaking Force and Twist Length of Developing SJ-2 Fibers

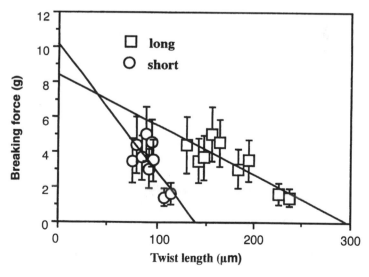

Source: Hsieh, Honic, and Hartzell (1995).

CRYSTALLINE STRUCTURE

The structure of cotton fibers is fibrillar and highly crystalline. The fine structure of cotton fibers, represented by structural parameters such as crystallinity, crystallite size, and orientation, plays important roles in fiber strength. Crystallinity and crystallite sizes of native and chemically treated cotton fibers with urea and sodium hydroxide are well documented (Khalifa, Abdel-Zaher, and Shoukr, 1991). Characterization of these fine structural parameters during fiber growth is not as well documented. Crystallinity of the hydrated fibers increased from 40 percent to 75 percent in fibers between 35 and 65 dpa. Drying increases the overall crystallinity at any of these stages (Kulshreshtha et al., 1973b).

We have investigated the crystalline structure of developing Acala fibers (Hu and Hsieh, 1996; Hsieh, Hu, and Nguyen, 1997) using wide-angle X-ray diffraction and a multipeak resolution method developed by Hindeleh, Johnson, and Montague (1980). This method takes overlapping into consideration and resolves the total scatter into a background and peaks. The area under the background is the noncrystalline scatter. Crystallinity is defined as the ratio of the summation of all resolved peaks to the total scatter under the unresolved trace, including the area under the background.

Calculation of crystallite dimensions normal to the hkl planes is based upon peak broadening, which requires an estimate of peak width at the half-maximum amplitude or the integral breadth of the peak (Elias, 1977). Following peak resolution and separation from background scatter, peak broadening (Br) caused by structural broadening ($\delta\beta$) is determined by separating the instrumental broadening (br). The instrumental broadening function is determined from a reference diffraction peak of a standard, with no broadening due to crystallite size, using the Fourier-Stokes method. The effects from crystal distortion, disorientation, and size distribution were not considered. Therefore, the average crystallite sizes reported are apparent crystallite size, or minimum crystallite size.

A typically resolved X-ray diffraction (WAXD) spectrum of Maxxa cotton fibers at 60 dpa is illustrated in Figure 6.6. The multiple peak resolution method yields seven peaks from the WAXD spectrum collected between 5° and 40°. The four peaks

FIGURE 6.6. Typical Wide-Angle X-Ray Diffraction (WAXD) Pattern of Matured Cotton Fibers

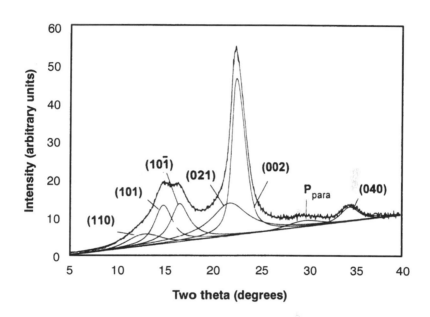

Source: Hu and Hsieh (1996).

Note: WAXD of cotton fibers ground in a Wiley mill and passed through a 20-mesh screen was performed using a Scintag XDS 2000 X-ray diffractometer. Diffraction intensities were counted at 0.05°-steps between a 5° to 40° two theta (2θ) angle range. The experimental data were corrected, normalized, and resolved by employing the X-ray diffraction data analysis program named XRAY. Corrections for each sample were made for air scattering, background radiation, polarization, and Lorentz factor, as well as for incoherent or Compton scattering. The corrected intensities were normalized into electron units and resolved into a background and peaks by the multipeak resolution method (Hindeleh, Johnson, and Montague, 1980). The total scatter was resolved into a background and peaks by taking overlapping into consideration. The area under the background is the noncrystalline scatter. Crystallinity is defined as the ratio of the summation of all resolved peaks to the total scatter under the unresolved trace, including the area under the background. The determination of apparent crystallite sizes of the 101, 10$\bar{1}$, and 002 reflection planes was based on the Sherrer equation. Further details on the experimental procedures and spectra analyses can be found in Hu and Hsieh (1996).

located near 2θ angles of 14.7°, 16.6°, 22.7,° and 34.4° are characteristic of the 101, 10$\overline{1}$, 002, and 040 reflections of cellulose I, respectively, and are used for structural analysis and comparison.

To correlate with strength data, structural characterization has been performed on fibers that are 21 dpa and older. The cellulose I crystalline structure is clearly evident at an early developmental stage of 21 dpa and has been confirmed on both the dried SJ-2 and Maxxa fibers collected at varying developmental stages and at maturity (Hsieh, Hu, and Ngyen, 1997). The crystal system remains unchanged during cotton fiber biosynthesis and at maturity. The degree of crystallinity increases with fiber development from about 30 percent at 21 dpa to 60 percent at 60 dpa (Hsieh, Hu, and Ngyen, 1997). The most significant increase in crystallinity, that is, from 30 percent to 55 percent, is observed between 21 and 34 dpa. The increasing crystallinity in fibers has been reported in the later stages of secondary cell wall development (35 to 65 dpa) of a field-grown Indian cotton variety (Kulshreshtha et al., 1973 a,b). Our data, thus far, have clearly shown that significant crystallinity is attained during the first half of secondary cell wall development of greenhouse-grown Acala cotton fibers.

The crystallite dimensions of the cotton fibers also increase with fiber development (see Figure 6.7), especially with crystallites perpendicular to the 002 direction. This observation indicates that the alignment of the glucosidic rings in respect to the 002 planes improves with fiber cell development. These increments in 002 crystallite dimensions occur also between 21 and 34 dpa, or during the first half of the secondary wall thickening process. These are consistent with the findings of Nelson and Mares (1965) that the lateral apparent crystallite sizes and their orientation increase during cellulose biosynthesis.

STRENGTH AND CRYSTALLINE STRUCTURE RELATIONSHIP

One intriguing question about the development of fiber strength is the contribution of structural factors. Although a positive relationship between the Stelometer bundle tenacities and crystallinity has been shown (Hindeleh, 1980), information on the strength-crys-

FIGURE 6.7. Crystallite Dimensions of Developing Acala Cotton Fibers

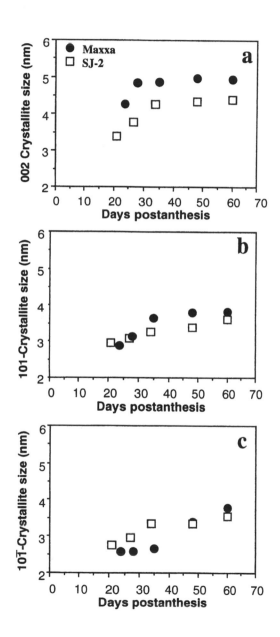

tallinity relationship in developing fibers is limited. Our work has shown positive relationships between single fiber breaking forces and overall crystallinity among developing fibers from the SJ-2 and Maxxa varieties (Hsieh, Hu, Nguyen, 1997). With increasing crystallinity from 30 to 58 percent, the forces required to break Maxxa fibers increase more than those for the SJ-2 fibers. In other words, single fiber breaking forces are higher for Maxxa fibers than for SJ-2 fibers when compared at the same crystallinity. Crystallinity of the Maxxa fibers is lower than for SJ-2 with the same amount of cellulose. This observation suggests that the positive relationship between single fiber breaking force and crystallinity may be variety dependent.

In general, the overall crystallinity, apparent crystallite sizes, and single fiber breaking forces of fibers increase with fiber development (Hsieh, Hu, Nguyen, 1997). The patterns by which these crystalline structure parameters and properties vary with age are different between these two varieties. The breaking forces of the developing Maxxa fibers increase more with crystallinity (between 30 and 50 percent) than do those of the SJ-2 counterpart. The increasing patterns in crystallite dimensions were different between the two varieties, resulting in different breaking force-crystallite dimension relationships, particularly with the $10\bar{1}$ and 002 crystallite dimensions (see Figure 6.8). Therefore, the dependency of single fiber breaking force on crystallite sizes is obviously different for these two varieties.

Although the overall crystallinity and apparent crystallite sizes increase with fiber development, the unit cell size decreases slightly, and thus, the crystal densities increase with fiber development. Among the crystal lattice planes, the alignment of the glucosidic rings in relation to the 002 planes improves most significantly with fiber development. The crystallinity and crystal density of SJ-2 fibers are higher than those of Maxxa fibers during 5 to 6 weeks of fiber development. The 002 and 101 crystallite dimensions of Maxxa fibers, on the other hand, are larger than those of the SJ-2 fibers. These increases coincide with the largest increase in force needed to break single fibers.

Within each variety, positive relationships are observed between single fiber breaking forces and the overall crystallinity, as well as

FIGURE 6.8. Relationships Between Single Fiber Breaking Force and Crystallite Dimensions of Acala Cotton Fibers

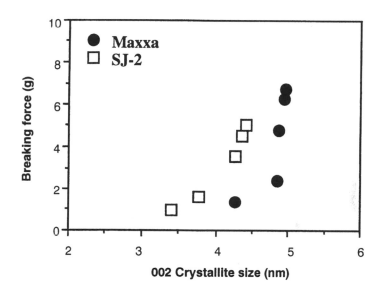

between single fiber breaking force and crystallite size. However, the relationships between tenacities and crystallinity are different between these two varieties (see Figure 6.9a). The single fiber breaking tenacities of the SJ-2 cotton fibers do not appear to vary from 21 dpa to maturity. For the Maxxa cotton fibers, however, breaking tenacities appear to be positively related to fiber development or thickening of the secondary cell wall. Additionally, negative relationships have been observed between breaking tenacities and crystal densities for both varieties (see Figure 6.9b). The increased crystallinity and crystallite sizes and perfection offer only partial explanation to the strength development of cotton fibers. At this point, some preliminary observations regarding the tenacity-age relationships and varietal differences in the tenacity-structure relationships of developing fibers have been made. Further analysis and data collection are necessary to fully establish these relationships. Also, the importance of structural parameters such as fibril

FIGURE 6.9. Relationships Between Single Fiber Breaking Tenacity and (a) Crystallinity and (b) Crystal Density of Acala Cotton Fibers

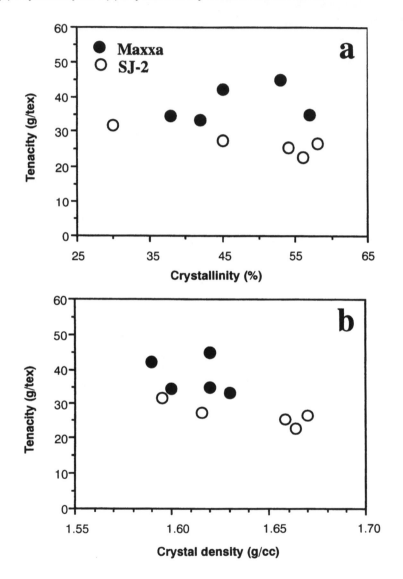

orientation and residual stress in cotton fiber strength should be considered in future work.

CONCLUSION

Fiber length, fineness, and strength are the most essential qualities of cotton fibers for textile processing. Structural characterization of cotton fibers has been extensively studied in their dried state and from a textile-processing viewpoint to enhance end-use performance. Biochemical and anatomical studies of cotton fibers during growth have been carried out from genotypic and environmental perspectives. Because of inevitable variations in fiber growth and development on a single ovule, within a boll, on a plant, and among plants, differences in the growth and development of fibers are major sources of quality variations. The linkages between fiber structure and growth and development are fundamental to the understanding of fiber quality and are crucial to any effort related to fiber quality improvement.

Of the three main fiber quality parameters, fiber length and fineness are determined in the early stages of growth and development. Fiber length is determined during the elongation stage, that is, during the first 20 to 25 dpa. The final thickness of fiber cells is developed at the base of the fibers by about 2 dpa. How cotton fibers gain strength has much to do with both primary and secondary cell wall development. Studies on fiber strength and its relationship to structural development in two Acala cotton fibers during primary wall formation (elongation) and secondary wall thickening (cellulose synthesis), as well as during desiccation (transition from mobile to highly hydrogen-bonded structure), has provided new information.

For both SJ-2 and Maxxa varieties, the single fiber breaking forces increase most significantly during the fourth week of fiber development and are similar between the two varieties. Beyond 30 dpa, both the single fiber breaking forces and tenacities of the Maxxa cotton fibers are higher than those of the SJ-2 cotton fibers. The linear densities and seed fiber weights of the Maxxa fibers are lower than those of the SJ-2 fiber when compared at the same developmental stages. The cellulose I crystalline structure is clearly evident at the early developmental stage of 21 dpa and remains unchanged throughout fiber development. The overall crystallinity and the apparent crystallite sizes increase with fiber development for both varieties. At the same developmental stages, Maxxa fibers have larger crystal sizes but lower crystallinity and lower crystal density than the SJ-2 fibers. Within each variety, the single fiber breaking forces are positively related to both the overall crystallinity and crystallite sizes.

Although positive relationships between single fiber breaking forces and overall crystallinity, as well as between single fiber breaking force and crystallite size, exist in both varieties, the relationships between tenacities and crystalline structure parameters are different between these two varieties. The tenacity-age relationships and varietal differences in tenacity-structure relationships need to be further studied. Additionally, experimentally determined fiber strengths are below those predicted theoretically. Better understanding of the structural factors for cotton fiber strength will help to explain these discrepancies and to improve fiber strength.

REFERENCES

Balls, W.L. (1928). *Studies of Quality in Cotton.* London: Macmillan.

Beasley, C.A. (1977). Ovule culture: Fundamental and pragmatic research for the cotton industry. In *Plant Cell, Tissue and Organ Culture*, Eds. J.R. Reinert and Y.P.S. Baja. Berlin: Springer-Verlag, pp. 160-178.

Benedict, C.R., R.J. Kohel, and G.M. Jividen (1994). Crystalline cellulose and cotton fiber strength. *Crop Science* 34: 147-151.

Berlin, J.D. (1986). The outer epidermis of the cottonseed. In *Cotton Physiology*, Eds. J.R. Mauney and J.M. Stewart. Memphis, TN: The Cotton Foundation, pp. 361-373.

Boylston, E.K., D.P. Thibodeaux, and J.P. Evans (1993). Applying microscopy to the development of a reference method for cotton fiber maturity. *Textile Research Journal* 63: 80-87.

deGruy, I.V., J.H. Carra, and W.R. Goynes (1973). Fine structure of cotton, an atlas of cotton microscopy. In *Fiber Science Series*, Volume 6, Ed. R.T. O'Conner. New York: Marcel Dekker, pp. 191-225.

Elias, H.G. (1977) *Macromolecules: Part 2—Synthesis and Materials.* New York: Plenum Press, p. 1081.

Fryxell, P.A. (1963). Morphology of the base of seed hairs of *Gossypium* I. Gross morphology. *Botanical Gazette* 124: 196-199.

Goldwaith, F.G. and J.D. Guthrie (1954). Matthew's textile fibers. Ed. H.R. Mauersberger. New York: Wiley Interscience.

Green, C.C. and T.W. Culp (1990). Simultaneous improvement of yield, fiber quality, and yarn strength in Upland cotton. *Crop Science* 30: 66-69.

Harig, H. (1992). Possibilities and problems in the further development of cotton as a raw material. *ITS Textile Leader* 10: 71-76.

Hebert, J.J. (1993). Strength of the primary wall of cotton fibers. *Textile Research Journal* 63: 695.

Hindeleh, A.M. (1980). Crystallinity, crystallite size, and physical properties of native Egyptian cotton. *Textile Research Journal* 50: 667-674.

Hindeleh, A.M., D.J. Johnson, and P.E. Montague (1980). Computational methods for profile resolution and crystallite size evaluation in fibrous polymers. *American Chemical Society Symposium* 141: 149-182.

Hornuff, G.V. and H. Richter (1964). Chemical composition of cotton fibers originating from various areas, Faserforsch. *Textiltechnik* 15: 115-126.

Hsieh, Y.-L., E. Honic, and M.M. Hartzell (1995). A developmental study of single fiber strength: Greenhouse grown SJ-2 Acala cotton. *Textile Research Journal* 65: 101-112.

Hsieh, Y.-L., X.P. Hu, and A. Nguyen (1997). Strength and crystalline structure of developing Acala cotton. *Textile Research Journal* 67: 529-536.

Hu, X.P. and Y.-L. Hsieh (1996). Crystalline structure of developing cotton fibers. *Journal of Polymer Science, Polymer Physics Edition* 34: 1451-1459.

Hu, X.P. and Y.-L. Hsieh (1997). Breaking elongation distributions of single fibers. *Journal of Materials Science* 32: 3905-3912.

Huwyler, H.R., G. Franz, and H. Meier (1979). Changes in the composition of cotton fiber cell walls during development. *Planta* 146: 635-642.

Ingram, P., D.K. Woods, A. Peterlin, and J.L. Williams (1974). Never-dried cotton, Part I: Morphology and transport properties. *Textile Research Journal* 44: 96-106.

Joshi, P.C., A.M. Wadhwani, and B.M. Johri (1967). Morphological and embryological studies of *Gossypium* L. *Proceedings of the National Institute of Science, India* B33: 37-93.

Kassenback, P. (1970). Bilateral structure of cotton fibers as revealed by enzymatic degradation. *Textile Research Journal* 40: 330-334.

Khalifa, B.A., N. Abdel-Zaher, and F.S. Shoukr (1991). Crystalline character of native and chemically treated Saudi Arabian cotton fibers. *Textile Research Journal* 61: 602-608.

Kulshreshtha, A.K., K.F. Patel, A.R. Patel, M.M. Patel, and N.T. Baddi (1973a). The fine structure and mechanical properties of cotton fibres at various stages of growth. *Cellulose Chemistry and Technology* 7: 307-314.

Kulshreshtha, A.K., K.F. Patel, A.R. Patel, M.M. Patel, and N.T. Baddi (1973b). A study of equatorial X-ray diffraction from fresh, undried cotton fibre at various stages of its growth. *Cellulose Chemistry and Technology* 7: 343-349.

Meinert, M.C. and D.P. Delmer (1977). Changes in biochemical composition of the cell wall of the cotton fiber during development. *Plant Physiology* 59: 1088-1097.

Muller, L.L. and M.L. Rollins (1972). Electron microscopical study of cotton fiber surfaces after purification treatment. *Microscope* 20: 143-152.

Muller, M. (1991). Precise measurement of important fiber properties in cotton. *International Textile Bulletin* 37: 73-88.

Nelson, M.L. and T. Mares (1965). Accessibility and lateral order distribution of the cellulose in the developing cotton fiber. *Textile Research Journal* 35: 592-603.

Patel G.S., P. Bhama Iyer, S. Sreenivasan, and K.R. Krishna Iyer (1990). Reversals in cotton: A study with scanning electron microscopy. *Textile Research Journal* 60: 771-774.

Peeters, M.-C. and E. DeLanghe (1986). Cellulose packing density in the secondary wall of never dried cotton fibers. *Textile Research Journal* 56: 755-758.

Peterlin, A. and P. Ingram (1970). Morphology of secondary wall fibrils in cotton. *Textile Research Journal* 40: 345-354.

Raes, G., T. Fransen, and L. Verschraege (1968). Study of the reversal phenomenon in the fibrillar structure of cotton fiber: Reversal distance distribution as origin of an extended hypothesis in the cotton fiber development. *Textile Research Journal* 38: 182-195.

Rebenfeld, L. (1977). Mechanical properties of native fibrous materials. In *Proceedings of the First Cleveland Symposium on Macromolecules,* Ed. A.G. Walton. Amsterdam: Elsevier, pp. 177-201.

Seagull, R.W. (1986). Changes in microtubule organization and wall microfibril orientation during in vitro cotton fiber development: An immunofluorescent study. *Canadian Journal of Botany* 64: 1373-1381.

Seagull, R.W. (1992). A quantitative electron microscopic study of changes in microtuble arrays and wall microfibril orientation during *in vitro* cotton fiber development. *Journal of Cell Science* 101: 561-577.

Stone, J.E. and A.M. Scallan (1965). In *Consolidation of the PaperWeb.* London, Technical Section of the British Paper and Board Makers' Association, pp. 145-174.

Timpa, J.D. and H.H. Ramey Jr. (1994). Relationship between cotton fiber strength and cellulose molecular weight distribution: HVI calibration standards. *Textile Research Journal* 64: 557-562.

Tripp, V.W., A.T. Moore, and M.L. Rollins (1957). The surface of cotton fibers, Part II: Modified fibers. *Textile Research Journal* 27: 427-436.

Tsuji, W., T. Nakao, M. Mori, and Y. Yamauchi (1990). Properties and structure of never-dried cotton fibers. Part II: American cotton cultivated in Japan. *Textile Research Journal* 60:738-743.

Vigil, E.L., W.S. Anthony, E. Columbus, E. Erbe, and W.P. Wergin (1996). Fine structure aspects of cotton fiber attachment to the seed coat: Morphological factors affecting saw ginning of lint cotton. *International Journal of Plant Sciences* 157: 92-102.

Warwicker, J.O., R. Jeffries, R.L. Colbran, and R.N. Robinson (1966). Shirley Institute Pamphlet, No. 93, Manchester, UK.

Yatsu, L.Y. and T.J. Jacks (1981). An ultrastructural study of the relationship between microtubules and microfibrils in cotton cell wall reversals. *American Journal of Botany* 68: 771-777.

Zeronian, S.H. (1991). The mechanical properties of cotton fibers. *Journal of Applied Polymer Science, Applied Polymer Symposium* 47: 445-461.

Chapter 7

Cotton Germplasm Resources and the Potential for Improved Fiber Productivity and Quality

Russell J. Kohel

INTRODUCTION

The identification of germplasm resources for improved fiber productivity and quality involves the identification of genes that confer these attributes. The inherent difficulty in this process is that wild cottons are perennial plants that possess no spinnable fibers for which we can measure fiber productivity and quality. Yet, we have cultivated cottons that possess abundant fiber productivity, those which have excellent fiber quality, and some that combine both attributes. Therefore, we must assume that genes exist in the primitive cottons that can be selected and recombined into new combinations to produce or enhance these attributes. The domesticated cotton species that are the source of raw materials for the textile industry are the product of man-directed evolution. The diploid species *Gossypium arboreum* and *G. herbaceum* and the allotetraploids *G. barbadense* and *G. hirsutum* have spinnable fibers. The more primitive members of these species have low productivity and poor fiber quality when compared to modern cultivars. Therefore, the influence of humans must be assumed to be the determining factor in improving these attributes. However, the genes must have existed in the primitive germplasm that allowed these improvements to take place.

The desire to improve fiber productivity and quality is inherent to the culture of cotton. Success in cotton improvement is readily observed by the existence of cottons that are highly productive and of others that have excellent quality fiber. The real challenge is to develop cottons that have both high fiber productivity and quality. The difficulty in combining these attributes reflects the history of cotton fiber improvement (Culp and Harrell, 1973; Webb, 1978; Niles and Feaster, 1984; Cooper, 1992; Culp, 1992; El-Zik and Thaxton, 1992; Gannaway and Dever, 1992; Meredith, 1992; Smith, 1992). A tenet of cotton breeding is that fiber yield and fiber quality are negatively associated (Miller and Rawlings, 1967; Abd Alla and Abo el-Zahab, 1975; Cateland and Schwendiman, 1976; Harrell and Culp, 1976; Scholl and Miller, 1976; Odemamedov, 1978). The history of cotton breeding success is not in the ability to change this negative relationship but in the ability to move the slope of the relationship to a higher level of intercept (Green and Culp, 1990; Culp and Green, 1992). As such, better yielding cottons are being produced with improved fiber quality, but at each new level of success, the same negative relationship exists.

Not only is cotton improvement faced with the negative relationship between fiber yield and quality, but practical considerations of cotton production confront the cotton breeder when decisions of how to improve cotton are made. The cotton producer is rewarded for productivity. Although economic considerations are given to fiber quality, the overwhelming consideration is yield of fiber. Thus, the product of the cotton breeder's successful efforts is the sale and use of cultivars because they are high yielding.

In contrast to the producer of cotton fiber, the textile industry consistently demands better fiber quality (Deussen, 1992). Fiber quality, in general, is important to the textile industry because it directly relates to improved processing performance and increased productivity. At the upper end of the fiber quality scale, high quality allows the spinning and weaving of higher value products. The high-value products represent only a small fraction of the total volume of production, but they are significant for their economic return. Even though the textile industry is noted for its demand of improved fiber quality, it is dependent also on the availability of abundant, cost-competitive raw cottons to supply a wide variety of products to meet consumer demands. Therefore, the entire cotton industry, from producer to processor, requires that cotton improvement provides highly productive cultivars.

FIBER PRODUCTIVITY

To address the issues of germplasm resources for improved fiber productivity and quality, let me first address productivity. First of all, let us be clear that when we say germplasm resources, we are, in reality, talking about the genetic variability of these resources. Fiber productivity seems to be an attribute that is easier to accomplish than quality. Often rapid success can be achieved in productivity, but at the expense of quality (Miller and Rawlings, 1967; Meredith and Bridge, 1973).

At the plant level, productivity can be defined simply as increased bolls per plant. At the field level, productivity is the optimization of plants per unit of land that maintains the maximum bolls per plant (Maner et al., 1971; Worley, Culp, and Harrell, 1974; Worley et al., 1976). Bolls per plant is undoubtedly under genetic control, and the optimum plant density must have a genetic component. However, the maximum number of bolls per unit of land has a large environmental component (Meredith and Bridge, 1973; Waddle, 1980). Fertility, moisture, soil, pests, and so forth have major roles in total productivity. These same factors can influence fiber quality, but more often than not, the result will be positively correlated to yield. That is to say, as these factors increase or decrease productivity, they will most often have a similar effect on fiber quality. Therefore, the real problem of improving fiber productivity and quality, that is, overcoming the negative relationship, must be addressed at the level of the individual cotton boll.

Since the cotton fiber is a single epidermal cell on the surface of the seed (DeLanghe, 1986; Graves and Stewart, 1988), productivity of fiber can result from more seeds per boll and more fibers per unit of seed surface area (Stewart, 1975; Ramsey and Berlin, 1976). Productivity can be further enhanced through increasing the amount of cellulose in the fiber by making longer fibers with thicker cell walls (Cheng, Kohel, and Benedict, 1985).

Having defined how to increase productivity, we might look at modern cultivars to see what characteristics they have that contribute to their productivity. Modern U.S. cottons certainly produce more bolls per unit of land. At the boll level, the bolls are smaller and the seeds are smaller. In general, fiber length has increased or been maintained at about the same level, and it does not appear as a major contributor to the yield increase. Although thicker cell walls would increase the total amount of cellulose produced, high-quality fiber requires a mature, fine fiber. As such, there is an optimum cell wall thickness, and thicker cell walls beyond this optimum are undesirable. Cell wall thickness has not increased. The main increase in productivity has been increased boll production (National Cotton Variety Test Program, 1996, and previous years).

With this introduction, we will now focus on what happens in the individual cotton boll to understand the potential of germplasm resources for improving cotton fiber productivity and quality. We will investigate the genetic variability for those factors which determine fiber initiation, growth, and development.

FIBER QUALITY

We must first define fiber quality. Fiber quality is the combination of those properties of the fiber that improve the spinnability of fiber. The first attempts to identify fiber quality were through visual observations and "feel" of the fibers. One could spread the fibers and observe their length. The ease of breaking a bundle of fibers by hand provided an estimate of fiber strength. The feel of a bundle of fibers helped estimate the ease of processing. Although these were subjective measures, they were determinations that practitioners could employ with great skill. As researchers were able to quantitatively measure fiber properties, they were able to identify those fiber properties which were most highly correlated with spinning performance. With the identification of fiber properties highly correlated with spinning performance, instruments were developed to routinely measure these properties, and fiber quality took on characteristics defined by the physical properties of the fiber (Hertel and Craven, 1951; Craven, 1952; Hertel, 1953, 1956; Craven, 1959).

Conventional fiber quality is measured primarily in terms of length, strength, and fineness. These fiber properties are a consequence of finding those fiber properties which best correlate to ring-spinning performance and of the development of instruments to measure them. With the development of new spinning technologies, these properties have remained important, but their relative importance differs with different spinning technologies. New properties may also be important. Some attempts have been made to develop new instruments to measure fiber properties, but most of these are refinements to the measurements of length, strength, and fineness. As new spinning technologies are developed, fiber properties not currently measured may become important in defining fiber quality for specific textile uses.

Biological Function of Fiber Quality

Wild cottons do not have spinnable fibers, but some have appressed or rudimentary fibers. Therefore, the genes for seed fiber formation are present. The role of these fibers is not obvious. The fibers on seeds of the species that have spinnable fibers have rudimentary fiber development compared to their modern derivatives. The biological function of these fibers is not known, but it has been postulated that they play a role in seed dispersal or seed germination and plant establishment. Dispersal can be through birds gathering nesting material or through adherence to animal fur. In modern cultivars, the seed fuzz and lint fibers initially insulate the seed from occasional moisture, but with continued exposure to water, the fibers become wet and retain moisture. Primitive cottons grow in xerophytic environments with distinct wet and dry seasons. Under these conditions, the fibers plus the hard seed coats could protect the seed from moisture until the rainy season arrives, and seeds could germinate at a time when the seedlings would have a greater probability of survival. We have to recognize that fiber quality requirements for industrial use do not relate to known biological function requirements.

Sources of Quality

Wild Species

There are approximately fifty wild species with worldwide distribution—Africa, Middle East, Asia, South America, Central

America, North America, Australia, and Hawaii (Percival and Kohel, 1990). There are the two domesticated diploid species (*G.
arboreum* and *G. herbaceum*) and the two domesticated allotetraploid species (*G. barbadense* and *G. hirsutum*). The diploid species
have been classified into seven genomic groups. The A and D
genomic groups are considered the donor genomes to the allotetraploids (Beasley, 1940). Since over 90 percent of the cotton produced, and closer to 100 percent of that processed by the textile
industry, is from the allotetraploids, my discussion will focus on
those species.

The wild species do not have spinnable fibers so we cannot make
a direct evaluation of their potential contribution to fiber quality.
However, introgression from the wild diploids to allotetraploids has
resulted in germplasm with enhanced fiber properties that have
been attributed to the donor diploid parent (Mansurov, Shebitchenko, and Bakhramov, 1973; Muramoto, 1973; Krishnaswai and Kothandaraman, 1975; Schwendiman, 1975; Burkalov, 1978; Bazhanova, 1979; Culp, Harrell, and Kerr, 1979; Egamberdiev, 1979;
Huang et al., 1981; Morozova, Krasilnikova, and Ruban, 1983;
Popova and Lavygina, 1984; Imamaliev and Livygina, 1985). It is
often assumed that the donor parent was the contributor of the
enhanced fiber properties, but seldom does direct evidence indicate
that introgression has taken place. It is possible also that introgression of fiber quality has not taken place, but rather, the disruption
caused by the interspecific hybridization resulted in greater recombination in the recurrent germplasm such that new gene combinations produced the enhanced fiber quality. This recombination of
genes would be similar to what apparently has taken place in conventional cotton improvement programs (Culp and Harrell, 1973;
Cooper, Hyer, and Turner, 1977; Webb, 1978; Niles and Feaster,
1984; Dever and Gannaway, 1987; Cooper, 1992; Culp, 1992;
El-Zik and Thaxton, 1992; Gannaway and Dever, 1992; Meredith,
1992; Smith, 1992).

Domesticated Species

Within the domesticated species, the most exotic center-of-origin
cottons do not exhibit fiber quality equal to modern cultivars
(Anonymous, 1974; Percival, 1987). Within the *G. hirsutum* germ-

plasm, there is a wide range of the individual fiber properties of length, strength, and fineness. Even though this wide range for fiber length exists, the cottons with the longest fibers approach the values of contemporary cultivars, but modal values are much shorter. A similar pattern is exhibited for fiber strength. There is a range of values, but the modal values are weak compared with contemporary cultivars. Micronaire exhibits a wide range in the exotic cottons from very fine to coarse. The modal values are not unlike contemporary cottons. The data for exotic cottons represent nonreplicated samples obtained primarily from winter nursery growth because they are predominantly photoperiodic. Even though they are not replicated values, the trend they show indicates that fiber quality cannot be enhanced by simple and direct selection of exotic germplasm accessions.

The fiber properties of the *G. barbadense* center-of-origin cottons do not differ greatly from *G. hirsutum*. The exotic *G. barbadense* are represented by fewer accessions than *G. hirsutum*. However, the major contemporary production of *G. barbadense* is of the extra-long-staple (ELS) cottons that are characterized by long fibers, strong fibers, and fine fibers. The exotic *G. barbadense* cottons exhibit a trend for longer fibers than *G. hirsutum;* the longest fibers are in the range of contemporary *G. hirsutum*, with the modal values much shorter. Fiber strength of exotic *G. barbadense* tends to exceed exotic *G. hirsutum*, but again, it does not exceed the values of contemporary *G. hirsutum*. Micronaire has a range similar to *G. hirsutum* and much coarser than modern ELS cottons.

The modern ELS cottons trace their origins through the Sea Island cottons from the center-of-origin *G. barbadense*. These *G. barbadense* cottons were unique in their exceptional high-quality fibers through the combination of long, strong, and fine fibers. They emerged early in the modern period of cotton production, but they precede the period of formal cotton breeding efforts (Kearney, 1943; Hutchinson and Manning, 1945; Turcotte and Percy, 1986; Roberts and Davis, 1991). It has been postulated that they are a product either of introgression with *G. hirsutum* and transgressive segregation or of selection from within *G. barbadense* germplasm (Hutchinson and Manning, 1945; Stephens, 1976). No direct evidence establishes a basis to strongly support one possibility over the other (Kohel, Park, and Slocum, 1994). These cottons differed from the center-of-origin germplasm in that they were day neutral, and that allowed their expansion over a wide range of latitudes. Despite their exceptional fiber quality, their production ceased due to their poor productivity and limited adaptation to other cotton-growing areas. However, they were direct contributors to modern ELS cottons, characterized by the Egyptian and Pima cottons through modern cotton improvement programs. From time to time, there has been renewed interest in their production because of their unique fiber.

Within the two dominant domesticated allotetraploid species, there are occurrences of high fiber quality that are unique. Within *G. barbadense,* this is represented by the Sea Island cottons (Hutchinson and Manning, 1945; Niles and Feaster, 1984), and within *G. hirsutum,* by the Acala cottons (Turner, 1974; Niles and Feaster, 1984). In both cases, these are unique events that are divergent from the modal characteristics of their primitive forms. These were derived in a human-directed evolutionary context in which changes arose through a selection process that was probably not characterized by the controlled hybridizations of modern cotton cultivar development. Selections must have been made on observed phenotypic variation.

Fiber Quality and Productivity

I have indicated that fiber quality and productivity are both controlled at the boll or seed level (Worley et al., 1976). We can divide the genetics of these characteristics into three main areas. The first is the area relating to boll and seed, factors that determine numbers and size. Second is the area relating to fiber initiation and elongation, factors that determine the density of fibers on the seed surface and the amount of elongation. Then there is the area of secondary wall formation, factors that determine the amount and quality of cellulose in the secondary wall. Underlying all of these is the genetics of cellulose synthesis that deals with the mechanisms of cellulose production and deposition.

Phenotypic Expression of Fiber Quality

The phenotypic expression of fiber quality is confined to the domesticated species. That genetic advances in fiber quality can be made indicates the existence of genes that contribute to fiber quality in germplasm that does not express the phenotype. The inability to observe the phenotypic expression of fiber quality means that to explore the potential of germplasm for fiber quality, we must develop methods to identify that trait. This ability takes on two dimensions. First, fiber quality is not a single trait. Rather, fiber quality is the result of specific combinations of fiber properties that interact together to produce the desired processing performance in the textile industry. Therefore, the phenotype of fiber quality is the combination of the phenotypes of fiber properties that we have identified as important to spinning performance. These properties are those which we have the means to quantify their phenotypic expression and which we have identified as important. The phenotypic measures of fiber quality and fiber properties are imprecise. These traits have been, and can be, genetically manipulated, but imprecisely. Molecular markers could provide the opportunity to use precision in identifying the phenotype of these traits. However, to explore with precision the genetic variability of the germplasm, for which there is no phenotypic expression, requires an understanding of the underlying genetic controls. Contemporary molecular attempts to identify fiber genes or genes expressed during fiber development may provide information helpful in identifying the complex process, but in and of themselves, they only provide a new dimension of phenotypic measurement (John and Crow, 1992; Dang, Heinen, and Allen, 1996).

Physical Properties of Fibers

The physical properties of cotton fiber cells are the result of a complex series of events for which we have fragmentary information and for which genetic control and interrelations are not well understood. First, we have the events that determine the initiation of epidermal seed cells that elongate into fiber cells. The number of fibers per unit of seed surface can be varied by increasing the number of epidermal cells that elongate to form fiber cells, as well as by increasing the relative number of cells that do elongate into fiber cells rather than the short fuzz fiber cells. Studies of in vitro and in vivo fiber elongation indicate that specific hormones influence the initiation of elongation of fiber from epidermal cells (Beasley, 1973; Beasley and Ting, 1973, 1974; Imamaliev and Uzilevskaia, 1982; Kosmidou-Dimitropoulou, 1986; Butenko and Azizkhodzaev, 1987; Zhang, Sun, and Pan, 1992; Davidonis, 1993a,b; Liu et al., 1999). The genetics of this regulation is not known, but if the genetic controls could be identified, it would have a major impact on fiber productivity and quality.

The rate and duration of elongation determines the length of the fiber cell. These events are associated with the deposition of cellulose in the primary cell wall. These are observable events, but the biochemical processes and the genetic controls are not fully known (Delmer et al., 1985; Dugger and Palmer, 1986; Delmer, 1990; Amor et al., 1991; Andrawis, Solomon, and Delmer, 1993).

Genetic control of fiber fineness may be defined at fiber initiation, or it may be a result of the secondary wall process. Fiber fineness might be controlled by the interaction of both processes of fiber initiation and secondary wall systems. Certainly, environmental inputs can influence both. Secondary wall deposition accounts for more than 90 percent of the cellulose in cotton fiber. The genetic controls of cellulose in the secondary wall are not known. We do not know whether cellulose deposition in the primary and secondary walls is under the same control and/or regulation. The secondary wall of the cotton fiber is a primary determinant of fiber productivity and quality.

Cellulose of the secondary wall has been characterized in many ways. Measurements of X-ray diffraction, microtubular orientation, molecular weight, crystallinity, and other traits have been characterized and related to traditional fiber properties (Duckett and Tripp, 1967; Muratov and Gesos, 1985; Moharir et al., 1986; Moharir, 1987; Gowarker, Viswanathan, and Sreedhar, 1987; Benedict, Kohel, and Jividen, 1991, 1992; Seagull, 1993). The role and interrelations of these properties are not well known, and the genetic control of cellulose deposition in the secondary walls is unknown.

CONCLUSIONS

Since what we describe as fiber quality is not observable in wild cottons, it is obvious that humans have selected genes in combinations that produce fibers with characteristics that allow spinning and higher quality selections. That the highest quality fibers are in improved cottons indicates that continued selection and recombination of fiber genes have produced new combinations of genes from the existing germplasm sources.

We must understand the genetics of fiber growth and development, which would be a major breakthrough in the ability to manipulate fiber productivity and quality and to explore the potential of germplasm resources. However, this information would only provide the tools to understand and to manipulate fiber properties. The ultimate task will be to understand which fiber properties and which combination of fiber properties allow us to improve fiber productivity and quality and to answer the question of whether, and to what extent, the observed negative relation can be broken.

Understanding the biochemical process of cellulose biosynthesis in the cotton fiber is the first step in being able to genetically manipulate fiber properties. Identifying the genes would open up another very important step. In many cases, the resultant properties may be greatly influenced by the timing and regulation of gene expression.

REFERENCES

Abd Alla, S.A. and A.A. Abo el-Zahab (1975). Estimates of genotypic variances for fiber properties and their correlation with agronomic characters in Egyptian cotton. *Zeitschrift Pflanzenzucht* 74: 162-167.

Amor, Y., R. Mayer, M. Benziman, and D. Delmer (1991). Evidence for a cyclic diguanylic acid-dependent cellulose synthase in plants. *Plant Cell* 3: 989-995.

Andrawis, A., M. Solomon, and D.P. Delmer (1993). Cotton fiber annexins: A potential role in the regulation of callose synthase. *Plant Journal* 3: 763-772.

Anonymous (1974). The regional collection of *Gossypium* germplasm. *United States Department of Agriculture—Agricultural Research Service-H2*, 105 pp.

Bazhanova, A.P. (1979). New variety of fine fiber cotton ASh I67B developed by the method of remote hybridization. *Izvestiya Akademii Naukowe* 2: 41-44.

Beasley, C.A. (1973). Hormone regulation of growth in unfertilized cotton ovules. *Science* 179: 1003-1005.

Beasley, C.A. and I.P. Ting (1973). The effects of plant growth substances on *in vitro* fiber development from fertilized cotton ovules. *American Journal of Botany* 60: 130-139.

Beasley, C.A. and I.P. Ting (1974). Effects of plant growth substances on *in vitro* fiber development from unfertilized cotton ovules. *American Journal of Botany* 61: 188-194.

Beasley, J.O. (1940). The origin of American tetraploid *Gossypium* species. *American Naturalist* 74: 285-286.

Benedict, C.R., R.J. Kohel, and G.M. Jividen (1991). A new concept of the polymeric nature of cellulose in secondary walls. In *Third International Congress of Plant Molecular Biology* (Abstract). Tucson, AZ.

Benedict, C.R., R.J. Kohel, and G.M. Jividen (1992). Cellulose polymers: Crystallinity and cotton fiber strength. In *Proceedings from Cotton Fiber Cellulose: Structure, Function and Utilization Conference.* Memphis, TN: National Cotton Council of America, pp. 227-246.

Burkalov, N. (1978). Inheritance of fiber length in the interspecific cotton hybridization. *Nauchni Trudove Vissh Selskostop Institut Vasil Kolarov. Plovdivski Khristo G. Danov* 23: 43-48.

Butenko, R.G. and A. Azizkhodzaev (1987). Fiber growth in the isolated cotton ovule culture. *Physiology and Biochemistry of Cultivated Plants* 19: 235-239.

Cateland, B. and J. Schwendiman (1976). Diallel crossing between American and African cotton plant varieties. Behavior of six characteristics of fiber, approach to genetic components and their possible implications for improvement. *Cotton Fibres Tropicales* 31: 349-367.

Cheng, H.L., R.J. Kohel, and C.R. Benedict (1985). Genetic improvement of upland cotton fiber quality. II. Genetic variation of the *in vitro* cellulose synthesis in developing bolls. *Journal of Agricultural Science,* Nanjing, China, 1: 33-37.

Cooper, H.B. (1992). Cotton for high fiber strength. In *Proceedings from Cotton Fiber Cellulose: Structure, Function and Utilization Conference.* Memphis, TN: National Cotton Council of America, pp. 303-314.

Cooper, H.B., A.H. Hyer Jr., and J.H. Turner Jr. (1977). Cotton germplasm development. *California Agriculture* 31: 14-15.

Craven, C.J. (1952). A physicist looks at cotton. *Tennessee Farm and Home Science* November 4.

Craven, C.J. (1959). Scientific evaluation of cotton. *Tennessee Engineer* 62: 24-26.

Culp, T.W. (1992). Simultaneous improvement in yield and fiber strength of Upland cotton. In *Proceedings from Cotton Fiber Cellulose: Structure, Function and Utilization Conference.* Memphis, TN: National Cotton Council of America, pp. 247-288.

Culp, T.W. and C.C. Green (1992). Performance of obsolete and current cultivars and Pee Dee germplasm lines of cotton. *Crop Science* 32: 35-41.

Culp, T.W. and D.C. Harrell (1973). Breeding methods for improving yield and fiber quality of upland cotton (*Gossypium hirsutum* L.). *Crop Science* 13: 686-689.

Culp, T.W., D.C. Harrell, and T. Kerr (1979). Some genetic implications in the transfer of high fiber strength genes to upland cotton. *Crop Science* 19: 481-484.

Dang, P.M., J.L. Heinen, and R.D. Allen (1996). Expression of a "cotton fiber specific" gene, Gh-1, in transgenic tobacco and cotton. *Plant Physiology* 111: 55.

Davidonis, G. (1993a). A comparison of cotton ovule and cotton cell suspension cultures: Response to gibberellic acid and 2-chloroethylphosphonic acid. *Journal of Plant Physiology* 141: 505-507.

Davidonis, G. (1993b). Cotton fiber growth and development *in vitro*: Effects of tunicamycin and monensin. *Plant Science* 88: 229-236.

DeLanghe, E.A.L. (1986). Lint development. In *Cotton Physiology*, Eds. J.R. Mauney and J.M. Stewart. Memphis, TN: The Cotton Foundation, pp. 325-350.

Delmer, D.P. (1990). Role of the plasma membrane in cellulose synthesis. In *The Plant Plasma Membrane: Structure, Function, and Molecular Biology*, Eds. C. Larsson and I.M. Moller. New York: Springer-Verlag, pp. 256-268.

Delmer, D.P., G. Cooper, D. Alexander, J. Cooper, T. Hayashi, C. Nitsche, and M. Thelen (1985). New approaches to the study of cellulose biosynthesis. *Journal of Cell Science* 2: 33-50.

Deussen, H. (1992). Improved cotton fiber properties: The textile industry's key to success in global competition. In *Proceedings from Cotton Fiber Cellulose: Structure, Function and Utilization Conference*. Memphis, TN: National Cotton Council of America, pp. 43-64.

Dever, J.K. and J.R. Gannaway (1987). Breeding for fiber quality on the Texas High Plains. In *Proceedings of the Beltwide Cotton Production Research Conferences*. Memphis, TN: National Cotton Council of America, pp. 111-112.

Duckett, K.E. and V.W. Tripp (1967). X-rays and optical orientation measurements on single cotton fibers. *Textile Research Journal* 37: 517-524.

Dugger, W.M. and R.L. Palmer (1986). Incorporation of UDP Glucose into cell wall glucans and lipids by intact cotton fibers. *Plant Physiology* 81: 464-470.

Egamberdiev, A.E. (1979). Wild cotton species, donors of fiber quality and resistance to wilt disease. *Doklady Akademii Nauk (Uzbekskoi SSR)* 8: 66-67.

El-Zik, K.M. and P.M. Thaxton (1992). Simultaneous genetic improvement of fiber strength, yield, and resistance to pests of MAR cottons. In *Proceedings from Cotton Fiber Cellulose: Structure, Function and Utilization Conference*. Memphis, TN: National Cotton Council of America, pp. 315-332.

Gannaway, J.R. and J.K. Dever (1992). Development of high quality cottons adapted to stripper harvested production areas. In *Proceedings from Cotton Fiber Cellulose: Structure, Function and Utilization Conference*. Memphis, TN: National Cotton Council of America, pp. 333-348.

Gowarker, V.R., N.V. Viswanathan, and J. Sreedhar (1987). Crystallinity in polymers. In *Polymer Science*. New Delhi, India: Wiley Eastern Limited.

Graves D.A. and J.M. Stewart (1988). Chronology of the differentiation of cotton (*Gossypium hirsutum* L.) fiber cells. *Planta* 175: 254-258.

Green, C.C. and T.W. Culp (1990). Simultaneous improvement of yield, fiber quality, and yarn strength in upland cotton. *Crop Science* 30: 66-69.

Harrell, D.C. and T.W. Culp (1976). Effects of yield components on lint yield of upland cotton with high fiber strength. *Crop Science* 16: 205-208.

Hertel, K.L. (1953). Fiber strength and extensibility as measured by the Stelometer. *The Cotton Research Clinic,* p. 10.

Hertel, K.L. (1956). Significance of fiber properties and the need for progress in the field of instrumentation. *Textile Bulletin* 82: 110-112.

Hertel, K.L. and C.J. Craven (1951). Cotton fineness and immaturity as measured by the Arealometer. *Textile Research Journal* 21: 765-774.

Huang, J.C., S.Y. Chien, K.L. Liu, C. Weng, I.S. Tseng, and K.Y. Chou (1981). Variations in the characters of upland cotton (*Gossypium hirsutum*) induced by exotic DNA of Sea Island cotton *(Gossypium barbadense)*. *Acta Genetica Sinica* 8: 56-62.

Hutchinson, J.B. and H.L. Manning (1945). The Sea Island Cottons. *Trinidad Cotton Research Station Memoirs, Genetics Series,* 25: 80-92.

Imamaliev, A.I. and N.E. Livygina (1985). Dynamics of monosaccharides and cellulose in interspecies cotton hybrids fiber. *Soviet Agricultural Sciences* 4: 14-17.

Imamaliev, A.I. and N.I. Uzilevskaia (1982). *In vitro* cultivation of different cotton varieties. Tissue and cell culture for studies of the regularities of fiber formation characters. *Uzbekshii Biologicheskii Zhurnal (Tashkent, USSR)* 2: 61-62.

John, M.E. and L.J. Crow (1992). Gene expression in cotton (*Gossypium hirsutum* L.) fiber: Cloning of the mRNAs. *Proceedings of the National Academy of Sciences USA* 89: 5769-5773.

Kearney, T.H. (1943). Egyptian-type cottons: Their origin and characteristics. Memo. Beltsville, MD: United States Department of Agriculture, Agricultural Research Administration, 23 pp.

Kohel, R.J., Y.H. Park, and M.K. Slocum (1994). Analysis of the origin of Extra Long Staple fiber in *Gossypium barbadense* L. In *Symposium: Biochemistry of Cotton.* Raleigh, NC: Cotton Incorporated, pp. 83-88.

Kosmidou-Dimitropoulou, K. (1986). Hormonal influences in fiber development. In *Cotton Physiology,* Eds. J.R. Mauney and J.M. Stewart. Memphis, TN: The Cotton Foundation, pp. 361-374.

Krishnaswai, R. and R. Kothandaraman (1975). Favorable influence of *Gossypium raimondii* genes on fibre properties of cultivated cotton: *Genetica Agraria* 29: 277-281.

Liu, K., T.Z. Zhang, J. Sun, J.J. Pan, and R.J. Kohel (1999). Relationship between plant growth substances and fiber initiation in Upland cotton. *Acta Gosypii Sinica* 11: 48-56.

Maner, B.A., S.D. Worley, D.C. Harrell, and T.W. Culp (1971). A geometrical approach to yield models in Upland cotton (*Gossypium hirsutum* L.). *Crop Science* 11: 904-906.

Mansurov, N.I., V.I.U. Shebitchenko, and A.M. Bakhramov (1973). Characteristics of indices of fiber output of interspecific cotton hybrids. *Doklady Akademii Nauk (Tadzhikskoi SSR)* 16: 64-66.

Meredith, W.R. Jr. (1992). Cotton breeding for fiber strength. In *Proceedings from Cotton Fiber Cellulose: Structure, Function and Utilization Conference.* Memphis, TN: National Cotton Council of America, pp. 289-302.

Meredith, W.R. and R.R. Bridge (1973). Yield, yield component and fiber property variation of cotton (*Gossypium hirsutum* L.) within and among environments. *Crop Science* 13: 307-312.

Miller, P.A. and J.O. Rawlings (1967). Selection for increased lint yield and correlated responses in Upland cotton, *Gossypium hirsutum* L. *Crop Science* 7: 637-640.

Moharir, A.V. (1987). Structure and strength-crystallite orientation relationship in native cotton fibres. *Indian Journal of Textile Research* 12: 106-119.

Moharir, A.V., K.M. Vijayraghavan, B.C. Panda, D.K. Suri, and K.C. Nagpal (1986). Crystallite orientation in some cotton varieties of *Gossypium herbaceum*. *Indian Journal of Textile Research* 11: 117-120.

Morozova, A.V., S.I. Krasilnikova, and N.Z. Ruban (1983). Increase of fiber length in *Gossypium hirsutum* and *Gossypium barbadense*. *Izvestiya Akademii Nauk (Turkmenskoi SSR)* 4: 11-14.

Muramoto, H. (1973.) Hexaploid cotton: Some fiber and spinning properties. *Crop Science* 13: 396-397.

Muratov, A. and K.F. Gesos (1985). Light and electron-microscopic investigations of the variability of supramolecular fiber structure of the interspecific *Gossypium thurberi* × *Gossypium raimondii* cotton hybrid. *Cytological Genetics* 19: 7-12.

National Cotton Variety Test Program (1996). *1995 National Cotton Variety Test.* Stoneville, MS: United States Department of Agriculture, 217 pp.

Niles, G.A. and C.V. Feaster (1984). Breeding. In *Cotton,* Eds. R.J. Kohel and C.F. Lewis. Madison, WI: American Society of Agronomy, Monograph No. 24, pp. 201-231.

Odemamedov, A. (1978). Correlation analysis of the characters of fine fiber cotton. *Khlopkovodstvo,* pp. 34-35.

Percival, A.E. (1987). The national collection of *Gossypium* germplasm. *Southern Cooperative Series Bulletin* 321: 362 pp.

Percival, A.E. and R.J. Kohel (1990). Distribution, collection, and evaluation of *Gossypium*. *Advances in Agronomy* 44: 225-256.

Popova, P.IA. and I.E. Lavygina (1984). Content of hemicellulose in developing fiber of interspecific hybrids of cotton. *Uzbekshii Biologicheskii Zhurnal (Tashkent, USSR)* 4: 66-68.

Ramsey, J.C. and J.D. Berlin (1976). Ultrastructure of early stages of cotton fiber differentiation. *Botanical Gazette* 137: 11-19.

Roberts, C. and D. Davis (1991). Variability in original and selected populations of Montserrat Sea Island cotton. In *Proceedings of the Beltwide Cotton Production Research Conferences.* Memphis, TN: National Cotton Council of America, pp. 538-539.

Scholl, R.L. and P.A. Miller (1976). Genetic association between yield and fiber strength in upland cotton. *Crop Science* 16: 780-783.

Schwendiman, J. (1975). Hybrid lines from the cross between *Gossypium hirsutum* L. and *Gossypium barbadense* L. V. Separation and relative importance of the genetic effects for the fiber yield and length. *Cotton Fibres et Tropicales* 30: 185-194.

Seagull, R.W. (1993). Cytoskeletal involvement in cotton fiber growth and development. *Micron* 24: 643-660.

Smith, C.W. (1992). Development of improved fiber quality for Central and South Texas. In *Proceedings from Cotton Fiber Cellulose: Structure, Function and Utilization Conference.* Memphis, TN: National Cotton Council of America, pp. 341-348.

Stephens, S.G. (1976). The origin of Sea Island cotton. *Agricultural History* 50: 391-399.

Stewart, J.M. (1975). Fiber initiation on the cotton ovule (*Gossypium hirsutum*). *American Journal of Botany* 62: 723-730.

Turcotte, E.L. and R.G. Percy (1986). Status and potential of exotic *Gossypium barbadense* L. germplasm. *Agronomy Abstracts,* p. 85.

Turner, J.H. (1974). History of Acala cotton varieties bred for San Joaquin Valley, California. *United States Department of Agriculture, Agriculture Research Service W-16,* 23 pp.

Waddle, B.A. (1980). Genetic and environmental effects on quality of fiber crops. Cotton, structural fibers. In *Crop Quality, Storage, and Utilization,* Ed. C.S. Hoveland. Madison, WI: American Society of Agronomy, pp. 169-183.

Webb, H.W. (1978). Goals and achievements in cotton breeding. In *Proceedings of the Beltwide Cotton Production Research Conferences.* Memphis, TN: National Cotton Council of America, pp. 18-20.

Worley, S., T.W. Culp, and D.C. Harrell (1974). The relative contributions of yield components to lint yield in upland cotton, *Gossypium hirsutum* L. *Euphytica* 23: 399-403.

Worley, S., H.H. Ramey, D.C. Harrell, and T.W. Culp (1976). Ontogenetic model of cotton yield. *Crop Science* 16: 30-34.

Zhang T.Z., J. Sun, and J.J. Pan (1992). *In vitro* induction of fiber initial development of a fuzzless lintless mutant in Upland cotton. *Acta Gossypii Sinica* 4: 84-88.

Chapter 8

Genetic Variation in Fiber Quality

O. Lloyd May

INTRODUCTION

Historical records suggest that cotton textile products have existed for over two and one-half millennia (Ramey, 1980). Improvement of cotton fiber quality to enhance its use as a textile fiber began when ancient peoples domesticated cotton, altering it from a wild, photoperiodic, short-staple, low-yielding plant into one that produces more and longer lint and is adapted to extratropical latitudes (Fryxell, 1979). With the advent of nonsubjective fiber quality evaluations, knowledge of how fiber properties contribute to textile performance, and expressed needs by textile manufacturers for improved fiber quality, breeders began to emphasize fiber quality in their genetic improvement programs. Breeders are charged also with simultaneously improving lint yield and traits such as plant type and host resistance to insects and pathogens. Applied breeders probably would generally not emphasize improvement of fiber quality ahead of lint yield in a list of priorities. The complexity of these issues is such, however, that this chapter will focus on the genetics of fiber quality for Upland (*Gossypium hirsutum* L.) cotton. Consideration will be given also to the future as textile industry priorities for cotton fiber properties change to reflect the requirements of more efficient yarn-spinning and fabric-manufacturing technologies.

Before examining genetic variation for fiber quality, it is instructive to review the brief history of breeding cotton for improved fiber quality and how fiber properties became recognized as important predictors of textile performance. This brief history lesson will demonstrate the relationship between fiber quality and a healthy cotton industry.

Brief History of Efforts to Improve Cotton Fiber Quality

Cotton became an important crop in the United States in the mid-1700s (Ware, 1936). Quality of cotton fiber at this time was assessed by fiber length because long-fibered types generally spun more efficiently and produced higher quality yarns on the spinning equipment of this era. Ware (1935) describes the names of U.S.-grown cottons that were recognized by foreign buyers as representing different levels of fiber quality. So-called "Benders"-type cottons were grown along the bends of the Mississippi River and were recognized as having excellent fiber quality. The next lower level of fiber quality was associated with "Rivers" types produced along the Mississippi River tributaries. Finally, there were "Creeks" and "Uplands," with creeks denoting a lower level of quality than "Rivers" and the term "Upland" denoting short-staple types. Although there were few recorded organized breeding efforts at improving fiber quality around this time, some effort was made at selection for long-fibered cottons. Mass selection within existing varieties, along with chance outcrosses, directed these initial variety development efforts (Calhoun, Bowman, and May, 1994). Late maturity was associated with long-fibered cottons of the time, and when the boll weevil (*Anthonomus grandis* Bohemen) invaded the U.S. Cotton Belt, these varieties were abandoned in favor of short-staple, earlier maturing cultivars (Ware, 1936). As a breeder, this author wonders how many genes conferring fiber quality were lost, possibly forever, when the long-fibered types were discarded. It was not long before market demand decreased for fiber produced by the earlier maturing cottons as textile manufacturers recognized the lower fiber quality of these types. Subsequently, the United States Department of Agriculture (USDA) and some private breeders began to emphasize fiber quality, in addition to earlier maturity of cultivars that could produce in the presence of the boll weevil (Ware, 1936).

Additional events occurred in the early twentieth century that caused geneticists and breeders in the United States to consider fiber quality as important as yield, plant type, or disease resistance. The enactment of the U.S. Cotton Futures Act of 1914 set rules by which the quality of cotton would be determined (Brown, 1938). The use of these standards for the classing of cotton became mandatory in 1923 with the passage of the U.S. Cotton Standards Act (Ramey, 1980). The terminology U.S. Cotton Standards was soon replaced by Universal Cotton Standards to reflect the input from international textile groups (Brown, 1938). Cotton was classed based on staple length and grade, the latter consisting of color, preparation, and nonlint content (Lewis and Richmond, 1968). The preparation component of grade refers to a visual assessment of orientation of fiber in the sample (Hake et al., 1990). Cotton "classers" assigned a grade to a bale of cotton based on visual comparison between a lint sample drawn from the bale and a standard comparison sample devised by the U.S. government through the Cotton Standards Act. Classers also could assign grades based on a mental image comparison with an official standard (Lewis and Richmond, 1968). Staple or fiber length was determined from parallelized fibers "pulled" from the bale sample and called to the nearest $1/32$ of an inch (Hake et al., 1990). Although still considered subjective measures of fiber quality, cotton classing was an improvement over assessing quality based on area of growth or species, such as Upland or Sea Island (Perkins, Ethridge, and Bragg, 1984). These subjective measures of fiber quality were all that were available until advances occurred in mechanical fiber property measurement. The fact that all cotton sold was assigned a grade and staple elevated the status of fiber quality as an important breeding objective.

Mechanization in the textile industry was another vehicle by which the fiber properties of a cotton variety came to be recognized as equally important as yield potential. Yarn spinners banded together in the mid-1920s to demand cotton with longer staple lengths due to its superior performance relative to that of the short-staple varieties (Brown, 1938). International demand for U.S. cotton declined in the early 1930s as other countries made concerted efforts to improve the staple length of their cotton varieties (Ware, 1936). These factors also contributed to the emphasis on fiber quality in U.S. cotton-breeding programs.

Early cotton-breeding efforts at improving fiber quality were not just limited to the United States. International demand for cotton with improved fiber quality was spurred by development in the nineteenth or early twentieth century of a mechanized spinning industry that replaced hand spinning. The then-new spinning technology could not efficiently manufacture yarn with short-staple cottons that typically had staple lengths of less than 25 mm (millimeters). The cultivation of cotton in Russia, Brazil, Peru, and Argentina was expanded due to shortages caused by the decline in U.S. production during the U.S. Civil War. This emphasis on cotton production was a stimulus for variety development, specifically varieties with improved fiber quality. A common theme among international breeding efforts was the introduction of American Upland germplasm and almost complete abandonment of short-staple Asiatic species (Ware, 1936).

Prior to the development of the U.S. cotton industry, India was the world leader in cotton production (Ware, 1936). India is likely the original place where cotton was used in textile products (Percival and Kohel, 1990). As in the United States, organized cotton breeding did not begin until the early 1900s. Publication of Mendel's work and the replacement of hand spinning methods with machine technology spurred efforts to improve the fiber quality of Indian cotton varieties. India had historically grown the diploid cotton species *G. arboreum* L. and *G. herbaceum* L., with staple lengths of around 13 mm or less, rather than Upland types. To improve the fiber quality of Indian cotton, Upland and Sea Island (*G. barbadense*) cottons were introduced. Also, organized selection for longer staple length within *G. herbaceum* resulted in several cotton varieties with sufficient fiber quality to meet mill needs. Organized breeding efforts with the introduced Upland types in the cotton-producing provinces of India, such as the Punjab area, resulted in varieties with improved fiber quality.

Similar to the situation in India, Chinese cotton production initially relied on short-staple Asiatic cotton species (Ware, 1936). Not until a modern textile industry developed in the 1890s was there recognition that cotton with better fiber properties was needed. Initially, this need was met by importing cotton from the United States and India. During the 1920s, breeding efforts were initiated at the University of Nanking. Seed of American cottons was provided by the USDA, and breeding efforts commenced. In the early 1920s, these efforts produced the cultivar "Million Dollar," with a level of fiber quality such that textile mills paid a premium for its fiber.

Russian efforts to improve cotton fiber quality coincided with development of a mechanized textile industry in the late 1800s. Introduction of American Uplands replaced the short-staple Asiatic types, thereby improving the textile performance of the Russian cotton. Breeders and agronomists were hired, and emphasis was placed, in part, on improvement of fiber quality. Ultimately, from an initial germplasm base comprised primarily of the American Uplands of the mid- to late-1800s era, Russian breeders developed their own locally adapted varieties (Ware, 1936).

Similar events in Brazil, Peru, Argentina, Korea, and Sudan led to efforts to breed cottons with improved fiber quality.

Modern Research into Cotton Fiber Quality

Improvement of cotton fiber quality accelerated after advances in several areas of science occurred in the early 1900s. First, the specific fiber properties that define quality had to be understood through studies relating raw fiber properties to textile performance (Moore, 1938). Concurrently, technological advances in instrumentation to accurately measure fiber properties such as length, strength, and fineness were made. The speed and accuracy of these instruments in measuring fiber properties allowed breeders to study the genetics of quality parameters. A brief synopsis of several key events is provided in the following material.

Although the instrumentation was crude by modern standards, Balls (1928) was one of the first researchers to measure fiber physical properties to facilitate breeding cottons that would benefit the textile industry. In the United States, fiber technology research progressed in the late 1920s with the opening at Clemson College, Clemson, South Carolina, of a fiber and spinning laboratory by the Bureau of Agricultural Economics, in cooperation with the Bureau of Plant Industry. It was recognized at the time that more precise measurement of raw fiber properties than that provided by cotton classers would benefit yarn spinners and, ultimately, breeders in improving fiber quality (Willis, 1926). Soon after, a laboratory with similar functions was created in Washington, DC, by the Department of Agriculture (Lewis and Richmond, 1968; Lee, 1984). At the same time, H. W. Barre, head of the Cotton Division, Bureau of Plant Industry, recognized the importance of synergy between cotton breeders and the developing fiber technology. He fostered cooperative ties with R. W. Webb, head of the fiber and spinning laboratory, in bringing fiber technology to cotton breeding. Under the leadership of Webb, investigations were begun to determine the importance to spinning of fiber length, strength, fineness, and maturity.

Fiber length was initially assessed by measuring the length of fiber samples from ginned lint or by combing fibers on seeds (Brown, 1938; Harrison, 1939). Fiber length uniformity was similarly measured through a painstaking process of manually collecting lint into classes representing $1/32$ of an inch categories and then weighing to determine relative fiber length uniformity. An early instrument method of assessing fiber length uniformity was the Suter-Webb sorter (Brown, 1938), which remains in use today. Although potentially useful to the cotton technologist, this method was too labor intensive to be applied to breeding for improved fiber length uniformity. Another device that allowed fiber length uniformity to be assessed with seed cotton was reported by McNamara and Stutts (1935). This instrument, although not requiring ginned lint, similarly operated by successively combing and removing by hand fibers of decreasing length. Apparently, the labor required in the process kept this instrument from being used to any extent in efforts to reduce fiber length variability through breeding.

Fiber strength was measured first by hand breaking small tufts of fibers, until spinners recognized that these breaking measurements often had little relationship to yarn strength (Brown, 1938). Early machine testers of fiber strength included the Dewey single fiber method (Dewey, 1913) and the Chandler bundle strength tester (Brown, 1938). Although these instruments provided nonsubjective measurements of fiber strength and were, in this manner, an improvement over hand breaking tests, they still were slow and not readily available to cotton breeders.

By the mid- to late-1930s, fiber technology research had progressed with the development of instruments providing raw fiber properties useful to spinners, cotton technologists, and breeders. The fibrograph, an instrument that rapidly measures fiber length by photoelectric sensor, was developed (Hertel and Zervigon, 1936; Hertel, 1940). Fiber length uniformity could be calculated from fibrograph measurements without resorting to the tedious process of sorting and weighing fibers into length groups. A more rapid measure of fiber strength was available with the Pressley tester (Pressley, 1942). The Arealometer was developed and provided estimates of fiber surface area measured at two air pressures from which fiber perimeter and wall thickness could be calculated (Hertel and Craven, 1951). Fiber measurements from the fibrograph, Pressley tester, and the Arealometer were shown to relate to yarn strength (Barker and Pope, 1948; Landstreet, 1954). The Stelometer (Hertel, 1953), an instrument that gives both fiber bundle strength and fiber elongation, provided another tool for breeders to evaluate their experimental cottons. The relative speed of measurement of these instruments accelerated breeding efforts at improving cotton fiber quality.

GENETIC VARIATION IN FIBER QUALITY

Genetic variation must exist for selection to be effective. Fortunately, sources of genetic variation for fiber quality exist within and among the cultivated species of cotton. Within Upland cotton, genetic variability for fiber quality exists among cultivars (USDA, 1995), among germplasm lines (Anonymous, 1974; Green, 1950), and in primitive germplasm converted to nonphotoperiodic flowering habit (McCarty and Jenkins, 1992; McCarty et al., 1996). The range of fiber properties available in several germplasm collections is extensive (Green, 1950; Anonymous, 1974). Breeders have access to a wealth of germplasm from the U.S. National Cotton Germplasm Collection, maintained by the Department of Agriculture, Agricultural Research Service, College Station, Texas, containing over 5,100 seed samples of *Gossypium* spp. (Percival and Kohel, 1990; see also Kohel, Chapter 7, this volume). A general categorization of *Gossypium* spp. with respect to fiber type would be the relatively short length and coarse diploid Asiatic species, Upland, with intermediate levels of length and fineness, followed by Egyptian, Pima, and Sea Island, with generally fine, long, and strong fiber (Ramey, 1980). The primary gene pool, or that most readily available to Upland breeders, includes the tetraploid species *G. hirsutum* L. (Upland) and *G. barbadense* L. (Pima, Egyptian, and Sea Island) (Stewart, 1988). The Asiatic diploid species *G. herbaceum* and *G. arboreum* can also be included if more complicated crossing schemes are considered (Fryxell, 1984). It should also be noted, however, that attempts at stable introgression of genes from a non-Upland source into a *G. hirsutum* background are not without difficulty, which could limit the primary gene pool for Upland improvement (Stephens, 1949; McKenzie, 1970). Although the genus *Gossypium* is comprised of a number of species (Stewart, 1988), those producing useable fiber would seem most appropriate for consideration as a germplasm source for a breeder interested in improving fiber quality. Because *G. hirsutum* is the most widely grown species of cotton, this chapter will concentrate on this species.

An understanding of those fiber quality traits which contribute to textile performance is a prerequisite for setting priorities in a breeding program. This chapter will examine genetic control of the common fiber properties, consisting of fiber length measures, strength, elongation, and fineness, plus short fiber content and properties measured by the Advanced Fiber Information System (Behery, 1993). Other properties exist, such as dust content (Deussen, 1992), that also define fiber quality; however, space limitations prevent their consideration in this chapter. Genetic versus nongenetic influences, type of gene action, heritability, and, where available, selection response are presented. To simplify tables, data (variances, mean squares) have been rounded from the authors' original values or expressed as ratios where appropriate. The final discussion is whether new yarn manufacturing systems will require breeders to emphasize different fiber properties than those known to benefit ring spinning systems.

MEASURES OF FIBER LENGTH

Since the development of mechanical spinning frames in the 1700s, fiber length has been recognized as a contributor to yarn strength and processing performance (Brown, 1938; Perkins, Ethridge, and Bragg, 1984). As we have seen, fiber length was the initial property used to assess cotton quality and its suitability for certain end uses. Knowledge of fiber length is critical to manufacturing yarn of a specific size on ring spinning systems (Rusca and Reaves, 1968). With respect to processing, certain measures of fiber length are used to set the distance between rolls in the drafting procedure during yarn manufacture (Ducket, 1974; Behery, 1993). As for yarn strength, the effect of fiber length on the maximum strength of yarns spun with optimum twist, according to Landstreet (1954), is secondary. Holding other fiber properties constant, longer fiber requires less twist to produce maximum yarn strength. In contrast, relatively short-staple cotton has reduced holding surface compared with longer staples and requires increased twist to produce maximum yarn strength. Increasing twist beyond this optimum reduces fiber strength, which then causes loss in yarn strength. Longer fiber length is desirable for the production of fine yarns and low twist yarns such as those used for knitting (Landstreet, 1954). Longer fiber requires less twist in the roving process of cotton destined to be ring spun. The minimum twist insertion necessary to produce roving from sliver is desirable to control yarn manufacturing costs (Perkins, Ethridge, and Bragg, 1984).

Fiber length is measured by classers staple, reviewed earlier, and instrument estimates by the fibrograph, High-Volume Instrument (HVI), or the new Advanced Fiber Information System (AFIS) (Behery, 1993). These measures include upper half mean (mean length of the longer 50 percent of the fiber by weight, as tested by the HVI), 2.5 percent and 50 percent span lengths (respectively, the distance spanned by the indicated percentage of the fibers in the beard tested on a fibrograph), and upper quartile length (length at which 25 percent of the fiber by weight is longer when measured with Suter-Webb array) (Behery, 1993). That there is a strong genetic basis for fiber length despite various methods of measurement is evident from the data in Table 8.1. These data were derived from a broad array of studies conducted throughout the U.S. Cotton Belt that included breeding populations and cultivar evaluations. Whether assessed by classers staple, Suter-Webb array, or fibrograph (upper half mean or span length), where genetic differences exist, the magnitude of genetic variance generally is greater than that of nongenetic influences. In a few instances, genotype × location, year, or higher-order interactions are noted, yet they are small in magnitude when compared with genetic variation. These findings indicate that extensive environmental replication is not necessary to evaluate and select breeding material on the basis of fiber length parameters. Note also that experimental error (see references E, F, J, and K—Table 8.1) was of similar or greater magnitude as genotypic variance in some studies. However, these nongenetic influences apparently are not related to effects of years or locations and, ultimately, should not preclude the identification of genotypes with desired fiber length. Most studies in which genetic variation for measures of fiber length has been broken down into components report additive variance to be more important than nonadditive types of genetic variance (see Table 8.2). Meredith and Bridge (1972) report one instance in which the expression of additive effects governing 2.5 percent and 50 percent span lengths varied over locations (data not shown). In their combined analysis over locations, dominance main effects were greater than additive main effects, but the additive × location and dominance × location interactions indicated that additive effects were more important than dominance effects in specific locations. Remaining studies that report additive and nonadditive genetic variance or effects over environments support data in Table 8.1 that environmental influences on length parameters are not of a magnitude that should hinder breeding efforts. Quisenberry (1975) and May and Green (1994) found that nonadditive genetic variance was larger than additive variance for fiber length (see Table 8.2). Within the Pee Dee germplasm, the finding of low additive genetic variance could reflect its exhaustion from over forty years of breeding for improved fiber properties.

Heterosis, whether expressed as deviations from the mid- or extreme parent, is another means by which genetic control of fiber length can be inferred (see Table 8.3). These data indicate that extreme parent heterosis for fiber length can occur in cotton, but generally, the magnitude of the transgressive expression is small. Miller and Lee (1965) reached similar conclusions. Thus, despite a few examples of dominant gene action for fiber length, the majority of the data are consistent with mainly additive genetic control of fiber length. A sample of heritability estimates for fiber length

TABLE 8.1. Genetic and Environmental Influences on Various Measures of Fiber Length

Reference	Genotype	Genotype × Location	Genotype × Year	Genotype × Location × Year	Residual
A	350	—	5	—	—
B	9	2	<1	2	1
C	0.6	<0.01	<0.01	<0.01	<0.01
D	1	<0.01	—	—	<0.01
E	0.001	0*	0	<0.001	0.001
F	5	0*	0.06	1	5
G	16	<0.1	<0.1	0.4	0.8
H	61	8	<1	15	32
I	5	<1	<1	1	5
J	0.9	0.05	0.05	0.2	1
K	0.6	0.02	0.04	0*	0.9
L	315	4	—	—	<1
M	0.5	0.03	—	—	—
N	24	8	—	—	16

* Analysis of variance estimate of the indicated variance was negative; thus, the most reasonable estimate is zero.

A. Classers staple length, twenty-four genotypes, four years, one location, analysis of variance F values. Neely (1940).

B. Upper half mean length, four cultivars, three locations, two years, mean squares. Hancock (1944).

C. Upper quartile length, sixteen cultivars, seven locations, three years, mean squares. Pearson (1944).

D. Upper quartile length, sixteen cultivars, nine locations, one year, mean squares. Pope and Ware (1945).

E. Upper half mean length, ninety-five $F_{2:4}$ or $F_{2:5}$ lines, two locations, two years, variance components. Miller et al. (1958).

F. Upper half mean length, fifteen cultivars, nine locations, three years, variance components ($\times 10^{-4}$). Miller, Robinson, and Williams (1959).

G. Classers staple length, sixteen cultivars, eleven locations, three years, variance components. Miller, Robinson, and Pope (1962).

H. Upper half mean length, four cultivars, 101 location × year combinations, variance components. Abouh-El-Fittouh, Rawlings, and Miller (1969).

I. 2.5 percent span length, eight cultivars, three locations, three years, variance components ($\times 10^{-4}$). Bridge, Meredith, and Chism (1969).

J. Upper half mean length, three cultivars, twenty-eight locations, three years, ratio of indicated source to error variance. El-Sourady, Worley, and Stith (1969).

K. 2.5 percent span length, sixty-two BC_2F_4 lines, two locations, two years, variance components. Murray and Verhalen (1969).

L. 2.5 percent span length, four cultivars, four environments (year × soil type combinations), mean squares. Meredith and Bridge (1973).

M. 2.5 percent span length, eighty-nine early generation families evaluated in three environments (year × location combinations), variance components. Scholl and Miller (1976).

N. 2.5 percent span length, eighteen genotypes, including advanced breeding lines and cultivars, seven locations, variance components expressed as percent of total variance. Meredith, Sasser, and Rayburn (1996).

TABLE 8.2. Additive and Nonadditive Genetic and Environmental Influences on Fiber Length

Reference	Additive	Additive x Loc	Additive x Yr	Additive x Loc x Yr	Nonadditive	Nonadditive x Loc	Nonadditive x Yr	Nonadditive x Loc x Yr	Residual
A	48	—	—	—	—	—	—	—	—
B	2	—	—	—	—	—	—	—	—
C	2	—	—	—	0	—	—	—	—
D	5	—	—	—	<1	—	—	—	—
E	670	3	10	0	10	0*	30	0*	1640
F	0.006	—	—	—	<0.01	—	—	—	<0.01
G	11	—	—	—	—	—	—	—	—
H	0.6	—	—	—	—	—	—	—	—
I	2	—	—	—	—	—	—	—	—
J	1	—	—	—	—	—	—	—	—
K	4	0.5	—	—	3	1	—	—	1
L	9	0.5	—	—	1	0.3	—	—	0.3
M	0.05	0.02	—	—	4	—	—	—	—

* Analysis of variance estimate of the indicated variance was negative; thus, the most reasonable estimate is zero.

A. Upper half mean length. Ratio of general combining ability to specific combining ability mean squares calculated from data in Barnes and Staten (1961).

B. Upper half mean length. Ratio of general combining ability to specific combining ability mean squares calculated from data in Barnes and Staten (1961).

C. Upper half mean length. General and specific combining ability variances ($\times\ 10^{-3}$). Miller and Marani (1963).

D. Upper half mean length. Additive and dominance genetic variances ($\times\ 10^{-4}$). Ramey and Miller (1966).

E. Upper half mean length. Additive and dominance genetic variances rounded. Lee, Miller, and Rawlings (1967).

F. 2.5 percent span length. Additive and nonadditive genetic and environmental variances. Al-Rawi and Kohel (1970).

G. 2.5 percent span length. Ratio of general and specific combining ability mean squares averaged from the F_2 and F_3 generation of diallel progenies. Meredith and Bridge (1973).

H,I. 2.5 percent span length. Ratio of additive to nonadditive genetic variance within High-Plains and Acala germplasm, respectively. Quisenberry (1975).

J. 2.5 percent span length. Ratio of additive genetic variance to total genetic variance. Wilson and Wilson (1975).

K. 2.5 percent span length. General and specific combining ability mean squares. Green and Culp (1990).

L. 2.5 percent span length. General and specific combining ability mean squares. Tang et al. (1993).

M. 2.5 percent span length. Additive, additive $\times$ additive, and dominance genetic variances ($\times\ 10^{-3}$). May and Green (1994).

TABLE 8.3. Heterosis for Fiber Length Expressed As Number of Hybrids or Average Millimeters by Which the Hybrids Exceeded Midparent or Parental Values

Reference	No. hybrids > longest parent	No. hybrids < shortest parent	Avg. hybrid deviation from midparent	Avg. hybrid deviation from extreme parent
A	2	0	—	—
B	2	1	—	—
C	—	—	+1.0	—
D	5	—	—	+0.9
E	5	—	—	+1.5
F	—	—	+0.3	—
G	—	—	+0.8	—
H	—	—	—	+1.5
I	1	—	—	+1.0

A. Twenty-two F_1 hybrids from crossing Acala 1517C or Acala 1517D with eleven other *G. hirsutum* cottons representing Acala, Mississippi Delta, and southeastern United States germplasm. Upper half mean (UHM). Barnes and Staten (1961).
B. Twenty-one F_1 hybrids from half-diallel among seven Acala germplasms. UHM length. Barnes and Staten (1961).
C. Twenty-two F_1 hybrids from crosses between two medium-staple Uplands and eleven long- and extra-long-staple Uplands. UHM length. Harrell (1961).
D. Seven F_1 hybrids between seven Uplands and one Acala. UHM length. Pate and Duncan (1961).
E. Ten F_1 hybrids among Upland and wild Upland. UHM length. White and Richmond (1963).
F,G. Four F_1 intraspecific hybrids, respectively, from crosses within *G. hirsutum* and *G. barbadense* cultivar groups. UHM length. Marani (1968a).
H. Nine inter-specific F_1 hybrids between *G. hirsutum* and *G. barbadense* cultivars. UHM length. Marani (1968b).
I. Three F_1 hybrids from Upland crosses. 2.5 percent span length. Meredith, Bridge, and Chism (1970).

measures from a divergent sample of cotton populations and various selection units suggests that selection for various length parameters should be effective (see Table 8.4). Experiments designed to measure response of length parameters to selection report responses of a magnitude typical for a quantitatively inherited trait (see Table 8.5).

There exists an upper limit to the need to increase length of medium-staple (about 25 to 30 mm) cottons to enhance their spinning performance, particularly for open-end rotor spinning systems (Deussen, 1992). Consequently, the textile industry would benefit more by concentrating breeding efforts on other quality factors

TABLE 8.4. Heritability Estimates for Fiber Length Measurements

Reference		Reference	
A	0.79	I	0.56
B	0.88	J	1.00
C	0.39	K	0.91
D	0.49	L	0.67
E	0.10	M	0.48
F	0.31	N	0.77
G	0.85	O	0.54
H	0.55		

A. Upper half mean (UHM), F_3 line selection unit, broad sense. Al-Jibouri, Miller, and Robinson (1958).
B. UHM, F_4 line selection unit, broad-sense. Miller et al. (1958).
C. UHM, F_2 plant selection unit, narrow-sense. Lewis (1957).
D. 2.5 percent span length, F_1 entry mean selection unit, narrow sense. Verhalen and Murray (1967).
E. 2.5 percent span length, F_2 plant selection unit, narrow sense. Murray and Verhalen (1969).
F. 2.5 percent span length, F_2 plant selection unit, narrow sense. Murray and Verhalen (1969).
G. 2.5 percent span length, BC_2F_4 line selection unit, broad sense. Murray and Verhalen (1969).
H. 2.5 percent span length, F_1 entry mean selection unit, narrow sense. Verhalen and Murray (1969).
I. 2.5 percent span length, F_1 entry mean selection unit, narrow sense. Al-Rawi and Kohel (1970).
J. 2.5 percent span length, F_1 entry mean selection unit, narrow sense. Wilson and Wilson (1975).
K. 2.5 percent span length, F_3 progeny mean selection unit, broad sense. Scholl and Miller (1976).
L. 2.5 percent span length, F_6 line mean selection unit, narrow sense. Keim and Quisenberry (1983).
M. 2.5 percent span length, F_2 plant selection unit, narrow sense. May and Green (1994).
N. 2.5 percent span length, F_2 population bulk selection unit, broad sense. May and Green (1994).
O. 2.5 percent span length, F_3 line selection unit, broad-sense. May and Green (1994).

related to fiber length distribution, specifically on improving length uniformity and reducing short-fiber content. By its nature, fibers in a sample of cotton are not all of the same length, varying even on the same seed (Richmond and Fulton, 1936). High length uniformity and low short-fiber content are desired by textile manufacturers because these traits are associated with reduced manufacturing

TABLE 8.5. Summary of Selection Experiments Toward Fiber Length Modification

Reference	Response of fiber length measure (mm)	
A	+2.3	—
B	+1.0	−0.8
C	+2.5	—
D	+1.5	—

A. Ten years of mass selection for increased upper half mean length with forced self-pollination within four originally open-pollinated cultivars. Mean response of three cultivars exhibiting response to selection. Simpson and Duncan (1953).
B. Single cycle of divergent mass selection for 2.5 percent span length. Murray and Verhalen (1969).
C. Three cycles of divergent selection for combination of 2.5 percent span length and micronaire. Mean difference between 2.5 percent span length groups presented. Quisenberry, Ray, and Jones (1975).
D. Single cycle of divergent mass selection within a cultivar. Average response from two methods of selection and two selection intensities. Verhalen, Baker, and McNew (1975).

waste, neps, and ends down during yarn production, along with improved yarn appearance and strength (Behery, 1993). Two measures of length uniformity are commonly defined. Length uniformity index (LUI) is the ratio of the mean length and the upper half mean length, whereas length uniformity ratio (UR) is the ratio of two span lengths, 50 percent/2.5 percent span length (Behery, 1993). Deussen (1992) indicates that for ring spinning, fiber length uniformity is an important determinant of yarn quality and spinning performance.

Less attention has been paid to fiber length uniformity than to the genetics of fiber length, as indicated by comparing the volume of data between Tables 8.1 through 8.5 and Table 8.6. Variation for fiber length uniformity exists and, in part, is attributable to genetic variation (see Table 8.6). Environmental influences on fiber length uniformity are present, but are not of a magnitude to preclude separation of genetic differences (Meredith et al., 1991; Meredith, Sasser, and Rayburn, 1996). Genetic variance for fiber length uniformity seems to be mostly of the fixable types for a self-pollinated crop, and reasonable heritability estimates exist. Hence, further improvement of fiber length uniformity seems possible and should receive additional attention in breeding programs. However, it is probably not realistic that all fiber length variation will be alleviated through breeding.

TABLE 8.6. Pertinent Data Describing the Genetic Control of Fiber Length Uniformity

Reference	Genotype	Genotype × Environment	Residual
A	20	2	—
B	3	1	15
	GCA/SCA	—	—
C	5	—	—
D	0.5	—	—
E	1.2	—	—
F	Add.	Dom.	A×A
	1.6	0	0.4
G	F_2 plant	F_2 bulk	F_3 row
	0.4	0.6	0.2

A. ANOVA F values (rounded to nearest whole number) of length uniformity index (LUI) from evaluation of nineteen cultivars and advanced breeding lines in two states in the United States. Meredith et al. (1991).
B. Percentage (rounded to whole numbers) of total variance of LUI accounted for by the indicated source of variation from the evaluation of eighteen genotypes at seven locations in the Upland Cotton Belt of the United States. Meredith, Sasser, and Rayburn (1996).
C. Ratio of GCA to SCA mean squares for LUI from a cross-classified mating design. Barnes and Staten (1961).
D. Average ratio of variances of additive to nonadditive effects for LUI averaged over seven parents crossed in a half-diallel. Barnes and Staten (1961).
E. Ratio of GCA to SCA mean squares for length uniformity ratio (UR) derived from a five-parent half-diallel. Green and Culp (1990).
F. Additive, dominance, and additive × additive genetic variance ($\times 10^{-4}$) for UR resolved from the F_2 and F_3 generations of a 4 × 4 design II mating. May and Green (1994).
G. Standard unit heritability (Frey and Horner, 1957) estimates for UR. May and Green (1994).

Short-fiber content (SFC) is defined as the percentage of fibers by weight with lengths less than 12.7 mm (Behery, 1993). Sources of short fibers include those inherent to the genotype and its reaction to the environment and those introduced by mechanical handling of the cotton. Breeders have relied mainly on indirect indicators of SFC, the LUI and UR, as direct measurements by Suter-Webb array are too slow. However, Behery (1993) and Meredith, Sasser, and Rayburn (1996) suggest that LUI and UR might not be acceptable measures of SFC. For more information on the limitations of estimating SFC from UR or LUI data, see Bargeron (1990) and Woo and Suh (1994). A relatively new instrument, the AFIS, provides a direct and relatively rapid measurement of SFC (Behery, 1993; Bragg and Shofner, 1993). The genetic control of SFC has not been extensively investigated because a suitably rapid measurement needed for breeding work was not available until the development of the AFIS. Behery (1993) suggests that the genotype causes short fibers only indirectly and that most are the result of fiber breakage related to fiber strength. Meredith, Sasser, and Rayburn (1996) report a correlation between fiber strength and AFIS SFC of only -0.02, suggesting that genetic factors other than those imparting fiber strength may be involved in causing SFC. Another reason why the correlation between fiber strength and SFC was low in this study could be related to how the cotton was processed. These data were derived from the High Quality Regional Cotton Variety Test (USDA, 1995) for which fiber samples are not ginned on commercial gins nor processed through one or more lint cleaners. Thus, as Meredith, Sasser, and Rayburn (1996) point out, this fiber may not be similar to that derived from commercial ginning with respect to the degree of stress and the associated fiber breakage from commercial processing. As a result, the effect of fiber strength on SFC may not have been realized and may partially explain their low correlation. Genotypic differences in SFC have been reported (Williford, Meredith, and Griffin, 1984; Meredith, Sasser, and Rayburn, 1996), and the genotypic component of SFC, as measured by AFIS, is greater than the genotype $\times$ environment variation (Meredith, Sasser, and Rayburn, 1996). Now that a reliable and relatively quick method of assessing SFC, the AFIS, is available, it should be possible to further investigate the genetic basis of SFC and the degree to which it has a separate genetic basis than that of fiber strength.

MEASURES OF FIBER STRENGTH

Fiber strength or tenacity is perhaps the most important fiber property other than length contributing to cotton's use as a textile fiber. Thus, its inheritance and efforts at improvement have been the subject of extensive study. Fiber strength translates almost directly into yarn strength (Landstreet, 1954; Meredith et al., 1991; Deussen, 1992) and is related to spinnability, defined as ends down during yarn manufacture (Deussen, 1992). In woven and knit fabric manufacturing, it affects production speed (Perkins, Ethridge, and Bragg, 1984) and is essential to the maintenance of cotton's natural qualities after chemical processing of fabric (Rowland et al., 1976). Among the common fiber properties, fiber strength ranks first in order of precedence in rotor spinning systems (Deussen, 1992), and it contributes to fiber durability from mechanical stresses in the harvest, ginning, and yarn-manufacturing processes (Perkins, Ethridge, and Bragg, 1984; Deussen, 1992). Fiber strength is more important to open-end spinning because of the different yarn structure of an open-end versus a ring-spun yarn (Konishi, 1975). Although fiber strength can be measured on single fibers (Sasser, 1992), it is most commonly measured on a bundle of fibers by the Pressley tester (Pressley, 1942), Stelometer (Hertel, 1953), or HVI instruments (Taylor, 1982).

Data in Table 8.7 indicate that fiber strength is very much a genetic property and that genotype × environment interactions are small relative to genetic influences. Similar to the data for fiber length, additive gene action predominates for fiber strength (see Table 8.8), with usually small, and not generally meaningful, amounts of heterosis (see Table 8.9). Additionally, heritability is generally high for selection units ranging from single plants to population bulks (see Table 8.10). Recurrent and mass selection has resulted in positive gains in strength (see Table 8.11). Other examples of strength improvement through plant selection have been documented by Singh and colleagues (1990, 1991). Estimates of gene number influencing fiber strength range from five (Self and Henderson, 1954) to as many as fourteen (Tipton et al., 1964a), typical for a quantitatively inherited trait. We should note that studies reporting the number of loci influencing strength do not account for the possible bias of duplicate loci, as Upland cotton is an allotetraploid (Endrizzi, Turcotte, and Kohel, 1984). Abdel-Nabi, Jones, and Tipton (1965) reported finding only a single transgressive segregate out of 1,731 F$_3$ plants from a strong-fibered Acala × a weak-fibered Upland cross, suggesting that the parents differed for many genes affecting fiber strength. However, other data suggest fiber strength may not always segregate in a quantitative manner. Richmond (1951) indicated that recovery of high strength segregates from small backcross populations during introgression of strength from the triple hybrid *G. thurberi* × *G. arboreum* × *G. hirsutum* (Beasley, 1940) was evidence for only a few major genes controlling strength. Meredith (1977) came to a

TABLE 8.7. Genetic and Environmental Influences on Various Measures of Fiber Strength

Reference	Genotype	Genotype × Location	Genotype × Year	Genotype × Location × Year	Residual
A	2.3	<0.1	<0.1	<0.1	<0.1
B	1870	30	—	—	7
C	0.06	0*	<0.01	0	0.09
D	29	0*	<.1	5	14
E	0.05	<0.01	<0.01	<0.01	<0.01
F	0.2	0*	0*	0.1	0.4
G	0.5	0*	0.1	0.3	1.0
H	0.2	<0.01	0.06	<0.01	1.7
I	58	<1	—	—	<1
J	44	0*	—	—	—

* Analysis of variance estimate of the indicated variance was negative; thus, most reasonable estimate is zero.
A. Pressley strength, four cultivars, three locations, two years, mean squares. Hancock (1944).
B. Chandler strength, sixteen cultivars, ten locations, mean squares. Pope and Ware (1945).
C. Pressley strength, ninety-two $F_{2:4}$ or $F_{2:5}$ lines, two locations, two years, variance components. Miller et al. (1958).
D. Pressley strength, fifteen cultivars, nine locations, three years, variance components. Miller, Robinson, and Williams (1959).
E. Stelometer strength, four cultivars, 101 location × year combinations, variance components. Abouh-El-Fittouh, Rawlings, and Miller (1969).
F. Stelometer strength, eight cultivars, three locations, three years, variance components. Bridge, Meredith, and Chism (1969).
G. Stelometer strength, three cultivars, twenty-eight locations, three years, ratio of indicated source to error variance. El-Sourady, Worley, and Stith (1969).
H. Stelometer strength, sixty-two BC_2F_4 lines, two locations, two years, variance components. Murray and Verhalen (1969).
I. Stelometer strength, four cultivars, four environments (year × soil type combinations), mean squares. Meredith and Bridge (1973).
J. Stelometer strength, eighty-nine early generation families evaluated in three environments (year × location combinations), variance components. Scholl and Miller (1976).

similar conclusion after several cycles of backcross breeding to improve fiber strength. Meredith (1992) reported the superior fiber strength of his cultivar, MD51ne, was conditioned by as few as two major genes, with the high strength being derived from the germplasm line FTA 263 (Culp and Harrell, (1980). The high fiber strength of FTA 263 results from a germplasm pool that includes

TABLE 8.8. Additive and Non-additive Genetic and Environmental Influences on Fiber Strength

Reference	Additive	Additive x Loc	Additive x Yr	Additive x Loc x Yr	Nonadditive	Nonadditive x Loc	Nonadditive x Yr	Nonadditive x Loc x Yr	Residual
A	20	—	—	—	—	—	—	—	—
B	4	—	—	—	3	—	—	—	—
C	0.03	—	—	—	<0.01	—	—	—	—
D	680	<10	<10	0*	0*	<10	<10	17	116
E	0.1	—	<0.01	—	—	—	—	—	<0.01
F	24	—	—	—	16	—	—	—	5
G	29	—	—	—	—	—	—	—	—
H	4	—	—	—	—	—	—	—	—
I	3	—	—	—	—	—	—	—	—
J	0.73	—	—	—	—	—	—	—	—
K	42	14	—	—	5	24	—	—	9
L	2900	141	—	—	200	90	—	—	70
M	0.05	—	—	—	2	—	—	—	—

* Analysis of variance estimate of the indicated variance was negative; thus, most reasonable estimate is zero.

A. Stelometer strength. Ratio of general combining ability to specific combining ability mean squares calculated from data in Barnes and Staten (1961).

B. Stelometer strength. General and specific combining ability variances ($\times\ 10^{-3}$). Miller and Marani (1963).

C. Stelometer strength. Additive and dominance genetic variances. Ramey and Miller (1966).

D. Stelometer strength. Additive and dominance genetic variances. Lee, Miller, and Rawlings (1967).

E. Stelometer strength. Mean squares of the additive component of genetic variation. Verhalen and Murray (1969).

F. Stelometer strength. Additive and nonadditive genetic and environmental variances. Al-Rawi and Kohel (1970).

G. Stelometer strength. Ratio of general and specific combining ability mean squares averaged from the F_2 and F_3 generation of diallel progenies. Meredith and Bridge (1973).

H,I. Stelometer strength. Ratio of additive to nonadditive genetic variance within High-Plains and Acala germplasm, respectively. Quisenberry (1975).

J. Stelometer strength. Ratio of additive genetic variance to total genetic variance. Wilson and Wilson (1975).

K. Stelometer strength. General and specific combining ability mean squares. Green and Culp (1990).

L. Stelometer strength. General and specific combining ability mean squares. Tang et al. (1993).

M. Stelometer strength. Additive and dominance genetic variances. May and Green (1994).

TABLE 8.9. Heterosis for Fiber Strength Expressed As Number of Hybrids or Average g/tex by Which the Hybrids Exceeded Parental or Midparent Values

Reference	No. hybrids> longest parent	No. hybrids < shortest parent	Avg. hybrid deviation from midparent	Avg. hybrid deviation from extreme parent
A	1	0	—	—
B	4	7	—	—
C	4	—	—	0.4
D	—	—	0.1	—
E	—	—	0.9	—
F	—	—	3.4	—
G	0	0	—	0

A. Twenty-two F_1 hybrids from crossing Acala 1517C or Acala 1517D with eleven other *G. hirsutum* cottons representing Acala, Mississippi Delta, and southeastern United States germplasm. Stelometer strength. Barnes and Staten (1961).
B. Twenty-one F_1 hybrids from half-diallel among seven Acala germplasms. Stelometer strength. Barnes and Staten (1961).
C. Seven F_1 hybrids between seven Uplands and one Acala. Stelometer strength. Pate and Duncan (1961).
D,E. Four F_1 intra-specific hybrids, respectively, from crosses within *G. hirsutum* and *G. barbadense* cultivar groups. Pressley strength converted to Stelometer units. Marani (1968a).
F. Nine interspecific F_1 hybrids between *G. hirsutum* and *G. barbadense* cultivars. Pressley strength converted to Stelometer units. Marani (1968b).
G. Three F_1 hybrids from Upland crosses. Stelometer strength. Meredith, Bridge, and Chism (1970).

Acala (AHA 6-1-4) and *G. barbadense* types, but is primarily thought to derive from introgression from triple hybrid origin (Culp, 1992). Some indirect evidence also suggests that fiber strength can be conditioned by only a few major genes. May and colleagues (1994) released F_4 germplasm combining brown lint color and relatively high fiber strength. In that material, the highest fiber strength is associated with plants heterozygous for lint color, where lint color is expressed in an incompletely dominant manner to produce a light-brown phenotype in the F_1. Upon selfing of light-brown types, the progeny segregates into parental dark-brown and weak-fiber types, light-brown color and high-strength heterozygotes, and normal white lint color and high-strength types (O.L. May, unpublished data). Apparently, the lint color locus is tightly linked to genes conditioning

TABLE 8.10. Heritability Estimates for Fiber Strength

Reference		Reference	
A	0.76	J	0.10
B	0.56	K	0.60
C	0.90	L	0.86
D	0.84	M	0.49
E	0.75	N	0.27
F	0.72	O	0.86
G	0.59	P	0.27
H	0.67	Q	0.64
I	0.56	R	0.15

A. Pressley strength, F_2 plant selection unit, narrow sense. Self and Henderson (1954).
B. Pressley strength, F_2 plant selection unit, narrow sense. Lewis (1957).
C. Stelometer strength, F_3 line selection unit, broad sense. Al-Jibouri, Miller, and Robinson (1958).
D. Pressley strength, F_4 line selection unit, broad sense, mean heritability for two populations calculated from data in Miller et al. (1958).
E. Pressley strength, F_2 plant selection unit, narrow sense. Worley (1958).
F. Stelometer strength, F_2 plant selection unit, narrow sense, mean heritability from two populations. Tipton et al. (1964a).
G. Stelometer strength, F_2 plant selection unit, narrow sense. Abdel-Nabi, Jones, and Tipton (1965).
H. Stelometer strength, F_1 entry mean selection unit, narrow sense. Verhalen and Murray (1967).
I. Stelometer strength, plot mean selection unit, broad sense. Murray and Verhalen (1969).
J. Zero gauge Stelometer strength, F_2 plant selection unit, narrow sense, mean of the heritabilities given in two years. Murray and Verhalen (1969).
K. Stelometer strength, plot mean selection unit, narrow sense, mean heritability from F_1 and F_2 data. Verhalen and Murray (1969).
L. Stelometer strength, plot mean selection unit, narrow sense. Al-Rawi and Kohel (1970).
M. Stelometer strength, plot mean selection unit, narrow sense. Baker and Verhalen (1973).
N. Stelometer strength, F_1 entry mean selection, narrow sense. Wilson and Wilson (1975).
O. Stelometer strength, F_3 line selection unit, broad sense. Scholl and Miller (1976).
P. Stelometer strength, F_2 plant selection unit, narrow sense. May and Green (1994).
Q. Stelometer strength, F_2 population bulk selection unit, broad sense. May and Green (1994).
R. Stelometer strength, F_3 line selection unit, broad sense. May and Green (1994).

fiber strength, again suggesting effects of a few major genes controlling fiber strength. The source of fiber strength from the normal white lint parent was the cultivar PD-3 (Culp et al., 1988), which similarly experienced triple hybrid introgression, albeit distant, in its ancestry. Is it possible then that fiber strength derived from a

TABLE 8.11. Summary of Selection Experiments Toward Fiber Strength Modification

Reference	Response of fiber strength measure (KN mkg^{-1})
A	24
B	31
C	60

A. Five cycles of recurrent selection from a Coker 100 strain/Acala 1517 strain population. Stelometer strength selected in last four cycles, with Pressley strength selected in Cycle 1. Response calculated as difference between Cycle 5 and Cycle 0 (base population). Parents of base population isolated from Coker 100 and Acala 1517, respectively, by at least five generations of selfing. Miller (1965).
B. Single cycle of divergent mass selection for Stelometer strength in a Texas Upland/Acala population. Response calculated as mean difference between high- and low-strength selected populations. Turner, Worley, and Ramey (1980).
C. Five cycles of divergent mass selection for Stelometer fiber strength accomplished with forced self-pollination. Base population was a composite F_2 derived from forty-five F_1s created with a ten-parent half-diallel. Response calculated as mean difference between fifth cycle high- and low-strength selected populations. McCall, Verhalen, and McNew (1986).

triple hybrid background is conditioned by only a few major genes, whereas that from an Acala background (other than Del Cerro, which has triple hybrid in its pedigree) (Calhoun, Bowman, and May, 1994) is controlled by numerous loci?

Meredith (1992) studied the components of bundle fiber strength, which include single fiber strength, number of fibers in the bundle, and fiber length. These data illustrate the nature of interaction of basic fiber properties. Significant contributors to bundle fiber strength were 50 percent span length and individual fiber strength. In a multiple regression model predicting bundle strength from 50 percent span length, individual fiber strength, and Arealometer perimeter, most of the sums of squares were accounted for by the three-factor interaction among these predictors. Ultimately, breeders might make more progress in improving bundle strength by selection for one or more of its components through multiple-trait selection strategies such as index selection or independent culling. The decision to conduct direct versus indirect selection for bundle strength by selection for correlated traits would have to consider magnitude of genetic variance and heritability among bundle strength and its components along with ease of measurement.

Overall, the improvement of strength would seem a relatively straightforward breeding goal. However, the challenge to the breeder is to develop a product that meets the needs of both the textile industry and the producer. The antagonistic relationship between yield and fiber strength has made their simultaneous improvement difficult (Culp, Harrell, and Kerr, 1979; Culp, 1992).

An unresolved issue in breeding cottons with higher fiber strength is which instrument, the Stelometer, Pressley tester, or HVI, should be used to select progenies. Each measures fiber bundle strength by different methods (Taylor, 1982; Taylor et al., 1995), which has led to reports that the various instruments may not evaluate the same genetic properties controlling fiber strength. The HVI instrument does not weigh the fiber sample to determine mass; rather, mass is indirectly estimated (Taylor and Godbey, 1992). In contrast, the operator of Pressley and Stelometer instruments measures the mass of fiber bundles of determined length (Taylor, 1982). Another variable between the methods is that fiber crimp is eliminated during operator sample preparation for Stelometer and Pressley, but not for HVI (Taylor, 1982). Taylor and Godbey (1992) indicate that high- and low-micronaire cottons are particularly subject to HVI strength measurement errors. Also, when measured with HVI, certain cottons exhibit unusually high bundle strength that is not reflected in higher yarn strength (Brown and Taylor, 1988). Where fiber strength measurements from Stelometer and HVI have been compared, only moderate (0.4 to 0.6) correlations have been found (Green and Culp, 1988). Generally, these findings suggest that each instrument may evaluate different components of bundle fiber strength, reflecting idiosyncrasies of each instrument and, perhaps, the sample preparation. Cooper, Oakley, and Dobbs (1988) and Green and Culp (1988) found that the HVI instrument was unable to separate small strength differences among experimental cottons. Such small differences in strength frequently represent the size of genetic gains that breeders have achieved over time. O. L. May (unpublished data) has found standard unit heritability of HVI strength in two populations to be lower than that of Stelometer strength. In contrast, Latimer, Wallace, and Calhoun (1996) found that heritability of HVI strength was similar to that determined with the Pressley instrument. Overall, their study showed that HVI fiber testing was sufficient for breeders to use in selecting for high fiber quality. There is no argument that the HVI technology has had a positive impact on the U.S. cotton industry (Chewning, 1992). The majority of the data, however, suggests that Stelometer and, perhaps, Pressley are more useful to breeders as measurement tools to select for improved fiber strength.

FIBER ELONGATION

Fiber elongation is a property of fiber that is measured during the determination of bundle strength (Hertel, 1953). The contribution of fiber elongation to spinning and textile performance occurs in several ways. Backe (1996) studied the effect of variation in fiber elongation on yarn and textile manufacturing. In this study, bales representing three levels of fiber elongation were grouped, while other fiber properties were held relatively constant between the elongation levels. Generally, increased elongation was associated with improved yarn quality of the open-end spun yarn, as measured by evenness, strength, and reduced hairiness, among other properties, and ability to withstand the demands of weaving. When the genetic association between elongation and yarn strength is examined, a different relationship is evident. Meredith and colleagues (1991) reported moderate negative phenotypic correlations between elongation and yarn strength of ring- and open-end spun yarns of various counts from a study of advanced breeding lines and cultivars. Though phenotypic correlations, the associations were deemed largely genetic, as non-genetic influences (interactions with environments and experimental error) were small. Green and Culp (1990) similarly found a low negative genetic correlation between elongation determined by Stelometer and skein strength of a 27 tex ring-spun yarn.

Because elongation is a property normally measured by Stelometer and reported along with strength and length parameters, its genetic parameters have been extensively studied. It is doubtful, however, that fiber elongation has ever been a selection criteria receiving much emphasis during breeding line or cultivar development, but as spinning and textile manufacturing technologies change, it may become a more important property. Similar to what we saw with measures of fiber length and strength where genetic differences exist, interactions of genotypes with locations and years are of minor importance and should not, in general, hinder the identification of superior types (see

Table 8.12). With a few exceptions (references H and L, see Table 8.13), the expression of elongation is most influenced by additive genetic variance. In the two studies reporting the magnitude of dominance genetic variance greater than additive variance (Quisenberry, 1975; May and Green, 1994), the additive genetic variance could simply have been exhausted in the germplasm studied. Heritability estimates for fiber elongation indicate that pedigree selection, or early generation testing schemes, should be effective breeding tools (see Table 8.14). Tipton and colleagues (1964b) found that four to five loci affected elongation in two single-cross cotton populations.

TABLE 8.12. Genetic and Environmental Influences on Fiber Elongation

Reference	Genotype	Genotype × Location	Genotype × Year	Genotype × Location × Year	Residual
A	0.7	<0.1	<0.1	0.1	0.2
B	1.0	<0.1	<0.1	<0.1	0.2
C	6	0.3	0.2	2	1
D	15	—	<1	—	<1
E	0.2	—	<0.01	—	—

A. Four cultivars, 101 location × year combinations, variance components. Abouh-El-Fittouh, Rawlings, and Miller (1969).
B. Eight cultivars, three locations, three years, variance components. Bridge, Meredith, and Chism (1969).
C. Three cultivars, twenty-eight locations, three years, ratio of indicated source to error variance. El-Sourady, Worley, and Stith (1969).
D. Four cultivars, four environments (year × soil type combinations), mean squares, genotype × year interaction is genotype × year interaction. Meredith and Bridge (1973).
E. Eighty-nine early generation families evaluated in three environments (year × location combinations), variance components, genotype × year interaction is the genotype × environment interaction. Scholl and Miller (1976).

MEASURES OF FINENESS/MATURITY

Fiber fineness determines the spin limit, defined by Faerber and Deussen (1994) as the finest yarn count that can be spun with an acceptable level of yarn quality and ends down. Deussen (1992) indicates that fineness contributes to yarn strength and spinnability,

TABLE 8.13. Additive and Nonadditive Genetic and Environmental Influences on Various Measures of Fiber Elongation

Reference	Additive	Additive x Loc	Additive x Yr	Additive x Loc x Yr	Nonadditive	Nonadditive x Loc	Nonadditive x Yr	Nonadditive x Loc x Yr	Residual
A	22	—	—	—	—	—	—	—	—
B	1.5	—	—	—	—	—	—	—	—
C	0.2	—	—	—	<0.01	—	—	—	—
D	950	15	23	9	2	35	0*	7	600
E	3	—	—	—	2	—	—	—	0.5
F	41	—	—	—	—	—	—	—	—
G	2.4	—	—	—	—	—	—	—	—
H	0.1	—	—	—	—	—	—	—	—
I	0.9	—	—	—	—	—	—	—	—
J	5	0.5	—	—	0.5	0.3	—	—	0.3
K	7	0.2	—	—	0.5	0.3	—	—	0.3
L	0.02	—	—	—	0.8	—	—	—	—

* Analysis of variance estimate of the indicated variance was negative; thus, most reasonable estimate is zero.

A. Ratio of general combining ability to specific combining ability mean squares calculated from data in Barnes and Staten (1961).

B. Ratio of general combining ability to specific combining ability mean squares from a diallel among seven Acala germplasms calculated from data in Barnes and Staten (1961).

C. Additive and dominance genetic variances. Ramey and Miller (1966).

D. Additive and dominance genetic variances. Lee, Miller, and Rawlings (1967).

E. Additive and nonadditive genetic and environmental variances. Al-Rawi and Kohel (1970).

F. Ratio of general and specific combining ability mean squares averaged from the F_2 and F_3 generations of diallel progenies. Meredith and Bridge (1973).

G,H. Ratio of additive to nonadditive genetic variance within High-Plains and Acala germplasm, respectively. Quisenberry (1975).

I. Ratio of additive genetic variance to total genetic variance. Wilson and Wilson (1975).

J. General and specific combining ability mean squares. Green and Culp (1990).

K. General and specific combining ability mean squares. Tang et al. (1993).

L. Additive and dominance genetic variances. May and Green (1994).

measured as number of ends down, particularly for open-end spinning systems. Increased levels of fiber fineness promote fiber-to-fiber cooperation in the yarn, permitting less yarn twist, which translates into a gain in productivity for the yarn manufacturer. Instrument measures of fineness/maturity include the Micronaire (Johnson, 1952), Shirley Fineness-Maturity Tester (American Soci-

TABLE 8.14. Heritability Estimates for Fiber Elongation

Reference		Reference	
A	0.90	F	0.80
B	0.36	G	0.36
C	0.80	H	0.77
D	0.77	I	0.21
E	0.43		

A. F_2 plant selection unit, narrow sense. Tipton et al. (1964b).
B. F_2 plant selection unit, narrow sense. Tipton et al. (1964b).
C. F_2 plant selection unit, narrow sense. Abdel-Nabi, Jones, and Tipton (1965).
D. Plot mean selection unit, narrow sense. Al-Rawi and Kohel (1970).
E. F_1 entry mean selection unit, narrow sense. Wilson and Wilson (1975).
F. F_3 line selection unit, broad sense. Scholl and Miller (1976).
G. F_2 plant selection unit, narrow sense. May and Green (1994).
H. F_2 population bulk selection unit, broad sense. May and Green (1994).
I. F_3 line selection unit, broad sense. May and Green (1994).

ety for Testing and Materials, 1993), Arealometer (Hertel and Craven, 1951), and the relatively new AFIS fineness and maturity module (Bradow et al., 1996).

Micronaire reading (MIC) is one of several properties textile mills use to make bale laydowns prior to yarn manufacture (Hake et al., 1990). High (>5.0) MIC fiber, usually indicating coarse fiber, does not spin efficiently into fine-count yarns, whereas low (<3.5) MIC cotton that is immature can cause neps and dye defects (Hake et al., 1990). MIC reading is used in combination with other fiber properties such as strength and span lengths to make a certain size yarn and to promote consistency of performance of a set of cotton bales in the yarn-manufacturing process (Perkins, Ethridge, and Bragg, 1984). Unfortunately, as MIC reading is a measure of resistance to airflow of a constant weight of fibers at one air pressure (Johnson, 1952), maturity and fineness can be confounded. With knowledge of fiber maturity, the degree to which the fiber lumen has filled in (Basra and Malik, 1984), MIC reading can be interpreted as a measure of fineness. Also, if fiber perimeter is known, MIC reading can indicate relative maturity (American Society for Testing and Materials, 1993). Without knowledge of fiber maturity or perimeter, low MIC cotton, for example, could result from immature fiber or genetically fine (e.g., small perimeter) fiber. Given this information, one question is how to interpret the genetic control of MIC. Meredith (1994) indicates that maturity and fineness account for 90 percent of the variation in MIC reading, with the remainder being experimental error. Genetic variation for MIC is due to nearly equal effects of maturity and perimeter (Meredith, 1991). Consequently, if we wish to investigate the genetic control of MIC, perhaps we should concentrate on the separate genetic control of fineness and maturity.

Despite these reservations, data in Table 8.15 indicate that breeders have relied extensively on MIC reading as a measure of fiber fineness. MIC reading is normally provided with length, strength, and elongation measurements for a nominal fee; this likely explains the prevalence of MIC reading as a measure of fineness. Extensive use of the Arealometer and Shirley Fineness-Maturity Tester in genetic studies has not occurred, but such use may reflect additional measurement costs over those of the common fiber properties. Of the seventeen studies of fiber fineness/maturity summarized in Tables 8.15 and 8.16, about half indicate that nongenetic influences are greater than genetic differences. Of the studies in which genetic differences were smaller than nongenetic variation, three involved MIC, and thus, we do not know the influence of immature fiber on MIC reading. Meredith, Sasser, and Rayburn (1996) note that AFIS fiber area and diameter measure large genetic as opposed to nongenetic influences. The AFIS measurements are not necessarily confounded with fiber maturity (because this estimate is provided) and would appear to provide breeders with a new tool to modify fiber fineness. MIC reading and Arealometer specific surface area are mostly influenced by additive genetic variance, though some studies report nonadditive variance (see Table 8.17). Reasonably high heritability estimates for MIC reading (see Table 8.18) and fiber shape parameters (see Table 8.19) suggest they can be modified through selection. Characterization of sources of variation for fiber fineness in *G. hirsutum* is needed if this trait is to be emphasized in breeding programs.

TABLE 8.15. Genetic and Environmental Influences on Various Measures of Fiber Fineness

Reference	Genotype	Genotype × Location	Genotype × Year	Genotype × Location × Year	Residual
A	0.4	<0.1	<0.1	<0.1	<0.1
B	1388	<10	13	<10	<10
C	14	<0.1	—	—	<0.1
D	540	30	100	80	990
E	0.06	<0.01	<0.01	<0.01	0.05
F	0.02	<0.01	<0.01	0.03	0.03
G	0.02	0*	0*	0.02	0.03
H	0.8	0.3	<0.01	0.4	1.0
I	0.5	0*	0.4	1	6
J	5	—	1	—	—
K	40	9	—	—	9

* Analysis of variance estimate of the indicated variance was negative; thus, most reasonable estimate is zero.

A. Arealometer specific surface area, four cultivars, three locations, two years, mean squares. Hancock (1944).

B. Mass per unit length, sixteen cultivars, seven locations, three years, mean squares. Pearson (1944).

C. Mass per unit length, sixteen cultivars, nine locations, one year, mean squares. Pope and Ware (1945).

D. Arealometer specific surface area, ninety-five breeding lines, two locations, two years, variance components. Miller et al. (1958).

E. Micronaire reading, fifteen cultivars, nine locations, three years, variance components. Miller, Robinson, and Williams (1959).

F. Micronaire reading, four cultivars, 101 year × location combinations, variance components. Abouh-El-Fittouh, Rawlings, and Miller (1969).

G. Micronaire reading, eight cultivars, three locations, three years, variance components. Bridge, Meredith, and Chism (1969).

H. Arealometer specific surface area, three cultivars, twenty-eight locations, three years, ratio of indicated source of variation to error variance. El-Sourady, Worley, and Stith (1969).

I. Micronaire reading, sixty-two BC_2F_4, two locations, two years, variance components ($\times 10^{-2}$). Murray and Verhalen (1969).

J. Shirley Fineness-Maturity Tester, nineteen cultivars and advanced breeding lines, two locations, one year, ANOVA F values. Meredith et al. (1991).

K. Advanced Fiber Information System area, eighteen genotypes, including advanced breeding lines and cultivars, seven locations, variance components expressed as percent of total variance. Meredith, Sasser, and Rayburn (1996).

TABLE 8.16. Genetic and Environmental Influences on Fiber Maturity and Related Properties

Reference	Genotype	Genotype × Location	Genotype × Year	Genotype × Location × Year	Residual
A	630	20	40	20	10
B	9	7	—	—	14
C	11	7	—	—	12
D	8	—	2	—	—
E	14	2	—	—	27
F	61	11	—	—	14

A. Percent thin-walled fibers, sixteen cultivars, seven locations, three years, mean squares. Pearson (1944).
B. Arealometer maturity, eighteen genotypes, including advanced breeding lines and cultivars, seven locations, variance components expressed as percent of total variance. Meredith, Sasser, and Rayburn (1996).
C. Advanced Fiber Information System maturity, eighteen genotypes, including advanced breeding lines and cultivars, seven locations, variance components expressed as percent of total variance. Meredith, Sasser, and Rayburn (1996).
D. Arealometer perimeter, nineteen cultivars and advanced breeding lines, two locations, one year, ANOVA F values. Meredith et al. (1991).
E. Arealometer perimeter, eighteen genotypes, including advanced breeding lines and cultivars, seven locations, variance components expressed as percent of total variance. Meredith, Sasser, and Rayburn (1996).
F. Advanced Fiber Information System diameter, eighteen genotypes, including advanced breeding lines and cultivars, seven locations, variance components expressed as percent of total variance. Meredith, Sasser, and Rayburn (1996).

WAX CONTENT

Another property of cotton fiber contributing to its ability to be spun into yarn is wax content (Perkins, Ethridge, and Bragg, 1984). Taylor (1996) reported that addition of wax content to models containing HVI fiber strength data improved prediction of fabric tear strength among bale cotton samples. Additionally, this study reported a rapid method of wax measurement, using near infrared reflectance as opposed to time-consuming wet chemistry (American Society for Testing and Materials, 1982). Perkins, Ethridge, and Bragg (1984) do not indicate what levels of wax are considered high or low, but they suggest that extreme values are detrimental to spinnability. Little genetic information exists about wax content in cotton. Conrad and Neely (1943) reported the inheritance of wax

TABLE 8.17. Additive and Nonadditive Genetic and Environmental Influences on Various Measures of Fiber Fineness

Reference	Additive	Additive x Loc	Additive x Yr	Additive x Loc x Yr	Nonadditive	Nonadditive x Loc	Nonadditive x Yr	Nonadditive x Loc x Yr	Residual
A	2.5	—	—	—	—	—	—	—	—
B	0.64	—	—	—	—	—	—	—	—
C	370	—	—	—	20	—	—	—	—
D	700	60	10	100	30	0*	40	0*	400
E	0.09	—	—	—	0.1	—	—	—	0.2
F	60	—	—	—	20	—	—	—	20
G	0.3	0.2	—	—	<0.1	0.2	—	—	—
H	12	—	—	—	—	—	—	—	—
I	0	—	—	—	—	—	—	—	—
J	0.06	—	—	—	—	—	—	—	—
K	0.96	—	—	—	—	—	—	—	—
L	1	0.4	—	—	0.2	0.3	—	—	0.3
M	1.5	0.1	—	—	0.1	<0.1	—	—	<0.1
N	<0.1	—	—	—	0.1	—	—	—	—

*   Analysis of variance estimate of the indicated variance was negative; thus, the most reasonable estimate is zero.

A. Micronaire reading, ratio of general combining ability to specific combining ability mean squares calculated from data in Barnes and Staten (1961).

B. Micronaire reading, ratio of variance due to additive versus nonadditive effects averaged over parents from a half-diallel among seven Acala germplasms. Barnes and Staten (1961).

C. Arealometer specific surface area, additive and dominance genetic variances. Ramey and Miller (1966).

D. Micronaire reading, additive and dominance genetic variances. Lee, Miller, and Rawlings (1967).

E. Micronaire reading, additive and nonadditive genetic and environmental variances. Al-Rawi and Kohel (1970).

F. Micronaire reading, general and specific combining ability mean squares. Thomson (1971).

G. Micronaire reading, generation means analysis of six crosses evaluated at three locations, data for one cross presented. Meredith and Bridge (1972).

H. Micronaire reading, ratio of general and specific combining ability mean squares averaged from the F_2 and F_3 generation of diallel progenies. Meredith and Bridge (1973).

I,J. Micronaire reading, ratio of additive to nonadditive genetic variance within High-Plains and Acala germplasm. Quisenberry (1975).

K. Micronaire reading, ratio of additive genetic variance to total genetic variance. Wilson and Wilson (1975).

L. Micronaire reading, general and specific combining ability mean squares. Green and Culp (1990).

M. Micronaire reading, general and specific combining ability mean squares. Tang et al. (1993).

N. Micronaire reading, additive × additive and dominance genetic variances, additive variance was not detected. May and Green (1994).

TABLE 8.18. Heritability Estimates for Micronaire Reading

Reference		Reference	
A	0.61	F	0.26
B	0.40	G	0.87
C	0.23	H	0.49
D	0.08	I	0.82
E	0.53	J	0.53

A. F_2 plant selection unit, broad sense. Bilbro (1961).
B. F_1 entry mean selection unit, narrow sense. Verhalen and Murray (1967).
C. F_2 plant selection unit, narrow sense. Murray and Verhalen (1969).
D. Plot mean selection unit, narrow sense. Al-Rawi and Kohel (1970).
E. Plot mean selection unit, narrow sense. Baker and Verhalen (1973).
F. F_1 entry mean selection unit, narrow sense. Wilson and Wilson (1975).
G. F_3 line selection unit, broad sense. Scholl and Miller (1976).
H. F_2 plant selection unit, narrow sense. May and Green (1994).
I. F_2 population bulk selection unit, broad sense. May and Green (1994).
J. F_3 line selection unit, broad sense. May and Green (1994).

TABLE 8.19. Heritability Estimates for Fiber Fineness-Related Traits

Reference		Reference	
A	0.51	E	0.58
B	0.43	F	0.68
C	0.55	G	0.72
D	0.07	H	0.61

A. Fiber perimeter, F_2 plant selection unit, narrow sense. Bishr (1954).
B. Fiber cell wall thickness, F_2 plant selection unit, narrow sense. Bishr (1954).
C. Arealometer fiber-specific area, F_2 plant selection unit, narrow sense. Lewis (1957).
D. Arealometer perimeter, F_2 plant selection unit, narrow sense. Lewis (1957).
E. Arealometer D value (measure of fiber shape), F_2 plant selection unit, narrow sense. Lewis (1957).
F. Arealometer fiber-specific area, F_3 line selection unit, broad sense. Al-Jibouri, Miller, and Robinson (1958).
G,H. Arealometer fiber-specific area, F_4 line selection unit, broad-sense, heritability from two populations calculated from data in Miller et al. (1958).

content in green lint, high wax × normal white lint, low wax crosses. The data indicated a pleiotropic effect of the green lint gene or tight linkage with the gene or genes affecting wax content. Further genetic study of the relationship of wax content and textile performance seems warranted.

GENETIC ASSOCIATIONS
AMONG FIBER QUALITY TRAITS

Thus far in discussing genetic variation for fiber properties, we have addressed each property, except MIC reading, as being an independent entity. We know that the fiber properties are not independent and that genetic correlations exist among them. The significance of genetic correlations to breeders is that selection for correlated traits results in their simultaneous modification. Fiber strength and length tend to be positively correlated (see Table 8.20), as are length and measures of fineness. The variation in direction of the genetic correlations between length and fineness in Table 8.20 reflects different scale measurements. Low MIC reading and high fiber-specific surface area in the absence of immature fiber indicate finer fiber and thus explain variation in sign of the correlation between length and fineness as longer fiber tends to have smaller perimeters. The positive association between fiber strength, length, and fineness would generally be considered advantageous, in that greater fineness, length, and strength are a desirable combination. Genes imparting fiber length and elongation appear to function independently, as their genetic correlations are low (see Table 8.20). The strongest genetic correlations existed between Stelometer strength and elongation, with the assumption that increased strength would occur at the expense of elongation. This association would hinder efforts to improve strength and elongation to benefit textile performance. Genetic correlations arise from pleiotropy or linkage, or they can be nongenetic, reflecting physiological relationships. Given that the major components of fiber bundle strength include 50 percent span length, fineness, and single fiber strength (Meredith, 1992), one wonders which component accounts for genetic gain in bundle strength. If the gain in bundle strength resulted from longer, and consequently generally finer, fiber as opposed to single fiber strength, then the association with length and strength might not be genetic. More fibers in the fiber bundle tested for strength might account for the increased strength. Data from the National Cotton Variety Tests (USDA, 1995) show that the strongest fiber also has the smallest perimeter. Were there a reasonably rapid method of measuring single fiber strength available to breeders, this might be an untapped source of genetic variation for the improvement of fiber strength.

TABLE 8.20. Summary of Genetic Correlations Among Fiber Quality Traits

Reference	Length			Strength	
	Strength	Fineness	Elongation	Fineness	Elongation
A	0.10	0.05	—	−0.02	—
B	−0.23	0.66	—	−0.25	—
C	0.25	0.70	—	0.06	—
D	0.33	0.16	—	−0.31	—
E	—	—	—	—	−0.57
F	—	—	—	—	−0.39
G	—	—	—	—	−0.84
H	0.36	0.12	0.15	0.14	−0.51
I	0.41	−0.42	0.07	−0.21	0.03
J	—	—	—	−0.11	—
K	0.36	−0.48	−0.17	−0.15	−0.16

A. Ninety-two F_3 progenies, upper half mean length, Stelometer strength, and Arealometer specific surface area. Al-Jibouri, Miller, and Robinson (1958).

B,C, Ninety-five, ninety-two, and eighty-one lines of separate populations evaluated in F_4 and
D. F_5, upper half mean length, Pressley strength, Arealometer specific surface area. Miller et al. (1958).

E,F. Two populations, phenotypic correlations, Stelometer strength/elongation. Tipton et al. (1964b).

G. Sixty-nine F_3 lines, Stelometer strength/elongation. Abdel-Nabi, Jones, and Tipton (1965).

H. Ninety-six F_3 lines, upper half mean length, Stelometer strength/elongation, and micronaire reading. Miller and Rawlings (1967).

I. Ninety-six F_3 lines, 2.5 percent span length, Stelometer strength/elongation, and micronaire reading. Meredith and Bridge (1971).

J. Forty-five F_1 populations, Pressley strength, micronaire reading. Thomson (1971).

K. Eighty-nine early generation families, 2.5 percent span length, Stelometer strength/elongation, and micronaire reading. Scholl and Miller (1976).

YARN STRENGTH

Yarn strength is a critical factor in efficient manufacturing of knit and woven fabrics (Faerber, 1995), and its improvement is necessary to maintain cotton's dominance as a textile fiber. Further demand by the textile industry for stronger yarns derives from rising consumer preference for wrinkle-resistant 100 percent cotton fabrics achieved by chemical treatment. Associated with the wrinkle-resistant treatment is a 30 to 50 percent reduction in strength of the treated product (Faerber, 1995). Data in Table 8.21 show that the strength of 12 to 42 tex ring- and open-end spun yarn is strongly

TABLE 8.21. Genetic and Environmental Influences on Yarn Strength

Reference	Genotype	Genotype × Location	Genotype × Year	Genotype × Location × Year	Residual
A	240	1	<1	10	20
B	0.5	<0.1	0*	0.3	1
C	3200	30	—	—	3
D	59	3	—	—	—
E	34	2	—	—	—
F	36	1	—	—	—
G	32	1	—	—	—
H	8	—	—	—	—
I	90	23	<20	<20	<20
J	0.74	—	—	—	—
K	50	7	—	—	12

* Analysis of variance estimate of the indicated variance was negative; thus, the most reasonable estimate is zero.

A. Twenty-seven tex, ring-spun yarn, four cultivars, 101 location × year combinations, variance components. Abouh-El-Fittouh, Rawlings, and Miller (1969).

B. Twenty-seven tex, ring spun yarn, three cultivars, twenty-eight locations, three years, ratio of indicated source to error variance. El-Sourady, Worley, and Stith (1969).

C. Twenty-seven tex, ring-spun yarn, four cultivars, four environments (year × soil type combinations), mean squares. Meredith and Bridge (1973).

D. Twenty-seven tex, rotor-spun yarn, nineteen cultivars and advanced breeding lines, two locations, one year, ANOVA F values. Meredith et al. (1991).

E. Forty-two tex, rotor-spun yarn, nineteen cultivars and advanced breeding lines, two locations, one year, ANOVA F values. Meredith et al. (1991).

F. Twelve tex, ring-spun yarn, nineteen cultivars and advanced breeding lines, two locations, one year, ANOVA F values. Meredith et al. (1991).

G. Thirty tex, ring-spun yarn, nineteen cultivars and advanced breeding lines, two locations, one year, ANOVA F values. Meredith et al. (1991).

H. Twenty-seven tex, ring-spun yarn, ratio between general and specific combining ability mean squares. Green and Culp (1990).

I,J. Twenty-seven tex, ring-spun yarn, twenty-five advanced breeding lines, two production systems, two years, and heritability with F_5 line selection unit. May and Bridges (1995).

K. Twenty-seven tex, ring-spun yarn, eighteen genotypes, including advanced breeding lines and cultivars, seven locations, variance components expressed as percent of total variance. Meredith, Sasser, and Rayburn (1996).

determined by genetics and has high heritability. These data reflect the contribution of individual fiber properties such as bundle strength and length, which have reasonable heritability. Extensive replication of experiments over locations and years to select for improved yarn strength does not seem warranted, as interactions with locations and years are small. Genetic gain in yarn strength requires knowledge of which fiber properties, when selected, will result in better yarn strength, as this trait is too expensive to select for directly, except in late generations of breeding.

Because of global competition, yarn and textile producers have been forced to adopt more efficient manufacturing technology (Deussen, 1992; Faerber, 1995). This technology requires stronger fiber to operate competitively in a global economy. Additionally, the open-end yarn spinning systems being adopted in the name of efficiency and at the expense of older ring spinning systems may require fiber with different profiles of length, strength, and fineness (Deussen, 1992). Breeders, therefore, are faced with meeting the fiber quality needs of both ring and open-end spinning. Since there is quite a lag time between initiation of breeding efforts and cultivar release, knowledge of the genetic association between fiber properties and yarns produced by the two spinning systems would be helpful. Also, breeders do not select for yarn strength in early generations because it is too expensive to measure on large populations. Breeders select fiber properties such as length and strength with the aim of improving yarn strength. Yarn manufacturers indicate that breeders should rank the fiber properties strength, fineness, and length in decreasing priority for rotor spinning in contrast to length, strength, and fineness for ring spinning (Deussen, 1992). Meredith and colleagues (1991) and Meredith and Price (1996) provide the only data available comparing the genetic association of the common fiber properties with various count yarns produced on ring and rotor spinning systems. These data do not disagree with Deussen (1992), based on simple correlations between Stelometer strength, length (2.5 percent span length or AFIS mean length), fineness (Shirley Fineness-Maturity Test or AFIS diameter), and yarn strengths of ring- and open-end spun 12 to 42 tex yarns. These studies, however, do not necessarily show that breeders should attempt divergent selection strategies to meet the fiber profile needs of the two spinning systems. We do not know if the same holds for finer yarns and higher rotor speeds. If breeders make progress for higher bundle strength through finer fiber and single fiber strength, then the resulting fiber should benefit both yarn-manufacturing systems. Again, breeders are faced with developing a germplasm that can produce economically sufficient amounts of lint, yet meet textile processing requirements. Culp, Harrell, and Kerr (1979) have shown that, although difficult, it is possible to simultaneously improve lint yield and ring-spun yarn strength. Meredith and Price (1996) show an antagonistic correlation between lint yield and rotor-spun yarn strength that suggests similar difficulty, though not impossibility, in achieving this goal.

CONCLUSION

The challenge facing breeders is to produce a cultivar which meets the needs of a textile industry in the midst of technological advancement and which also produces enough lint for growers to make a profit. Common fiber properties such as length and strength tend to be moderately to highly heritable for various selection units, with additive genetic variance playing a major role in their expression. Thus, their continued improvement is expected. New tools will facilitate breeding for fiber quality traits such as wax content, short-fiber content, and fineness. The AFIS provides breeders with a direct measure of short-fiber content and separate measures of fiber fineness and maturity. Progress in reducing short-fiber content and achieving greater fiber fineness and length uniformity should be possible. Although yarn manufacturers are demanding fiber with greater strength and fineness for open-end spinning and better textile performance, the data so far indicate that breeders do not necessarily need to alter their fiber quality objectives to meet the needs of different yarn-spinning systems. Biotechnology will provide genes conferring specific fiber properties (John, 1992; see John, Chapter 10, this volume), and possibly, molecular markers will allow direct selection for the genotype, thereby providing a more efficient means of selecting for fiber properties. The incentive to genetically improve fiber quality must come from a realization that the long-term health of the cotton industry depends on it. A cotton marketing system that recognizes quality and adequately compensates growers for its delivery would be beneficial, but should not be a prerequisite to moving ahead with efforts to improve fiber quality.

REFERENCES

Abdel-Nabi, H., J.E. Jones, and K.W. Tipton (1965). Studies on the inheritance of fiber strength and fiber elongation in the F_3 generation of a cross between two varieties of Upland cotton. In *Proceedings of the Seventeenth Annual Cotton Improvement Conference*, Ed. H.H. Ramey. Memphis, TN: National Cotton Council, pp. 80-89.

Abouh-El-Fittouh, H.A., J.O. Rawlings, and P.A. Miller (1969). Genotype by environment interactions in cotton—Their nature and related environmental variables. *Crop Science* 9: 377-381.

Al-Jibouri, H.A., P.A. Miller, and H.F. Robinson (1958). Genotypic and environmental variances and covariances in an Upland cotton cross of interspecific origin. *Agronomy Journal* 50: 623-636.

Al-Rawi, K.M. and R.J. Kohel (1970). Gene action in the inheritance of fiber properties in intervarietal diallel crosses of Upland cotton, *Gossypium hirsutum* L. *Crop Science* 10: 82-85.

American Society for Testing and Materials (1982). *Quantitative Analysis of Textiles*. ASTM D629-77.

American Society for Testing and Materials (1993). *Standard Test Method for Linear Density and Maturity Index of Cotton Fibers (IIC-Shirley Fineness/Maturity Tester)*. ASTM D3818-92, pp. 127-130.

Anonymous (1974). The regional collection of *Gossypium* germplasm. *United States Department of Agriculture Report ARS-H-2*, Washington, DC, pp. 1-105.

Backe, E.E. (1996). The importance of cotton fiber elongation on yarn quality and weaving performance. In *Proceedings of the 9th Annual Engineered Fiber Selection System Conference*, Ed. C. Chewning. Raleigh, NC: Cotton Incorporated, pp. 1-13.

Baker, J.L. and L.M. Verhalen (1973). The inheritance of several agronomic and fiber properties among selected lines of Upland cotton, *Gossypium hirsutum* L. *Crop Science* 13: 444-450.

Balls, W.L. (1928). *Studies of Quality in Cotton*. London: McMillan and Company, Ltd.

Bargeron, J.D. III (1990). Cotton length uniformity and short fiber. Paper Number 901026. Columbus, OH: American Society of Agricultural Engineers International Summer Meeting.

Barker, H.D. and O.A. Pope (1948). Fiber and spinning properties of cotton: A correlation study of the effect of variety and environment. *United States Department of Agriculture Technical Bulletin 970*.

Barnes, C.E. and G. Staten (1961). Combining ability of some varieties and strains of *Gossypium hirsutum*. *New Mexico Agricultural Experiment Station Bulletin 457*.

Basra, A.S. and C.P. Malik (1984). Development of the cotton fiber. *International Review of Cytology* 89: 65-113.

Beasley, J.O. (1940). The origin of American tetraploid *Gossypium* species. *American Naturalist* 74: 285-286.

Behery, H.M. (1993). Short fiber content and uniformity index in cotton. *International Cotton Advisory Committee and Center for Agriculture and Biosciences Review Article 4*. Washington, DC, pp. 1-40.

Bilbro, J.D. Jr. (1961). Comparative effectiveness of three breeding methods in modifying coarseness of cotton fiber. *Crop Science* 1: 313-316.

Bishr, M.A. (1954). Inheritance of perimeter and wall thickness of fiber in a cross between two varieties of Upland cotton. PhD Dissertation, Louisiana State University, Baton Rouge, Louisiana.

Bradow, J.M., O. Hinojosa, L.H. Wartelle, and G. Davidonis (1996). Application of AFIS fineness and maturity module and X-ray fluorescence spectroscopy in fiber maturity evaluation. *Textile Research Journal* 66: 545-554.

Bragg, C.K. and F.M. Shofner (1993). A rapid, direct measurement of short fiber content. *Textile Research Journal* 63: 171-176.

Bridge, R.R., W.R. Meredith Jr., and J.F. Chism (1969). Variety × environment interactions in cotton variety tests in the delta of Mississippi. *Crop Science* 9: 837-838.

Brown, H.B. (1938). *Cotton*. New York: McGraw-Hill Book Company, Inc.

Brown, R.S. and R.A. Taylor (1988). Investigations on HVI strength values for Deltapine 90 cottons. In *Proceedings of the Beltwide Cotton Production Research Conferences*, Eds. J.M. Brown and D.A. Richter. Memphis, TN: National Cotton Council, pp. 608-610.

Calhoun, D.S., D.T. Bowman, and O.L. May (1994). Pedigrees of Upland and pima cotton cultivars released between 1970 and 1990. *Mississippi Agricultural and Forestry Experiment Station Technical Bulletin 1017*. Mississippi State, Mississippi, pp. 1-53.

Chewning, C.H. (1992). Cotton fiber management using high volume instrument testing and Cotton Incorporated's engineered fiber selection system. In *Proceedings from Cotton Fiber Cellulose: Structure, Function and Utilization Conference*, Eds. C.R. Benedict and G.M. Jividen. Memphis, TN: National Cotton Council, pp. 29-42.

Conrad, C.M. and J.W. Neely (1943). Heritable relation of wax content and green pigmentation of lint in Upland cotton. *Journal of Agricultural Research* 66: 307-312.

Cooper, H.B., S.R. Oakley, and J. Dobbs (1988). Fiber strength by different test methods. In *Proceedings of the Beltwide Cotton Production Research Conferences*, Eds. J.M. Brown and D.A. Richter. Memphis, TN: National Cotton Council, pp. 138-139.

Culp, T.W. (1992). Simultaneous improvement of lint yield and fiber quality in Upland cotton. In *Proceedings from Cotton Fiber Cellulose: Structure, Function and Utilization Conference*, Eds. C.R. Benedict and G.M. Jividen. Memphis, TN: National Cotton Council, pp. 247-287.

Culp, T.W. and D.C. Harrell (1980). Registration of extra-long staple cotton germplasm. *Crop Science* 20: 289.

Culp, T.W., D.C. Harrell, and T. Kerr (1979). Some genetic implications in the transfer of high fiber strength genes to Upland cotton. *Crop Science* 19: 481-484.

Culp, T.W., R.F. Moore, L.H. Harvey, and J.B. Pitner (1988). Registration of "PD-3" cotton. *Crop Science* 28: 190.

Deussen, H. (1992). Improved cotton fiber properties—The textile industry's key to success in global competition. In *Proceedings from Cotton Fiber Cellulose: Structure, Function and Utilization Conference*, Eds. C.R. Benedict and G.M. Jividen. Memphis, TN: National Cotton Council, pp. 43-63.

Dewey, L.H. (1913). The strength of textile fibers. *United States Department of Agriculture Bureau of Plant Industry Circular 128*, Washington, DC.

Ducket, K.E. (1974). Cotton fiber instrumentation research in Tennessee. *University of Tennessee Agriculture Experiment Station Bulletin 536*.

El-Sourady, A.S., S.W. Worley Jr., and L.S. Stith (1969). The relative varietal stability for fiber properties and yarn strength in Upland cotton. In *Proceedings of the Beltwide Cotton Production Research Conferences*, Ed. J.M. Brown. Memphis, TN: National Cotton Council, pp. 83-86.

Endrizzi, J.E., E.L. Turcotte, and R.J. Kohel (1984). Qualitative genetics, cytology, and cytogenetics. In *Cotton*, Eds. R.J. Kohel and C.F. Lewis. Madison, WI: American Society of Agronomy, Crop Science Society of America, pp. 81-129.

Faerber, C. (1995). Future demands on cotton fiber quality in the textile industry, technology-quality-cost. In *Proceedings of the Beltwide Cotton Production Research Conferences*, Eds. D.A. Richter and J. Armour. Memphis, TN: National Cotton Council, pp. 1449-1454.

Faerber, C. and H. Deussen (1994). Improved cotton fiber quality and improved spinning technology—a profitable marriage. Part I. Progress in rotor spinning and progress in the quality profile of U.S. Upland cotton. Part II. The contributions of improved cotton quality and those of rotor spinning developments to higher profits in cotton production and in spinning. In *Proceedings of the Beltwide Cotton Production Research Conferences*, Eds. D.J. Herber and D.A. Richter. Memphis, TN: National Cotton Council, pp. 1615-1621.

Frey, K.J. and T. Horner (1957). Heritability in standard units. *Agronomy Journal* 49: 59-62.

Fryxell, P.A. (1979). *The Natural History of the Cotton Tribe*. College Station, TX: Texas A & M University Press.

Fryxell, P.A. (1984). Taxonomy and germplasm resources. In *Cotton*, Eds. R.J. Kohel and C.F. Lewis. Madison, WI: American Society of Agronomy, Crop Science Society of America, pp. 27-57.

Green, C.C. and T.W. Culp (1988). Utilization of fiber strength measurements in the development of high fiber strength cottons. In *Proceedings of the Beltwide Cotton Production Research Conferences*, Ed. J.M. Brown and D.A. Richter. Memphis, TN: National Cotton Council, pp. 613-614.

Green, C.C. and T.W. Culp (1990). Simultaneous improvement of yield, fiber quality, and yarn strength in Upland cotton. *Crop Science* 30: 66-69.

Green, J.M. (1950). Variability in the properties of lint of Upland cotton. *Agronomy Journal* 42: 338-341.

Hake, K., B. Mayfield, H. Ramey, and P. Sasser (1990). *Producing Quality Cotton*. Memphis, TN: National Cotton Council of America.

Hancock, N.I. (1944). Length, fineness, and strength of cotton lint as related to heredity and environment. *Agronomy Journal* 44: 530-536.

Harrell, D.C. (1961). Yield and fiber quality of intraspecific hybrids involving long and extra-long staple Upland cottons. In *Proceedings of the Thirteenth Cotton Improvement Conference*, Ed. W.P. Sappenfield. Memphis, TN: National Cotton Council, pp. 29-37.

Harrison, G.J. (1939). Breeding and seed distribution of Acala cotton in California. In *Proceedings of the Southern Agricultural Workers Conference*. New Orleans, LA. Geneva, NY: American Society of Agronomy, pp. 1-9.

Hertel, K.L. (1940). A method of fibre-length analysis using the fibrograph. *Textile Research* 10: 510-525.

Hertel, K.L. (1953). The Stelometer, it measures fiber strength and elongation. *Textile World* 103: 97-260.

Hertel, K.L. and C.J. Craven (1951). Cotton fineness and immaturity as measured by the Arealometer. *Textile Research Journal* 21: 765-774.

Hertel, K.L. and M.G. Zervigon (1936). An optical method for the length analysis of cotton fibres. *Textile Research* VI: 331-339.

John, M.E. (1992). Genetic engineering of cotton for fiber modification. In *Proceedings from Cotton Fiber Cellulose: Structure, Function and Utilization Conference*, Eds. C.R. Benedict and G.M. Jividen. Memphis, TN: National Cotton Council, pp. 91-105.

Johnson, B. (1952). *Use and Application of Fiber and Spinning Tests*. Memphis, TN: National Cotton Council of America.

Keim, K.R. and J.E. Quisenberry (1983). Inheritance of fiber quality in a semi-dwarf composite population of cotton (*Gossypium hirsutum* L.). In *Proceedings of the Beltwide Cotton Production Research Conferences*, Ed. J.M. Brown. Memphis, TN: National Cotton Council, p. 108.

Konishi, T. (1975). Russell corporation experiences in open end spinning. In *Proceedings of the Technical Seminar on Open End Spinning and its Implications for Cotton*, Ed. R.H. McRae. Memphis, TN: National Cotton Council, pp. 40-44.

Landstreet, C.B. (1954). The relation of cotton fiber properties to spinning performance. In *Proceedings of the Seventh Cotton Improvement Conference*, Ed. D.M. Simpson. Memphis, TN: National Cotton Council, pp. 1-21.

Latimer, S.L., T.P. Wallace, and D.S. Calhoun (1996). Cotton breeding: High volume instrument versus conventional fiber quality testing. In *Proceedings of the Beltwide Cotton Production Research Conferences*, Eds. P. Dugger and D.A. Richter. Memphis, TN: National Cotton Council, p. 1681.

Lee, J.A. (1984). Cotton as a world crop. In *Cotton*, Eds. R.J. Kohel and C.F. Lewis. Madison, WI: American Society of Agronomy, Crop Science Society of America, pp. 1-25.

Lee, J.A., P.A. Miller, and J.O. Rawlings (1967). Interaction of combining ability effects with environments in diallel crosses of Upland cotton (*Gossypium hirsutum* L.). *Crop Science* 7: 477-481.

Lewis, C.F. (1957). Genetic recombination in a hybrid involving three species of *Gossypium*. *Agronomy Journal* 49: 455-460.

Lewis, C.F. and T.R. Richmond (1968). Cotton as a crop. In *Advances in Production and Utilization of Quality Cotton: Principles and Practices,* Eds. F.C. Elliot, M. Hoover, and W.K. Porter Jr. Ames, IA: The Iowa State University Press, pp. 1-21.

Marani, A. (1968a). Inheritance of lint quality characteristics in intraspecific crosses among varieties of *Gossypium hirsutum* L. and of *Gossypium barbadense* L. *Crop Science* 8: 36-38.

Marani, A. (1968b). Inheritance of lint quality characteristics in interspecific crosses of cotton. *Crop Science* 8: 653-657.

May, O.L. and B.C. Bridges Jr. (1995). Breeding cottons for conventional and late-planted production systems. *Crop Science* 35: 132-136.

May, O.L. and C.C. Green (1994). Genetic variation for fiber properties in elite Pee Dee cotton populations. *Crop Science* 34: 684-690.

May, O.L., C.C. Green, S.H. Roach, and B.U. Kittrell (1994). Registration of PD 93001, PD 93002, PD 93003, and PD 93004 germplasm lines of Upland cotton with brown lint and high fiber quality. *Crop Science* 34: 542.

McCall, L.L., L.M. Verhalen, and R.W. McNew (1986). Multidirectional selection for fiber strength in Upland cotton. *Crop Science* 26: 744-750.

McCarty, J.C. Jr. and J.N. Jenkins (1992). Cotton germplasm, characteristics of 79 day-neutral primitive race accessions. *Mississippi Agricultural and Forestry Experiment Station Technical Bulletin 184,* Mississippi State, Mississippi.

McCarty, J.C. Jr., J.N. Jenkins, B. Tang, and C.E. Watson (1996). Genetic analysis of primitive cotton germplasm accessions. *Crop Science* 36: 581-585.

McKenzie, W.H. (1970). Fertility relationships among interspecific hybrid progenies of *Gossypium. Crop Science* 10: 571-574.

McNamara, H.C. and R.T. Stutts (1935). A device for separating different lengths of fibers from seed cotton. *United States Department of Agriculture Circular 360,* Washington, DC, pp. 1-45.

Meredith, W.R. Jr. (1977). Backcross breeding to increase fiber strength of cotton. *Crop Science* 17: 172-175.

Meredith, W.R. Jr. (1991). Associations of maturity and perimeter with micronaire. In *Proceedings of the Beltwide Cotton Production Research Conferences,* Ed. J.M. Brown. Memphis, TN: National Cotton Council, p. 569.

Meredith, W.R. Jr. (1992). Improving fiber strength through genetics and breeding. In *Proceedings from Cotton Fiber Cellulose: Structure, Function and Utilization Conference,* Eds. C.R. Benedict and G.M. Jividen. Memphis, TN: National Cotton Council, pp. 289-302.

Meredith, W.R. Jr. (1994). Genetic and management factors influencing textile fiber quality. In *Proceedings of the Seventh Annual Cotton Incorporated Engineered Fiber Selection System Research Forum,* Ed. C. Chewning. Raleigh, NC: Cotton Incorporated, pp. 256-261.

Meredith, W.R. Jr. and R.R. Bridge (1971). Breakup of linkage blocks in cotton, *Gossypium hirsutum* L. *Crop Science* 11: 695-698.

Meredith, W.R. Jr. and R.R. Bridge (1972). Heterosis and gene action in cotton, *Gossypium hirsutum* L. *Crop Science* 12: 304-310.

Meredith, W.R. Jr. and R.R. Bridge. (1973). The relationship between F_2 and selected F_3 progenies in cotton *(Gossypium hirsutum* L.). *Crop Science* 13: 354-356.

Meredith, W.R. Jr., R.R. Bridge, and J.F. Chism (1970). Relative performance of F_1 and F_2 hybrids from doubled haploids and their parent varieties in Upland cotton *Gossypium hirsutum* L. *Crop Science* 10: 295-298.

Meredith, W.R. Jr., T.W. Culp, K.Q. Robert, G.F. Ruppenicker, W.S. Anthony, and J.R. Williford (1991). Determining future cotton variety fiber quality objectives. *Textile Research Journal* 61: 715-720.

Meredith, W.R. Jr. and J.B. Price (1996). Genetic association of fiber traits with high speed rotor yarn strength. In *Proceedings of the Ninth Annual Cotton Incorporated Engineered Fiber Selection System Research Forum*, Ed. C. Chewning. Raleigh, NC: Cotton Incorporated, pp. 226-240.

Meredith, W.R. Jr., P.E. Sasser, and S.T. Rayburn (1996). Regional high quality fiber properties as measured by conventional and AFIS methods. In *Proceedings of the Beltwide Cotton Production Research Conference*, Eds. P. Dugger and D.A. Richter. Memphis, TN: National Cotton Council, pp. 1681-1684.

Miller, P.A. (1965). Correlated responses to selection for increased yield and fiber tensile strength in cotton. In *Proceedings of the Seventeenth Cotton Improvement Conference*, Ed. H.H. Ramey. Atlanta, GA: National Cotton Council, pp. 29-37.

Miller, P.A. and J.A. Lee (1965). Heterosis and combining ability in varietal top crosses of Upland cotton, *Gossypium hirsutum* L. *Crop Science* 5: 646-649.

Miller, P.A. and A. Marani (1963). Heterosis and combining ability in diallel crosses of Upland cotton, *Gossypium hirsutum* L. *Crop Science* 3: 441-444.

Miller, P.A. and J.O. Rawlings (1967). Breakup of initial linkage blocks through intermating in a cotton breeding population. *Crop Science* 7: 199-204.

Miller, P.A., H.F. Robinson, and O.A. Pope (1962). Cotton variety testing: Additional information on variety $\times$ environment interactions. *Crop Science* 2: 349-352.

Miller, P.A., H.F. Robinson, and J.C. Williams (1959). Variety $\times$ environment interactions in cotton variety tests and their implications in testing methods. *Agronomy Journal* 51: 132-134.

Miller, P.A., H.F. Robinson, J.C. Williams, and R.E. Comstock (1958). Estimates of genotypic and environmental variances and covariances in Upland cotton and their implications in selection. *Agronomy Journal* 50: 126-131.

Moore, J.H. (1938). The relation of certain physical fiber properties in improved cotton varieties to spinning quality. *North Carolina Agricultural Experiment Station Bulletin* 58: 1-54.

Murray, J.C. and L.M. Verhalen (1969). Genetic studies of earliness, yield, and fiber properties in cotton *(Gossypium hirsutum* L.). *Crop Science* 9: 752-755.

Neely, J.W. (1940). The effect of genetical factors, seasonal differences and soil variations upon certain characteristics of Upland cotton in the Yazoo-Mississippi delta. *Mississippi Agricultural Experiment Station Technical Bulletin 28.* Mississippi State, MI, pp. 1-44.

Pate, J.B. and E.N. Duncan (1961). Yield and other characteristics of experimental cotton hybrids. In *Proceedings of the Thirteenth Cotton Improvement Conference*, Ed. W.P. Sappenfield. Memphis, TN: National Cotton Council, pp. 51-56.

Pearson, N.L. (1944). Neps in cotton yarns as related to variety, location, and season of growth. *United States Department of Agriculture Technical Bulletin 878.* Washington, DC.

Percival, A.E. and R.J. Kohel (1990). Distribution, collection, and evaluation of *Gossypium. Advances in Agronomy* 44: 225-256.

Perkins, H.H. Jr., D.E. Ethridge, and C.K. Bragg (1984). Fiber. In *Cotton*, Eds. R.J. Kohel and C.F. Lewis. Madison, WI: American Society of Agronomy, Crop Science Society of America, pp. 437-509.

Pope, O.A. and J.O. Ware (1945). Effect of variety, location, and season on oil, protein, and fuzz of cottonseed and on fiber properties of lint. *United States Department of Agriculture Technical Bulletin Number 903.* Washington, DC, pp. 1-45.

Pressley, E.H. (1942). A cotton fiber strength tester. *American Society for Testing and Materials Bulletin 118.* Philadelphia, PA, pp. 13-18.

Quisenberry, J.E. (1975). Inheritance of fiber properties among crosses of Acala and High Plains cultivars of Upland cotton. *Crop Science* 15: 202-204.

Quisenberry, J.E., L.L. Ray, and D.L. Jones (1975). Response of Upland cotton to selection for fiber length and fineness in a non-irrigated semi-arid environment. *Crop Science* 15: 407-409.

Ramey, H.H. Jr. (1980). Fiber crops. In *Crop Quality, Storage, and Utilization*, Ed. C.S. Hoveland. Madison, WI: American Society of Agronomy, pp. 35-58.

Ramey, H.H. Jr. and P.A. Miller (1966). Partitioned genetic variances for several characters in a cotton population of interspecific origin. *Crop Science* 6: 123-125.

Richmond, T.R. (1951). Procedures and methods of cotton breeding with special reference to American cultivated species. *Advances in Genetics* 4: 213-245.

Richmond, T.R. and H.J. Fulton (1936). Variability of fiber length in a relatively uniform strain of cotton. *Journal of Agricultural Research* 53: 749-763.

Rowland, S.P., M.L. Nelson, C.M. Welch, and J.J. Hebert (1976). Cotton fiber morphology and textile performance properties. *Textile Research Journal* 46: 194-214.

Rusca, R.A. and W.A. Reaves (1968). Utilization developments. In *Advances in Production and Utilization of Quality Cotton: Principles and Practices,* Eds. F.C. Elliot, M. Hoover, and W.K. Porter Jr. Ames, IA: The Iowa State University Press, pp. 487-525.

Sasser, P.E. (1992). The physics of fiber strength. In *Proceedings from Cotton Fiber Cellulose: Structure, Function and Utilization Conference,* Eds. C.R. Benedict and G.M. Jividen. Memphis, TN: National Cotton Council, pp. 19-27.

Scholl, R.L. and P.A. Miller (1976). Genetic association between yield and fiber strength in Upland cotton. *Crop Science* 16: 780-783.

Self, F.W. and M.T. Henderson (1954). Inheritance of fiber strength in a cross between the Upland cotton varieties AHA 50 and Half and Half. *Agronomy Journal* 46: 151-154.

Simpson, D.M. and E.N. Duncan (1953). Effect of selecting within selfed lines on the yield and other characters of cotton. *Agronomy Journal* 45: 275-279.

Singh, H., V.P. Singh, N.B. Patil, and B.M. Petkar (1991). Improvement of yield and fibre strength in medium and superior-medium staple Upland cotton (*Gossypium hirsutum*). *Indian Journal of Agricultural Sciences* 61: 11-15.

Singh, M., V.P. Singh, C.B. Lal, and K. Paul (1990). Breeding for high fiber strength in Upland cotton (*Gossypium hirsutum*). *Indian Journal of Agricultural Sciences* 60: 137-138.

Stephens, S.G. (1949). The cytogenetics of speciation in *Gossypium*. I. Selective elimination of the donor parent genotype in interspecific backcrosses. *Genetics* 34: 627-637.

Stewart, J.McD. (1988). Update on the taxonomy of *Gossypium*. In *Proceedings of the Beltwide Cotton Production Research Conferences*, Eds. J.M. Brown and D.A. Richter. Memphis, TN: National Cotton Council, pp. 95-97.

Tang, B., J.N. Jenkins, J.C. McCarty, and C.E. Watson (1993). F_2 hybrids of host plant germplasm and cotton cultivars. II. Heterosis and combining ability for fiber properties. *Crop Science* 33: 706-710.

Taylor, R.A. (1982). Measurement of cotton fiber tenacity on $^1/_8$ gage HVI tapered bundles. *Journal of Engineering for Industry* 104: 169-174.

Taylor, R.A. (1996). Natural waxes on cotton contribute to yarn and fabric quality. *Journal of Textile Chemists and Colorists* 29: 1-14.

Taylor, R.A. and L.C. Godbey (1992). Influence of micronaire on HVI bundle mass and strength measurements. In *Proceedings of the Beltwide Cotton Production Research Conferences*, Eds. D.J. Herber and D.A. Richter. Memphis, TN: National Cotton Council, pp. 1000-1005.

Taylor, R.A., L.C. Godbey, D.S. Howle, and O.L. May (1995). Why we need a standard strength test for cotton variety selection. In *Proceedings of the Beltwide Cotton Production Research Conferences*, Eds. D.A. Richter and J. Armour. Memphis, TN: National Cotton Council, pp. 1175-1178.

Thomson, N.J. (1971). Heterosis and combining ability of American and African cotton cultivars in a low latitude under high yield conditions. *Australian Journal of Agricultural Research* 22: 759-770.

Tipton, K.W., M.A.A. El-Sharkawy, B.M. Thomas, J.E. Jones, and M.T. Henderson (1964a). Inheritance of fiber strength in two separate crosses of Upland cotton having a common parent. In *Proceedings of the Sixteenth Annual Cotton Improvement Conference*, Ed. J.B. Pate. Memphis, TN: National Cotton Council, pp. 20-27.

Tipton, K.W., M.A.A. El-Sharkawy, B.M. Thomas, J.E. Jones, and M.T. Henderson (1964b). Inheritance of fiber elongation in two separate crosses of Upland cotton having a common parent. In *Proceedings of the Sixteenth Annual Cotton Improvement Conference*, Ed. J.B. Pate. Memphis, TN: National Cotton Council, pp. 13-20.

Turner, J.H. Jr., S. Worley Jr., and H.H. Ramey Jr. (1980). Response to selective pressure in early generation progenies of Upland cotton (*Gossypium hirsutum* L.). *Euphytica* 29: 615-624.

United States Department of Agriculture (USDA) (1995). *National Cotton Variety Tests,* Revised. Washington, DC: USDA, Cotton Physiology and Genetics Research Unit.

Verhalen, J.M., J.L. Baker, and R.W. McNew (1975). Gardner's grid system and plant selection efficiency in cotton. *Crop Science* 15: 588-591.

Verhalen, J.M. and J.C. Murray (1967). A diallel analysis of several fiber property traits in Upland cotton (*Gossypium hirsutum* L.). *Crop Science* 7: 501-505.

Verhalen, J.M. and J.C. Murray (1969). A diallel analysis of several fiber property traits in Upland cotton (*Gossypium hirsutum* L.). II. *Crop Science* 9: 311-315.

Ware, J.O. (1935). Opportunities for improving the quality of cotton. *Commercial Fertilizer* 3: 1-8.

Ware, J.O. (1936). Plant breeding and the cotton industry. In *Yearbook of Agriculture.* Washington, DC: United States Department of Agriculture, pp. 657-744.

White, T.G. and T.R. Richmond (1963). Heterosis and combining ability in top and diallel crosses among primitive foreign and cultivated American Upland cottons. *Crop Science* 3: 58-63.

Williford, J.R., W.R. Meredith, and A.C. Griffin Jr. (1984). Effect of variety, harvest method and lint cleaners on cotton quality and value in 1983. In *Proceedings of the Beltwide Cotton Production Research Conferences,* Ed. J.M. Brown. Memphis, TN: National Cotton Council, pp. 114-115.

Willis, H.H. (1926). Cotton lint research. In *Yearbook of Agriculture 1926.* Washington, DC: United States Department of Agriculture, pp. 267-271.

Wilson, F.D. and R.L. Wilson (1975). Breeding potentials of noncultivated cottons. I. Some agronomic and fiber properties of selected parents and their F_1 hybrids. *Crop Science* 15: 763-766.

Woo, J.L. and M.W. Suh (1994). HVI's potential in the estimation of SFC in cotton. In *Proceedings of the Seventh Annual Cotton Incorporated Engineered Fiber Selection System,* Ed. C.H. Chewning. Raleigh, NC: Cotton Incorporated, pp. 173-176.

Worley, S. Jr. (1958). Inheritance of $1/8$ gauge fiber strength in an interspecific cotton hybrid. In *Proceedings of the Eleventh Cotton Improvement Conference,* Ed. C.F. Lewis. Memphis, TN: National Cotton Council, pp. 28-33.

Chapter 9

Molecular Genetics
of Developing Cotton Fibers

Thea A. Wilkins
Judith A. Jernstedt

COTTON FIBER DEVELOPMENT

To appreciate the contribution a particular gene makes to cotton fiber structure and morphology, it is important to keep in mind the developmental events in surrounding tissues and organs that may directly affect fiber development. For this reason, key stages of cotton flower and boll development in relation to seed and fiber development are summarized in Table 9.1. Unquestionably, the size, shape, and structure of the fiber and, thus, to a large degree, the economic properties of the mature fiber are linked to the development of the boll and flowers, as well as to embryo and seed development. Thus, it is clearly evident from the stages and events listed in Table 9.1 that fiber development is only one aspect of a highly complex, integrated developmental process. Whether it will be feasible in the future to uncouple fiber growth from other developmental processes for the purpose of controlling fiber properties remains an open question.

A major step in understanding fiber biology in recent years has been profoundly impacted by the characterization of dynamic

The authors extend their thanks to the members of the Wilkins lab, past and present, who contributed to the research discussed in this chapter. Support from the U.S. Department of Energy Grant DE-FGO3-92ER20067 (T.A.W.) and Cotton Incorporated Grants 92-815 (T.A.W.) and 94-983 (J.A.J.) is gratefully acknowledged.

TABLE 9.1. Cotton Flower and Boll Development

DPA	Flower/Boll	Ovule/Embryo	Fibers
−40	Floral stimulus		
−32	Carpel and anther differenti- ation established **Flower bud development begins** Linear elongation of boll		
−23		Number of ovules established	
−7	Exponential expansion of corolla	Stomata development begins	
−3	Exponential elongation of boll	Number of stomata increases	**Fiber differentiation** Preexpansion preparation
−1	Rapid expansion of corolla Elongation of style and anthers		**Fiber expansion** Stage I —1 to 3 dpa
0	Anthesis		Appearance of fiber initials/ Onset of isodiametric cell expansion
1	Petal senescence **Boll enlargement begins**	Fertilization of egg and polar nuclei	
2		Liquid endosperm development begins	
3-4		First division of zygote	Stage II—3 to 5 dpa Transition stage—preparation for polar expansion
5-6		Embryo at globular stage	Stage III—5 to 20 dpa Rapid polar expansion of fibers Fuzz fibers initiated
12-13		Endosperm becomes cellular Embryo differentiation begins	Peak rate of fiber expansion
14-16	Maximum volume of boll established	Embryo starts to elongate	**Secondary cell wall synthesis** Fiber expansion declines
20-21		Ovules attain maximum length Cotyledons enlarge rapidly	Fiber expansion completed
24-27		Embryos attain maximum length	Peak rate of cellulose synthesis
30-32	**Filling phase***	Embryos fill the embryo sacs	Maximum rate of cellulose deposition
40-45			**Fiber maturation starts**
42	**Maturation phase**	Seed maturation	Cellulose deposition stops
50-60	Boll dehiscence	Seed coat turns black	Desication of fibers

Sources: Data adapted from J. McD. Stewart (1986) and Wilkins (1996).

* Fresh weight plateaus, whereas dry weight continues to increase.

stage-specific cellular events revealed by structural studies in conjunction with molecular analyses. The shifts in the metabolic and cellular processes underlying fiber growth and development are, of course, predictably accompanied by major changes in gene expression and polypeptide synthesis. A summary of known major cellular processes or events associated with particular subcellular compartments, which are key to understanding fiber biology at the molecular level, as discussed in the section on gene expression, are depicted in Table 9.2.

REGULATION OF GENE EXPRESSION IN DEVELOPING COTTON FIBERS

A general desire to unravel the genetic mechanisms regulating fiber growth and development, and, hence, fiber yield and quality, has been the driving force for the focus on the isolation of cotton genes. In the past five years, the number of isolated cotton fiber genes has risen dramatically and continues to rise, although the functions of only a handful of genes have been reported as yet. Despite initial expectations, the expression of only a very few of the isolated cotton genes is, in truth, fiber specific, compared to the vast majority of identified genes that are simply fiber enriched. This is not entirely unexpected, of course, since most, if not all, of the cellular processes associated with developing cotton fibers are also inherent to other cell types at one stage of development or another. As a result, the complexity of models for gene expression in developing cotton fibers is increased accordingly to account for stage-specific spatial and temporal regulation of fiber genes. A detailed discussion of cotton fiber genes, their regulation, expression patterns, and effects on fiber development is organized in the following section by developmental stage.

Fiber Differentiation

For more than twenty years, the in vitro cultured ovule experiments conducted by Beasley's group (Beasley and Ting, 1973, 1974) provided the only evidence for fiber differentiation in the days preceding anthesis (see Table 9.1). Recently, however, advanced cryo-

TABLE 9.2. Major Cellular Events Related to Cotton Fiber Development

Developmental Stage[1]	Golgi[2]	Vacuole	Nucleus[3]	Cytoskeleton[4]	Cell Wall
Differentiation	Morphologically distinct Golgi populations	Formation of two vacuole populations[2]		Random orientation of cortical MTs	Amorphous MEP distributed evenly in cell wall[5]
Expansion Stage I	Increased number and complexity of Golgi stacks	Fusion of vacuole populations; de novo synthesis of ER-derived provacuoles; formation of large central vacuole[2]	Enlargement		Formation of MEP matrix into helical fibrillar network transverse to axis of elongation[5]
Stage II		Change in tonoplast composition[6]		MT arrays shift to transverse orientation	Reorientation of MEP fibrillar arrays to steeply pitched helices[5]
Stage III		Change in tonoplast composition after termination of expansion >20dpa[6]	Migration to maintain fixed distance to tip	Shallow-pitched MT arrays stabilized by 7 dpa and maintained to 22 dpa; $\geq$ threefold increase in tubulin subunits.[7]	Incorporation of structural cell wall proteins 15-20 dpa;[8] peak accumulation of xyloglucan 15 dpa[9]
Secondary Cell Wall Synthesis				MT arrays shift orientation to steeply pitched helices beginning at 24 dpa; MT reorientation associated with change in MT number, length, and proximity to plasmalemma	Dramatic increase in rate of cellulose synthesis and cellulose content $\geq$21 dpa;[10] increase in fiber strength $\geq$ 21 dpa[11]
Maturation (cell death)					Flat, twisted ribbons formed upon desiccation

MT = microtubules; MEP = methylesterified pectins; ER = endoplasmic reticulum; dpa = days postanthesis
1 Each developmental stage is defined in terms of dpa as follows: differentiation, −3 to −1 dpa; expansion, −1 dpa to 20 dpa (Stage I: −1 to 3 dpa; Stage II: 3 to 5 dpa; Stage III: 5 to 20 dpa) (Wilkins, 1996); secondary cell wall synthesis, 15 to 40 dpa; desiccation, ~50 to 60 dpa.
2 S.C. Tiwari and T.A. Wilkins, unpublished data; Wilkins and Tiwari (1995).
3 Ramsey and Berlin (1976).
4 Seagull (1992).
5 Wilkins and Tiwari (1994); Wilkins (1996).
6 C.-Y. Wan and T.A. Wilkins, unpublished data.
7 Kloth (1989); Dixon, Seagull, and Triplett (1994).
8 John (1995).
9 Shimizu et al. (1997).
10 Meinhert and Delmer (1977); Shimizu et al. (1997).
11 Hsieh, Honic, and Hartzell (1995).

techniques, which offer vastly improved ultrastructural preservation, revealed that fiber differentiation is accompanied by the biogenesis of two morphologically and biochemically distinct vacuole and Golgi populations (see Table 9.2). Yet, despite these developmentally regulated subcellular changes, the gene or genes responsible for the differentiation of fiber cells continue to elude identification.

Many groups have proposed that *Arabidopsis* leaf trichomes, which require at least twenty genes for normal development (Hülskamp, Misera, and Jürgens, 1994), could potentially serve as a model for elucidating the underlying genetic mechanism controlling cotton fiber differentiation. *Glabrous1* (*GL1*), one of the first genes to be characterized, encodes a member of the *MYB* family of transcription factors involved in the initiation of leaf trichomes in *Arabidopsis* (Oppenheimer et al., 1991). In an unsuccessful attempt to isolate a *GL1* homolog from developing cotton fibers, an exhaustive search of cotton ovule cDNA libraries using a PCR (polymerase chain reaction)-based screening method resulted in the recovery of cDNAs encoding six new *MYB*-domain genes not previously identified in plants (Loguercio, Zhang, and Wilkins, 1999). Although none of the cotton clones encode a *GL1* homolog per se, the differential expression patterns produced by the cotton *MYB* genes warrant more detailed discussion.

Apart from the highly conserved DNA-binding domain (DBD) common to MYBs, each of the six unique cotton MYB polypeptides, designated as A, D, G, J, N, and O, contains distinguishing structural features within the variable transactivation domain (TAD). For instance, the cotton MYBs, with the notable exception of MYBD, contain a short basic TAD subdomain of forty amino acids that separates the DBD and the acidic TAD. In MYBD, the basic TAD subdomain is missing in its entirety. Interestingly, however, of the five MYBs containing the TAD basic subdomain (A, G, J, N, O), only MYBA and MYBO include the conserved peptide motif GIDPxxH, initially identified in a number of other plant MYBs. Similarly, variants related to a second conserved MYB motif, CPDLNLxISPP, are present in the TAD acidic subdomains encoded not only by *MYBA* and *MYBO*, but by *MYBJ* as well. The function of these conserved protein motifs is unknown, although it is probably safe to assume that each motif is important for regulating transcriptional activity. Yet another interesting feature of the cotton *MYB* clones is the presence of MYB-binding consensus sequences scattered along the length of the clones, suggesting a mechanism of feedback regulation via the binding of its own gene product or other MYB proteins to RNA transcripts to regulate the amount of MYB synthesized at both the transcriptional and translational levels.

The spatial expression patterns produced by cotton *MYB* genes fall into two general categories, Type I and II, that have little or no apparent relationship to the structural similarities or differences among the deduced polypeptides. Using semiquantitative reverse-transcription PCR (RT-PCR) to determine the relative abundance of individual *MYB* transcripts, Type I cotton *MYB* genes (*MYBA, MYBD* and *MYBG*) are ubiquitously expressed in all tissues examined, although *MYB* transcripts are much less abundant in pollen and stigmatic tissue. The expression of Type II cotton *MYB* genes (*MYBJ, MYBN, MYBO*), on the other hand, is differentially regulated to a much greater degree than that of Type I genes. In addition to the general absence of transcripts for Type II *MYB* genes from stigmatic tissue, the relative abundance of mRNAs varies significantly from tissue to tissue. Consistent with MYBJ as the most distantly related cotton MYB, the expression pattern of *MYBJ* shows the greatest tissue specificity relative to the other *MYB* genes. In fact, the expression of *MYBJ* in developing cotton fibers is quite distinct, in that *MYBJ* transcripts are absent or barely detectable, although this observation does not necessarily preclude a contributing role for MYBJ in fiber development. These results support the current model, suggesting that plants have evolved a plethora of diverse MYB transcription factors that serve to regulate gene expression in a broad spectrum of cellular functions, most notable of which are plant-specific functions (Erich Grotewold, personal communication, August 1997).

RT-PCR analysis of developing cotton ovules from −9 days preanthesis to 35 dpa (days postanthesis), inclusive of fibers, revealed that both Type I and Type II cotton *MYB* genes are temporally regulated. Expression of Type I *MYB* genes is fairly consistent throughout ovule and fiber development, although transcript levels show a marked decrease after 20 dpa, coincident with the end of fiber expansion. However, several minor peaks in expression levels of *MYBA* were observed to closely coincide with key stages of fiber development, as follows: -9 dpa (Predifferentiation), -1 dpa (Stage I Expansion), 5 dpa (Stage II Expansion), and 15 dpa (Stage III Expansion). Type II *MYBN* and *MYBO* genes are expressed at lower levels but are more highly modulated compared to Type I genes. For instance, *MYBN* transcripts, initially detected at -9 dpa, decrease by at least two orders of magnitude between 3 to 5 dpa, coincident with Stage II of fiber expansion, and cannot be detected in 35 dpa ovules/fibers. *MYBO* expression in developing ovules and fibers may be up-regulated slightly during fiber differentiation, more moderately at the onset of rapid polar expansion, and again at secondary cell wall synthesis in developing fibers. In Li_2, a fiber mutant defective in elongation, only Type II *MYBO* showed an altered pattern of expression that was more consistent with the levels and pattern of Type I *MYB* genes. Further experiments are obviously required to determine if the altered expression of *MYBO* is directly linked to the Li_2 lesion itself. However, determining what role, if any, these transcription factors may potentially play in regulating the expression of gene cascades during fiber development will require the identification of downstream genes targeted for transcriptional activation by cotton MYBs during fiber development.

Fiber Expansion

Fiber expansion is the net result of the complex interplay between cell turgor, the driving force of cell expansion, and cell wall extensibility. These two cellular processes, in turn, require the extensive involvement of numerous metabolic and biosynthetic pathways to provide the many components necessary to sustain prolonged polar growth of developing fibers. Not surprisingly, cell turgor and relaxation of cell structure are coordinately regulated and subject to developmental control. The major cellular events that characterize fiber expansion, as it is currently defined from -1 dpa to 20 dpa, are summarized in Table 9.2. Moreover, fiber expansion can be subdivided further into three discrete stages on the basis of ultrastructural and molecular studies (Wilkins, 1996). In general terms, the stages of expansion are defined as follows: Stage I (-1

to 3 dpa)—mobilization of cellular machinery, Stage II (3 to 5 dpa)—transition period, and Stage III—(5 to 20 dpa)—rapid polar elongation (see Tables 9.1 and 9.2).

The overwhelming majority of genes shown to exhibit stage-specific expression in developing cotton fibers are associated with Stage III of expansion, the period of rapid polar elongation (see Table 9.2). However, it is also important to keep in mind that determination of the developmental expression of fiber expansion-specific genes is most often based on RNA extracted from fibers harvested as early as 5 dpa. Thus, there is a decided lack of evidence establishing whether expansion genes are or are not expressed during Stages I and II of expansion. A few notable exceptions, described in the next section, indicate that a few of the expansion genes, at least, are differentially expressed in a stage-specific manner that is apparently directly related to the events taking place within the developing fiber cell (see Table 9.2).

Interestingly, Stage III expansion genes have been categorized into two discrete groups, designated as *primary* or *constitutive*, on the basis of differences in developmentally regulated expression patterns (see Table 9.3). The so-called primary expansion genes appear to be regulated, for the most part, at the transcriptional level, whereas posttranslational regulation of a few constitutively expressed expansion genes has also been documented. The primary expansion genes can be readily divided into major subclasses (see Table 9.3) corresponding to the major cellular processes—cell turgor, cytoskeleton dynamics, and loosening of the cell wall (see Table 9.2). Accordingly, the regulation and functional role of the primary expansion genes identified to date will be discussed in relation to cellular events (see Table 9.2) in the following section.

Primary Fiber Expansion Genes—Turgor

The primary expansion genes (see Table 9.3) exhibit similar expression during polar expansion that closely parallels the *rate* of expansion during fiber development. The relative abundance of expansion mRNAs detected at the onset of polar elongation (5 dpa) increases sharply to peak levels around 12 dpa, coincident with peak expansion rates. Almost immediately, developmental triggers signal the down-regulation of fiber expansion, resulting in a rapid

TABLE 9.3. Genes Developmentally Regulated in Cotton Fibers

	Expansion—Primary Cell Wall Synthesis		Secondary Cell Wall Synthesis
	Primary	Constitutive	
Turgor	acyl carrier protein[13] vacuolar H[+]-ATPase subunits A,B,c[1] plasma membrane H[+]-ATPase[2] tonoplast intrinsic protein[2,14] phosphoenolpyruvate carboxylase[3] sucrose synthase[15]	25S rRNA[1] vacuolar H[+]-pyrophosphatase[2]	rac9, rac13[9] H6[11] celA1, celA2[12] sucrose synthase[5,16] annexins[10] FbL2A[7] endo-1,3-β-glucanase[5]
Cytoskeleton	α-tubulin[2] (annexins)[10]	actin[5]	
Cell Wall	endo-1,4-β-glucanase[5] expansin[5] lipid transfer protein[4] E6[6] Fb-B6[8]	xyloglucan endotransglycosylase[5]	

[1] Wilkins (1993); Wan and Wilkins (1994); Hasenfratz, Tsou, and Wilkins (1995).
[2] Smart et al. (1998).
[3] Vojdani, Kim, and Wilkins (1997).
[4] Ma et al. (1995); Ma et al. (1997).
[5] Shimizu et al. (1997).
[6] John and Crow (1992); John (1996).
[7] Rinehart, Petersen, and John (1996).
[8] John (1995).
[9] Delmer et al. (1995).
[10] Andrawis, Solomon, and Delmer (1993); Potikha and Delmer (1997).
[11] John and Keller (1995).,
[12] Pear et al. (1996).
[13] Song and Allen (1997).
[14] Ferguson, Turley, and Kloth (1997).
[15] Nolte et al. (1995).
[16] Amor et al. (1995).

239

decline in both the rate of expansion and expansion transcript levels, so that by 20 dpa, expansion, for all practical purposes, is terminated. After termination of expansion, basal levels of expansion transcripts may or may not persist through 25 dpa, but not usually much beyond this point in development. There is some minor variability among the primary genes with respect to the rate at which individual transcripts increase and decrease during expansion. As is often the case, expression studies using the more sensitive RT-PCR method also tend to detect low levels of target messages at stages that are not readily discernible via conventional approaches for RNA analysis. For instance, carefully controlled RT-PCR experiments revealed that the primary expansion genes encoding phosphoenolpyruvate carboxylase (PEPCase) and subunit A of the vacuolar H^+-ATPase are expressed at low levels in 3 dpa fibers. These results suggest that primary expansion genes are, in fact, transcriptionally activated during the transition period (Stage II) preceding the onset of polar elongation, resulting in the detection of already appreciable amounts of primary expansion transcripts by 5 dpa of Stage III.

The requirement for increased de novo synthesis of membranes during the earliest stages of cell expansion is part of the general process involving the mobilization of cellular machinery that is required to sustain cellular structures in anticipation of rapid polar elongation (Wilkins, 1996). In keeping with the demand for membranes in the proliferation and enlargement of organelles, the expression of a putative fiber-specific gene encoding an acyl carrier protein (ACP) (Song and Allen, 1997), which functions in the synthesis of membrane lipids, is induced during all three stages of expansion (see Table 9.2). Initially detected by 2 dpa (Stage I), ACP transcripts increase rapidly through Stage II and into early Stage III of expansion, reaching peak accumulation in 5 to 8 dpa fibers. The fiber-specific *acp* gene is a member of a small gene family in cultivated cotton species and is currently the sole candidate for classification as a tentative "early" primary expansion gene, since peak accumulation of ACP transcripts reportedly precedes that of other primary expansion genes by at least 4 days.

Two membrane-bound electrogenic proton pumps, the vacuolar H^+-ATPase (V-ATPase) and the plasma membrane H^+-ATPase

(PM-ATPase), play a pivotal role in regulating cell turgor, the driving force of cell expansion. Both enzymes hydrolyze ATP (adenosine triphosphate) to pump H^+ ions against a concentration gradient, thereby generating both an electrochemical and a pH gradient across the membrane. The electrochemical potential also provides the energy necessary to activate secondary transport systems, including osmoregulatory transport mechanisms important to regulating cell turgor. Changes in RNA accumulation during the course of expansion are followed closely by corresponding changes in the levels of protein and enzymatic activity, which lag behind RNA levels by approximately 2 to 3 days, to attain peak levels by 15 dpa (Smart et al., 1998). RNA analysis revealed that the developmental regulation of the V-ATPase observed in isolated fibers could be reproduced in ovules with attached fibers indicating that highly accentuated gene expression in fibers is *not* significantly diluted by activity in the ovule, at least during rapid polar expansion of fibers. However, it is unknown what is occurring at the molecular level during early stages of expansion, between 0 and 5 dpa, especially since this period is characterized by at least a fiftyfold increase in fiber length and dynamic changes at the cellular level (see Table 9.2).

Molecular and biochemical analyses performed with excised ovules spanning -3 to 3 dpa revealed that the levels of V-ATPase and PM-ATPase proteins are significantly higher during fiber differentiation and/or Stage I of expansion relative to Stage III levels. However, these increased protein levels are *not* accompanied by an increase in enzyme activity (Smart et al., 1998), indicating that the regulation of these proton pumps is also subject to posttranscriptional regulation during the early stages of fiber development. In the case of the PM-ATPase, it is presumed that the protein has not yet been posttranslationally phosphorylated since it is well established that PM-ATPase activity is regulated by its phosphorylation status (Sussman, 1994). The situation for the V-ATPase differs in that the assembly and transport of the V-ATPase to subcellular compartments plays a major role in the posttranslational regulation of this proton pump. It appears that, early on, the active V-ATPase holoenzyme has not yet been assembled since a considerable proportion of the peripheral catalytic and regulatory subunits, subunits A and B, respectively, is present in the soluble pool. Assembled membrane-bound V-ATPase holoenzymes are primarily associated with organelles other than the vacuole. However, this situation is reversed as the fiber begins to expand, resulting in a decrease of soluble subunits and a corresponding increase in membrane-bound holoenzymes, which are almost exclusively associated with the vacuole during Stage III of expansion. These studies indicate that some, but not all, of the primary expansion genes are differentially regulated during fiber differentiation and all three stages of expansion and that transcriptional activation is, to some degree, cyclic. One interpretation of such a scenario would suggest that fiber primordia accumulate sufficient resources in reserve to support limited growth during Stages I and II of expansion, a time when partial or complete suppression of primary gene expression occurs.

At least two cotton clones encoding each of the major subunits (A, B, and c) have been isolated and characterized for the V-ATPase, a multimeric enzyme found in compartments of the endomembrane system (Wilkins, 1993; Wan and Wilkins, 1994; Hasenfratz, Tsou, and Wilkins, 1995). The principle subunits of the peripheral V_1 sector, located on the cytoplasmic face of the membrane, include subunits A and B, which are present in a stoichiometric ratio of A_3B_3. Subunit A is a 69 kDa (kilodalton) catalytic polypeptide responsible for the hydrolysis of ATP, whereas the 57 kDa subunit B, which is evolutionarily related to subunit A, serves only in a regulatory capacity. Six copies of subunit c, a 16 kDa proteolipid of the membrane-bound V_0 sector, form a proton pore that opens a passage for the transport of cytoplasmic H^+ across the membrane to the lumen of the organelle. Invariably, one of the clones for each V-ATPase subunit represents transcripts that are consistently more prevalent. Subunits B and c belong to small gene families including at least three to six family members (Wan and Wilkins, 1994; Hasenfratz, Tsou, and Wilkins, 1995). However, the genome organization of subunit A is decidedly more complex in cotton, encompassing a gene superfamily consisting of two discrete multigene families with approximately six or more members per family (Wilkins, Wan, and Lu, 1994). Despite expectations to the contrary, this organization of subunit A genes is not even remotely related to the polyploid nature of cultivated cottons, since A and D diploids exhibit a similar genome organization, albeit one less complex than in the allotetraploid species. Another interesting family feature is that the A and D gene members within a family are more closely related to each other than A or D genes between families, indicating that each family may perform a related, but unique, function. Estimates of nucleotide substitution in A and D genes from both diploid and allotetraploid cotton revealed that the two gene families evolved as early as 50 million years ago and are, therefore, as old as the genus *Gossypium* itself (Wilkins, Wan, and Lu, 1994). However, we contend that the evolution of the subunit A gene superfamily predates plant speciation and is not necessarily unique to *Gossypium* and is therefore likely to be a common feature of many plant families.

More recently, we established that the two V-ATPase subunit A gene families are differentially regulated in a physiologically relevant context. The deduced polypeptides of cDNA clones CVA69.75 (T. Wilkins et al., unpublished) and CVA69.24 (Wilkins, 1993), which are representative members of gene families *vat69A* and *vat69B*, respectively, differ by as few as four amino acid residues. To examine the expression of these two V-ATPase gene families, gene-specific primers were generated for quantitative RT-PCR analysis during fiber development (T. Wilkins et al., unpublished). We concluded that the low levels of expression associated with *vat69B* genes are consistent with a "housekeeping" function. The small elevation in CVA69.75 transcripts observed during Stage III of fiber expansion is believed to reflect the cell's need for additional housekeeping isoforms of V-ATPase subunit A to maintain the structural integrity of the central vacuole by preserving the same number of V-ATPase enzymes per unit membrane during vacuole enlargement in expanding fibers. The stable level of the Vat69B isoform detected throughout fiber development on immunoblots of two-dimensional protein gels provides compelling evidence in support of a housekeeping role. In contrast, not only are *Vat69A* transcripts more prevalent, but induction of *Vat69A* gene expression during polar expansion of fibers is greater than that of *Vat69B* genes by several orders of magnitude. A sizable increase in the abundance of the Vat69B isoform during fiber expansion significantly changes the ratio between the two V-ATPase subunit A isoforms. In our view, the Vat69B isoform performs tasks related to both housekeeping and expansion-associated functions. In fact, evidence suggests that it is the increase in the number of V-ATPase holoenzymes carrying the "expansion" isoform that, in turn, results in the dramatic increase in V-ATPase-specific activity of the tonoplast during Stage III of expansion. Developmental signals believed to trigger the termination of expansion also cue the disassembly and turnover of expansion-specific V-ATPases. It is quite possible that the events accompanying the termination of fiber expansion, including a dramatic change in the protein composition of the tonoplast, drastically alter the permeability characteristics of the tonoplast in the process.

The tonoplast intrinsic protein (TIP) is yet another turgor-related gene belonging to an extensive family of water channel proteins called aquaporins. A subfamily of TIP homologs localized to the plasma membrane (PM) are termed PIPs. Recently, clusters of PIPs were shown to occur in high concentrations in PM invaginations called plasmalemmasomes, structures which protrude into the cytoplasm to maintain intimate contact with the vacuole (Robinson et al., 1996). So far, γ-TIP and one member of the PIP1 subfamily are known to be preferentially expressed during cell expansion (Ludevid et al., 1992; Kaldenhoff et al., 1995). TIPs and plasmalemmosome PIPs are believed to function in the rapid exchange of water between the apoplast and vacuole in the maintenance of cell turgor, especially during cell expansion (Robinson et al., 1996). Two cotton cDNA clones encoding the same δ-TIP have been isolated from developing fibers (Ferguson, Turley, and Kloth, 1997; Smart et al., 1998). Ribonuclease protection assays (RPAs) revealed that steady-state δ-TIP RNA transepts were most abundant during the earliest stages of rapid fiber expansion, between 5 to 10 dpa, but then declined rapidly after 15 dpa as expansion rates decreased. The 26 kDa TIP proteins detected in tonoplasts by anti-TIP serum on immunoblots were fairly abundant throughout fiber expansion and exhibited only a minor peak at 15 dpa. The presence of TIP proteins decreased slightly overall after 30 dpa. The disparity between the abundance of δ-TIP RNAs and TIP proteins observed during fiber development indicates that the δ-TIP cDNA clone represents a relatively minor isoform and is not the predominant *TIP* gene expressed in developing fibers (Smart et al., 1998), although post-translational phosphorylation may play a role in modulating aquaporin activity and, hence, water transport (Maurel et al., 1995). Using a PCR-based strategy, Ferguson and co-workers (1997) reported that cotton *TIP* genes are organized as two families related to the A or D subgenomes.

In conjunction with increased water uptake, the sequestration of osmoregulatory solutes by the vacuole is an important aspect of turgor-driven expansion in developing cotton fibers. The active transport of osmoregulatory solutes into the vacuole lumen through secondary transport systems is facilitated by the membrane potential generated by the V-ATPase. The synthesis of malate, a key osmoregulatory solute in expanding fibers, is initially synthesized in the cytoplasm via a process involving dark CO_2 metabolism and the enzyme PEPCase. Indeed, in keeping with the increase of malate in vacuoles of expanding fibers, the induced expression of at least one major gene, accompanied by a corresponding increase in PEPCase and enzymatic activity, occurs coincident with increasing expansion (Smart et al., 1998). Moreover, the expansion of fiber initials during Stage I is also associated with a second peak in PEPCase and PEPCase activity. It is especially interesting to note that the PEPCase clone isolated from a fiber library does, in fact, encode a dark CO_2 isoform of PEPCase since RT-PCR experiments using gene-specific primers show that this gene is differentially expressed solely in tissues (embryos, roots, fibers) never exposed to light (Vojdani, Kim, and Wilkins, 1997). Thus, the PEPCase dark CO_2 isoform is encoded by a cotton gene that is expressed in a tissue-specific manner, as well as developmentally regulated in expanding fibers (Smart et al., 1998). It is likely that PEPCase activity is modulated posttranslationally during fiber expansion via phosphorylation of a conserved site located proximal to the amino-terminus (Vojdani, Kim, and Wilkins, 1997). At the other end of the scale, decreased V-ATPase activity during the termination of fiber expansion may produce a change in membrane permeability of the tonoplast that allows malate to "leak," or diffuse, out of the vacuole, resulting in feedback inhibition of PEPCase activity by increasing malate concentrations in the cytoplasm. This scenario clearly points to the importance of the vacuole and V-ATPases in turgor-driven expansion of developing fibers.

In summary, the pattern of several turgor-related genes (V-ATPase, PM, ATPase, PEPC, TIP, and V-PPase [vacuolar H$^+$-pyrophosphatase]) features cyclic, successive waves of expression that decrease in intensity from differentiation through expansion. It is clear that these genes also play important roles in the early stages of fiber expansion, and this is one of the reasons for recognizing the start of Stage I of fiber expansion as -1 dpa (see Table 9.2). It also makes sense that events at the molecular and cellular levels should precede the formation of fiber initials, the appearance of which provides the first morphological evidence that expansion is underway. The wealth of information to be derived from investigating early stages of fiber expansion may provide a more holistic perspective of fiber development, particularly from a molecular standpoint. One proviso that cannot be ignored, however, is the need to confirm that such results are truly fiber specific during early stages of ovule development via in situ hybridization experiments and immunocytochemistry.

Primary Fiber Expansion Genes—Cytoskeleton

Cortical microtubule (MT) arrays have been long regarded as the principle cytoskeletal component with a pivotal role in fiber development, based almost exclusively on three stage-specific changes in MT structural dynamics (Seagull, 1992) (see Table 9.2). The observed differential sensitivity of MT arrays to pharmacological MT-disrupting agents suggests that the MT cytoskeleton is, in fact, a heterogeneous population comprised of "drug-labile" and "drug-stable" subpopulations (Seagull, 1990). The cellular mechanisms regulating the differential stability of the two MT subpopulations are not known, but are likely to be subject to qualitative and/or quantitative changes in abundance of MT α- and ß-tubulin subunits, MT-associated proteins (MAPs), posttranslational modifications, or any combination thereof. For instance, in other plant systems, changes observed in the dynamic behavior of MTs were related to the differential accumulation of tubulin isotypes (Ludueña, 1993). A threefold increase in tubulin subunits (Kloth, 1989) is associated, at least in part, with a significant pool of monomeric tubulin subunits during fiber expansion *and* secondary cell wall synthesis (Seagull, 1990). Consistent with the isotype scenario, fiber-specific tubulin isotypes are temporally regulated in developing cotton fibers

(Dixon, Seagull, and Triplett, 1994). The α-tubulin isotypes (8 and 9) are associated with rapidly expanding 10 dpa fibers, whereas ß-isotypes (6 and 7) in 20 dpa fibers may, in fact, presage the shift in microtubule orientation to steeply pitched arrays associated with secondary cell wall synthesis (see Table 9.2). It would be especially interesting to determine if the α- and ß-tubulin isotypes are differentially associated with the labile and stabile MT subpopulations, respectively.

Given the number of cotton genes isolated to date (reviewed in L. Dirk and T. Wilkins, in press), it is somewhat surprising that the microtubule cytoskeleton has not been the target of molecular studies. Expression studies conducted in our lab revealed that α-tubulin mRNAs in isolated fiber cells increase sharply to peak levels around 20 dpa, during the "transition" period between expansion and secondary cell wall synthesis. Although there is a very moderate increase in α-tubulin subunits between 5 to 15 dpa, the major protein peak occurs at 25 dpa, *after* the termination of cell expansion. By 30 dpa, α-tubulin rapidly declines to previous levels, before falling to basal levels within the next few days (Smart et al., submitted). The RNA and protein results concur, yet these data clearly do not fit our definition of a primary expansion gene. However, the tubulin work was performed alongside studies of a battery of other cotton genes differentially expressed in developing fibers, indicating that the tubulin profiles are valid and not due to experimental variability. Kloth's (1989) developmental protein profile of tubulin subunits in fibers is not consistent with our results, although the method of extracting fiber proteins and/or antibody specificity may be contributing factors to this apparent discrepancy. It is obvious from these conflicting results that more in-depth studies of the MT cytoskeleton are sorely needed, and this further emphasizes the importance of including other developmental markers as points of reference in molecular studies.

In our current working model, we propose that the microtubule cytoskeleton performs a dual role related to expansion and secondary cell wall synthesis, suggesting that functional roles of MTs in developing cotton fibers are regulated both spatially and temporally. In light of the relatively small increase of tubulin mRNAs and proteins in expanding fibers, the preponderance of MTs are most likely of the labile class, such that rapid turnover of MTs is regulated posttranscriptionally via a mechanism of dynamic instability. Such dynamic instability allows the cell greater flexibility in mediating both antero- and retrograde vesicular transport, a major cellular function in delivering materials to the cell wall and subcellular destinations in rapidly expanding fibers. This scenario is also consistent with the location of microtubules deep within the cytoplasm of expanding fibers and the frequency of MT-associated secretory vesicles (Tiwari and Wilkins, 1995). The MT helices oriented more or less perpendicular to the axis of elongation, which in some way control the direction of cellulose microfibril deposition and thereby determine the direction of cell expansion, may be a stable MT subpopulation. In contrast, the increase in MT number, length, and proximity to the plasmalemma (Seagull, 1992) (see Table 9.2) denotes a shift in equilibrium dynamics toward the stable class of MTs during secondary cell wall synthesis. In this case, a predominantly stable MT population is regulated primarily at the transcriptional level to provide a sufficient pool of monomeric subunits for MT assembly. It is possible that fiber-specific ß-tubulin isotypes also play a role in stabilizing MTs. The location of stable MT subpopulations in close proximity to the plasma membrane (Seagull, 1992) supports a functional role in positioning the massive deposition of cellulose microfibrils during secondary cell wall synthesis.

Primary Fiber Expansion Genes—Cell Wall

The mechanistic basis for fiber expansion is a fine balance between cell turgor and relaxation of primary cell wall structure; an unfavorable shift in the equilibrium in either direction would profoundly affect the size and shape of the mature fiber cell. The rate of cell expansion is dictated by the dynamic interactions between the cellulose microfibrils and the noncellulosic polysaccharide matrices. In fact, *the* rate-limiting step to expansion is considered to be the degree to which xyloglucan (XG) polymers cross-link cellulose microfibrils (Carpita and Gibeaut, 1993). In this connection, it is therefore especially interesting to note that the xyloglucan content in the primary cell walls of expanding cotton fibers parallels primary expansion gene expression (Smart et al., 1998). The xyloglucan content increases to peak levels during rapid expansion before gradually decreasing by more than threefold to previous levels by the termination of fiber expansion (Shimizu et al., 1997).

The functional role of enzymes responsible for modulating cell wall dynamics via regulation of inter- and intramolecular interactions between structural components of the cell wall has become the subject of intense scrutiny in recent years. Yet, similar studies are limited to a few gene sequences, and there is very little expression data reported for developing cotton fibers. So far, the induction of at least two genes encoding cell wall enzymes that participate in regulating the intra- and intermolecular bonds of the cellulose-XG framework during the cell wall-loosening process in expanding fibers has been reported (Shimizu et al., 1997), although virtually nothing more is known about these two enzymes beyond the accumulation of steady-state mRNAs during fiber development. One partial cotton cDNA encodes an endo-ß-1,4-glucanase (EGase) closely related to a small subset of fruit-ripening EGases induced by auxin during cell expansion (Wu et al., 1996; Nakamura et al., 1995). At present, it is generally accepted that EGases are involved in regulating the cellulose-XG framework via cleavage of internal ß-1,4-linkages of the xyloglucan cellulosic backbone. Apart from their role in cleaving xyloglucan chains, auxin-induced EGases may generate xyloglucan oligosaccharides that also participate in the regulation of cell expansion (Darvill et al., 1992). A second partial cotton cDNA encodes expansin, an enzyme that disrupts hydrogen bonds between both cellulose microfibrils and cellulose-XGs, which allows the structural components to physically separate for deposition of newly synthesized cell wall material in a very controlled manner (McQueen-Mason and Cosgrove, 1994).

The cotton genes *LTP3* and *LTP6* encode two related isoforms of a lipid transfer protein (LTP) that share about 67 percent amino acid identity (Ma et al., 1995, 1997). Both *LTP* genes are preferentially expressed in rapidly expanding fibers between 10 to 15 dpa, although *LTP3* transcripts are by far the most abundant of the LTP messages. *LTP6* is apparently a fiber-specific gene, whereas the *LTP3* gene is only fiber enriched since a low level of *LTP3* message is also detected in leaves, although expression in leaves may be specific to epidermal cells. Synthesized initially as proproteins, mature LTPs are small, basic extracellular proteins that presumably function in cutin synthesis, according to current models (Thoma et al., 1994; Pyee and Kolattukudy, 1995). This hypothesis is consistent with three-dimensional structural analysis of plant LTPs, indicating that a hydrophobic core forms a cavity capable of transporting long fatty acid chains (Shin et al., 1995). The tertiary structure of the mature LTP polypeptide is determined, in part, by the presence of eight conserved cysteine residues that form four disulfide bridges (Shin et al., 1995). Unlike LTP3, which contains all eight cysteines, LTP6 lacks two essential cysteine residues and is therefore capable of forming only two disulfide bonds (Ma et al., 1997). Whether structural differences in the LTP6 isoform reflect a unique functional role in expanding cotton fibers remains to be seen.

The *E6* gene family, one of the first fiber-enriched genes to be isolated and characterized from cotton, encodes a unique asparagine- and glutamic acid-rich polypeptide that may be specific to fiber-producing plant species (John and Crow, 1992; John, 1996). Several *E6* genes are actively transcribed in fibers, producing 30 to 32 kDa proteins that range in length from 238 to 246 amino acids. The protein heterogeneity is due to the varying number of serine-glycine repeats, ranging between 1 to 5 copies, present in the polypeptide (John, 1996). Based upon computer analysis, the hydrophobic N-terminal region of E6 is a common characteristic of secretory proteins bearing a signal sequence for entry into the secretory pathway, a feature that belies statements by several groups referring to E6 as a cytoplasmic protein. Expression of chimeric gene constructs under the transcriptional control of the *E6* promoter in transgenic cotton fibers demonstrated that the *E6* promoter is most active between 5 to 15 dpa (Rinehart, Petersen, and John, 1996), an observation that is in keeping with the E6 protein profile (John and Crow, 1992). Antisense suppression of endogenous messages reduced E6 protein levels by 60 to 98 percent in transgenic plants, yet transgenic fiber properties did not differ significantly from elite cultivars, indicating E6 is not essential to the growth or structure of developing fibers (John, 1996). It has not been unequivocally established that E6 is, in fact, a structural cell wall protein. However, many cell wall structural proteins contain repeating peptide motifs in a pattern analogous to E6, which has, therefore, been placed in the cell wall category until evidence to the contrary becomes available.

The cotton *Fb-B6* gene is also preferentially expressed in developing cotton fibers, although very little data are currently available for this particular gene (John, 1995). *Fb-B6* is preferentially expressed in 15 to 22 dpa fibers and may be a potential candidate for a "late" class of primary expansion genes because B6 transcripts reach peak levels *after* the period of maximum expansion and do not persist well into secondary cell wall synthesis. One of the most interesting attributes of the 28 kDa Fb-B6 protein is the presence of hydrophobic domains at both the N- and C-termini (John, 1995). The N-terminal domain has all the classic features of a signal sequence, indicating that Fb-B6 is, indeed, a secretory protein. In most cases, the mature protein is produced following the cleavage of the N-terminal propeptide in the lumen of the endoplasmic reticulum (ER). Computer projections indicate that the hydrophobic C-terminal is a transmembrane domain, and as a consequence, Fb-B6 is presumably an integral membrane protein. In addition, two proline-rich domains, in which every other amino acid residue is proline, reside within the Fb-B6 mature polypeptide. Interestingly, the proline-rich domains are located in close proximity to the hydrophobic termini. It is unknown whether these proline-rich motifs are critical to the topology or function of Fb-B6 or whether hydroxylation of proline residues is an important consideration. Although it has been suggested that E6 and Fb-B6 are primary cell wall structural proteins, it is clear that more concrete evidence is required, especially localization of the gene products at the subcellular level, to firmly establish a putative role for E6 and Fb-B6 in fiber development.

Constitutive Fiber Expansion Genes

The recent discovery of several constitutively expressed genes clearly indicates that not all cotton genes are necessarily regulated at the transcriptional level in developing cotton fibers (Shimizu et al., 1997; Smart et al., 1998). Cytological evidence documenting the enlargement of the nucleolar organizing region (see Table 9.2) presumably indicates an increase in the synthesis and/or assembly of ribosomes to sustain demands for translation of mRNA templates during the early stages of fiber expansion. Yet, the 25S and 18S ribosomal RNA (rRNA) genes are apparently constitutively expressed throughout fiber development, or at least through 30 dpa (Smart et al., 1998), provided that the two major rRNA transcripts for both the 25S and 18S structural RNAs are totaled (T. Wilkins et al., unpublished). However, it is interesting to note that the relative abundance of the two transcripts shifts dramatically during fiber development in a stage-specific manner. For example, coincident with fiber expansion, the higher molecular weight 25S transcript is the more prevalent of the two 25S rRNAs from -1 to 10 dpa (T. Wilkins et al., unpublished), although it is not known if this developmental change influences translational activity in any way.

The V-PPase is yet another electrogenic proton pump associated with the tonoplast in developing cotton fibers. Unlike the V-ATPase, the V-PPase is composed of a single 66 kDa subunit and utilizes inorganic pyrophosphate as an energy source for H^+-translocation. RPAs demonstrated that the *V-PPase* gene is also constitutively expressed in developing fibers from 5 to 30 dpa (Smart et al., 1998). Between -3 to 3 dpa, however, immunoblots detected a sharp peak in the relative abundance of the V-PPase polypeptide that subsequently dropped by more than tenfold by 10 dpa, before stabilizing at low levels throughout the remainder of expansion and secondary cell wall synthesis. Nonetheless, starting at about 10 dpa, the decline in V-PPase polypeptides is accompanied by a rapid rise in V-PPase enzyme activity, which continues to increase by about 70 percent to peak at 20 dpa before rapidly decreasing to preexpansion activity levels by 35 dpa. The disparity between protein and activity levels strongly supports a posttranslational mechanism for regulating V-PPase activity in developing fibers, probably through changes in the phosphorylation state of the enzyme. One other interesting feature regarding the V-PPase is the 5-day lag in reaching peak activity levels relative to the V-ATPase or other primary expansion enzymes. Since V-PPase activity is highest at the *conclusion* of expansion and immediately precedes cellulose synthesis, it is increasingly unlikely that this proton pump is involved in turgor-related functions of the vacuole during fiber expansion. Since the V-PPase does not appear to be a turgor-related gene, does it have a major role in secondary cell wall synthesis, and if so, is the differential regulation of V-PPase in the tonoplast at all related to changes in concentrations of PP_i (pyrophosphate) as a function of development? These are certainly important issues to address to ascertain the primary function of V-PPase in fiber development.

Actin and xyloglucan endotransglycosylase (XET) genes are also constitutively expressed during cotton fiber development, although in both of these cases, the corresponding mRNAs decline after 25 dpa and can no longer be detected by 30 dpa in sensitive RT-PCR experiments (Shimizu et al., 1997). This expression pattern concurs with a role in fiber development, but also suggests that these genes are nonessential to fiber maturity. Although the actin cytoskeleton is important to the structure and function of a cell, treatment of cultured ovules with agents that disrupt actin microfilaments (e.g., cytochalasin) do *not* inhibit fiber growth during the early stages of expansion (Tiwari and Wilkins, 1995). However, the precocious reorientation of MT (microtubules) arrays to steeply pitched helices by cytochalasin during fiber expansion, suggests that actin microfilaments play an important, yet indirect, role in cellulose synthesis by regulating MT during the onset of secondary cell wall synthesis (Seagull, 1992). In contrast, XETs may play a pivotal role in a developmental process that is highly correlated with cell wall extensibility during cell expansion. XETs catalyze both the cleavage and endotransglycosylation of XG chains and, together with endo-ß-1,4-glucanases (see "Primary Fiber Expansion Genes—Cell Wall"), are responsible for modulating the degree of cross-linking between carbohydrate polymers of the cellulose-XG framework and, hence, the rate of cell expansion. Yet, the role of XET in cell wall loosening during expansion remains debatable, considering that XET, in and of itself, is not sufficient or necessary for sustained expansion of isolated cell walls (McQueen-Mason et al., 1993). On the other hand, xyloglucan oligosaccharides (XGOs) promote growth, presumably by acting as competitive substrates for free ends of cleaved high molecular weight XG polymers during transglycosylation. The result is a reduction in the mean length of XG polymers and temporary cell wall loosening (Fry, Smith, Hetherington, and Potter, 1992; Fry, Smith, Renwick, et al., 1992). The XGOs, in turn, may be generated in vivo by the action of hydrolases such as endo-1,4-ß-glucanase, a cellulase that catalyzes the irreversible cleavage of XGs. Similar to XETs, however, cellulase alone cannot support prolonged expansion of isolated cell walls (Cosgrove and Durachko, 1994). Unfortunately, RNA analysis alone is insufficient to determine the importance of actin and XET to fiber development, especially since potential translational or posttranslational regulation of gene expression is not readily apparent.

Transition from Primary to Secondary Cell Wall Synthesis

The period between 16 to 21 dpa represents a switch in emphasis from primary to secondary cell wall synthesis during fiber development, exemplified by the transition in gene expression from ß-1, 4-endoglucanase to ß-1,3-endoglucanase (Shimizu et al., 1997). This window of development is especially challenging to study because it is not necessarily a simple task to assign putative functions to cotton genes preferentially expressed during this transition period (see Tables 9.2 and 9.3). The majority of gene transcripts (*Fb-B6, H6*) that accumulate during the latter stage of fiber expansion, but disappear before peak expression of genes involved in secondary cell wall synthesis, are likely associated with expansion and primary cell wall synthesis, whereas other transcripts (e.g., *Rac, α-tubulin*) represent genes encoding functions related to secondary cell wall synthesis. Likewise, the existence of structural genes displaying an expression pattern characteristic of secondary cell wall genes does not necessarily preclude a functional role in primary cell wall architecture in postexpansion fibers, though this situation may arise infrequently. Thus, gene expression patterns alone are not sufficient to allow an accurate portrayal of gene function. Sorting of cotton genes preferentially expressed during the transition phase based on putative functions requires careful deliberation and, at present, is tentative at best, given the amount of information currently available.

Although there is a tremendous increase in the rate of cellulose synthesis (Meinert and Delmer, 1977), fiber expansion and synthesis of primary cell wall continues, albeit at a rapidly decreasing level, during the transition period. It has been a long-held belief that continued expansion of developing fibers is limited by a shift in MT orientation (see Table 9.2) and secondary cell wall deposition (Beasley, 1979). If so, at least two conditions must be met to satisfy this possible scenario. First, incorporation of newly synthesized secondary wall would need to initiate in the midregion of the fiber as proposed (Marx-Figini, 1966). Second, since initial fiber expansion occurs by a mechanism of diffuse growth, with a decided bias for deposition at the fiber tip (Tiwari and Wilkins, 1995), secretory vesicles carrying the last of the primary cell wall constituents could continue to deposit material preferentially at the tip of the cell, thereby allowing expansion to occur on a limited basis. However, discovery of the *immature* (*im*) fiber mutant, which is defective in secondary cell wall synthesis (Kohel, Quisenberry, and Benedict, 1974), casts serious doubt on the existence of a causal relationship between fiber expansion/primary cell wall synthesis and secondary cell wall synthesis. These two developmental processes may simply be independent cellular events under similar developmental control that, by happenstance, overlap for a brief period of time in developing fibers.

The nature of developmental signals that trigger the simultaneous termination of auxin-induced expansion and the onset of secondary cell wall synthesis in cotton fibers remains purely speculative. Nonetheless, a dramatic shift in the auxin/abscisic acid (IAA/ABA) ratio coincident with the decrease in the rate of fiber expansion is indicative of a signal-transducing role for the phytohormones in Upland (*Gossypium hirsutum* L.) cotton (W. Kin, E. Sutter, and T. Wilkins, in preparation). A similar proposal has been put forward for diploid cotton species (Nayyar, Kaur, Basra, et al., 1989; Nayyar, Kaur, Malik, et al., 1989). Although a changing IAA/ABA ratio is the only potential candidate at present for a regulatory role in the molecular switch between developmental stages, a number of important questions need to be answered before this notion can be fully accepted. Areas remaining open to investigation include (1) identification of components in the signal transduction pathway that spur hormonal changes in the first place; (2) determination of whether hormonal changes directly affect expansion and/or secondary cell wall synthesis processes; and (3) characterization of developmentally regulated genes that are differentially responsive to hormones in a stage-specific manner. Preliminary work from our lab, thus far, shows that only primary cell wall-related genes are induced in response to auxin during peak expansion of fibers, although secondary cell wall-related genes were not included in this study (W. Kin, E. Sutter, and T. Wilkins, in preparation).

As alluded to earlier in this section, the uncoupling of expansion and secondary cell wall synthesis in *im* mutant fibers points to independent, yet slightly overlapping, developmental processes. It is therefore increasingly likely that different signal transduction pathways are involved in regulating these processes, although a common global regulatory factor may be responsible for the simultaneous triggering of distinct signal cascades that suppress expansion on one hand, while initiating the onset of secondary cell wall synthesis on the other, during the transition period. In fact, a proposal put forth by Delmer and co-workers (Delmer et al., 1995) states that an oxidative burst, elicited by oligogalacturonides released from the primary cell wall (Meinert and Delmer, 1977), is involved in signaling the onset of secondary cell wall synthesis. The NADPH oxidase system located in the plasma membrane, which is responsible for the synthesis of reactive oxygen radicals, is activated by a Rac GTP (guanine triphosphate)-binding protein (Diekmann et al., 1994). The stimulus behind this proposal was the fortuitous finding that two cotton genes (*Rac9* and *Rac13;* see Table 9.3) are differentially expressed during the onset of secondary cell wall synthesis and encode basic 21 kDa plant homologs belonging to the *ras* superfamily of small signal-transducing GTP-binding proteins (Delmer et al., 1995). Of the two closely related *Rac* genes, *Rac13* is the most highly expressed and fiber specific, while *Rac9* is expressed at relatively low levels in both fibers and roots. Since the cotton Rac homologs are also members of the *rho* subfamily of *Ras* proteins, it is possible that one, or both, of the cotton Rac proteins functions in regulating the organization of actin microfilaments during the realignment of the cortical microtubule cytoskeleton and deposition of cellulose microfibrils with the onset of secondary cell wall synthesis (see Table 9.2).

It is presumed that in the final days preceding the termination of fiber expansion, around 20 to 21 dpa, the primary cell wall architecture undergoes a developmental metamorphosis that ultimately determines the size and shape of the mature fiber. The decline in the rate of expansion is accompanied by a corresponding decrease in extensibility and an increase in rigidity of the primary cell wall, resulting from gross changes in cross-linking relationships between cellulose microfibrils and polysaccharide polymers in the noncellulosic matrices (Carpita and Gibeaut, 1993). In support of molecular changes in structural relationships that "fix" the final shape of the fiber, the loss of approximately 36 percent of the xyloglucans synthesized during peak fiber expansion parallels cessation of expansion (Shimizu et al., 1997). A similar loss of xyloglucans has been reported during auxin-induced expansion of pea hypocotyls (Hayashi, Wong, and Maclachlan, 1984). With respect to the pectin fraction of the noncellulosic matrix, circumstantial evidence suggests that Ca^{2+} homeostasis influences the pectin macromolecular structure, resulting in increased rigidity of the cotton primary cell wall during the transition period (T. Wilkins, unpublished).

The incorporation of structural proteins into the primary cell wall also plays an important role in determining fiber length and shape. For instance, extensins become increasingly associated with the cell wall in response to a decrease in expansion (Sadava, Walker, and Chrispeels, 1973) and may serve as a key component in the "locking" mechanism that "fixes" the structure of the mature cell wall (reviewed in Carpita and Gibeaut, 1993). Yet, the locking function ascribed to extensins in fixing cell shape may, in actuality, be attributed to the hydroxyproline residues rather than to extensins per se because a strong correlation exists between the amount of hydroxyproline in the wall and final length of the cell (Iraki et al., 1989), at least in some systems (Ye and Varner, 1991). Strangely enough, extensins appear to be noticeably lacking from cotton fibers (John and Crow, 1992; John and Keller, 1995), and thus, other proline-rich proteins are likely to perform a structural role in primary cell wall architecture.

One of the most promising candidates for such a structural role is the proline-rich, arabinogalactan-like H6 protein of cotton fibers (John and Keller, 1995). H6 is a basic protein containing an N-terminal signal sequence and 17 repeats of the proline-rich pentameric motif A/ST/SPPP, features which are typical of secretory cell wall structural proteins (reviewed in Showalter, 1993). The deviation in molecular weight of H6 proteins synthesized in vitro and in vivo is highly suggestive of posttranslational modifications, which, in the case of proline-rich proteins, is oftentimes the hydroxylation of proline residues to hydroxyproline (Hyp). In addition, carbohydrate side chains are frequently added to serine and Hyp residues via O-linked glycosylation. RNA transcripts corresponding to the *H6* gene are present at very low levels during Stages I and II of expansion, but gradually increase through Stage III, peaking at 22 dpa before declining in 24 dpa fibers (John and Keller, 1995). The 65 kDa H6 protein, detected on immunoblots in 15 to 40 dpa developing cotton fibers, peaked around 30 dpa. The authors (John and Keller, 1995) contend that the H6 developmental profile is consistent with a functional role in assembly of the secondary cell wall matrix. However, the developmental profile alone does not preclude the possibility that H6 functions as a primary cell wall structural protein, especially when considering that xyloglucans can be replaced in the cell wall by a protein framework *without* changing the shape of the cell *after* expansion is complete (Carpita and Gibeaut, 1993). The simultaneous increase in H6 and a further 37 percent decline in XGs between 21 to 27 dpa is probably more than coincidence and provides compelling evidence in support of an integral role for H6 in primary cell wall structure. Determining the spatial distribution of H6 within cell walls of developing fibers would doubtless address the likelihood of either possible scenario.

During the final days of the transition period, the structure of the primary cell wall is essentially fixed around 20 dpa, which signals the termination of fiber expansion. The termination of expansion immediately precedes the first significant increase in fiber strength in 21 dpa fibers (Hsieh, Honic, and Hartzell, 1995), thereby lending credence to the notion that fixation of primary cell wall structure is a contributing factor to fiber strength. Moreover, it has also been proposed that it is the cellulose-XG framework in particular that is responsible for determining the strength of the primary cell wall (Shedletzky et al., 1990).

Fiber Secondary Cell Wall Synthesis

Remarkably, the thick secondary cell wall of cotton fibers is comprised almost exclusively of cellulose. The fiber cell's entry into the stage of active secondary cell wall synthesis around 21 dpa is signaled by (1) a decline in the transient synthesis of ß-1, 3-glucans (callose) and (2) a sharp increase in expression of cellulose synthase catalytic subunit genes (Pear et al., 1996), cellulose synthesis (Meinert and Delmer, 1977), and cellulose content (Shimizu et al., 1997). The pronounced differences in the structure and composition of the primary and secondary cell walls of cotton fibers, especially in terms of cellulose, are summarized in Seagull (1990). However, since cellulose biosynthesis is discussed in detail in another chapter of this book (see Delmer, Chapter 4), this section will focus primarily on the expression and developmental regulation of cotton genes selectively involved in secondary cell wall synthesis. Cotton genes that exhibit an atypical expression pattern (e.g., *Rac* GTPases, α-tubulin, and actin) relative to other secondary cell wall genes (see Table 9.2), which may be involved in the synthesis of cellulose only indirectly, are dealt with elsewhere in this chapter.

The not so subtle differences between primary and secondary cell wall cellulose provide ample support for the involvement of discrete biosynthetic mechanisms in developing fibers. Our understanding of the mechanistic basis of cellulose biosynthesis in plants has gained little ground during the past decade or so, at least until very recently. The long-awaited isolation of plant cDNA clones encoding cotton homologs of the bacterial cellulose synthase catalytic subunit (Pear et al., 1996) is a crowning achievement that promises to profoundly impact elucidation of the cellulose biosynthetic pathway. The expression of two cotton fiber-enriched genes, *CelA-1* and *CelA-2*, initially detected early in the transition period (17 dpa), are induced to high levels of expression throughout the period of active secondary cell wall synthesis (21 to 27 dpa) in developing fibers. The greatest degree of amino acid similarity (50 to 60 percent) between cotton and bacterial subunits resides within three conserved domains, designated as H-1, H-2, and H-3. Hydrophilic stretches in each H-domain also contain conserved subdomains (U-1 through U-4) involved in binding the substrate UDP (uridine diphosphate)-glucose and/or in the catalytic reaction itself. Overexpressed in *Escherichia coli*, a truncated fusion polypeptide spanning the conserved domains was shown to be fully capable of binding UDP-glucose in a Mg^{2+}-dependent manner analogous to bacterial cellulose synthase complexes. Interestingly, CelA proteins do not carry an N-terminal signal sequence and are therefore not bona fide secretory proteins. However, consistent with the location of cellulose synthase in the plasmalemma, computer algorithms predict that CelA-1 and CelA-2 contain a total of six putative transmembrane domains, two of which are retained within the N-terminal region, while the remainder are clustered in the C-terminus. However, plant CelA proteins also contain two unique plant-specific regions, termed P-CR (plant-conserved region) and HVR (hypervariable region), positioned at the junctions between the H-1/H-2 and H-2/H-3 domains, respectively. P-CR domains are well conserved among plant species, although the hypervariable region is not, leading to a speculative role for HVRs in bestowing the synthase complex with some sort of functional specificity (Pear et al., 1996). Clearly, identification of CelA in plants is the first step toward the characterization of other cellulose synthase subunits and accessory proteins.

Sucrose synthase (SuSy) is a degradative, multifunctional enzyme that utilizes sucrose as a substrate to channel carbon to several growth-related processes paramount to fiber growth and development, including increased respiratory demands, cell wall polysaccharide and starch biosynthesis, and the synthesis of osmoregulatory solutes in turgor-driven cell expansion. It is therefore not surprising that SuSy mRNA is detected through fiber development, from 0 to 27 dpa, before disappearing in 30 dpa fibers (Nolte et al., 1995; Shimizu et al., 1997). Of special interest is the marked elevation in SuSy transcript levels during key stages of fiber development, including (1) the onset of fiber expansion at 0 dpa (Nolte et al., 1995), (2) peak polar elongation of fibers at around 15 dpa, and (3) the active phase of secondary cell wall synthesis between 21 to 27 dpa (Shimizu et al., 1997). The recent identification of a 91 kDa plasma membrane-bound SuSy isoform in developing fibers undergoing active secondary cell wall synthesis (Amor et al., 1995) is a significant discovery that has revolutionized our thinking about the biosynthesis of both callose and cellulose. Not only is carbon from sucrose converted at high rates into callose and cellulose, but the spatial distribution of SuSy parallels the orientation of cellulose microfibrils. These data strongly support a model in which SuSy is an integral component of ß-glucan synthases and provides carbon in the form of UDP-glucose to facilitate the synthesis of celluose and/or callose as an energy-efficient process (Amor et al., 1995). Although $\geq$ 50 percent of SuSy is associated with the plasma membrane during secondary cell wall synthesis, a sizable pool of SuSy remains in the cytoplasm. In other plant systems, the majority of studies have focused primarily on the cytoplasmic form of SuSy. It is clear that the diverse functional roles ascribed to SuSy are performed by at least two distinct enzymes represented by cytoplasmic and membrane-associated isoforms. The number and type of SuSy isoforms potentially involved in each of the cellular processes requiring SuSy activity is yet to be determined. Likewise, the mechanism by which SuSy is associated with membranes and/or ß-glucan synthase complexes is a mystery that needs to be thoroughly investigated because of its monumental promise for elucidating the structure and function of cellulose synthase in plants.

Annexins are another group of multifunctional proteins that have been implicated in the regulation and/or assembly of ß-glucan synthases during secondary cell wall synthesis in developing cotton fibers (Andrawis, Solomon, and Delmer, 1993). In general, annexins are Ca^{2+}-binding proteins that reversibly associate with membranes in response to Ca^{2+} and in a phospholipid-dependent manner (reviewed in Moss, 1997). All eukaryotes characteristically contain a fair number of related annexins, although some organisms, including plants, encode organism-specific homologs as well. In animal systems, convincing evidence now links annexins to roles in important cellular processes, such as vesicle fusion and trafficking, Ca^{2+} homeostasis, and cell proliferation. Recently, one or more of three cotton annexin isoforms, purified from plasma membranes of developing fibers actively engaged in secondary cell wall synthesis, were found to bind plasma membranes in a Ca^{2+}-dependent manner and to inhibit callose synthase activity in in vitro assays (Andrawis, Solomon, and Delmer, 1993). Interestingly, radiolabeling studies suggest that each of the 34 kDa cotton fiber annexins exhibits distinct phosphorylation/dephosphorylation kinetics. The phosphorylation site(s) probably reside within the variable N-terminal region, which serves to define the function of a particular annexin, depending on the number and type of sites within the domain that, in turn, determine the nature of the interactions with cellular ligands (Moss, 1997).

Together, the existence of annexins, shown to interact with the cytoskeleton in both plants (Calvert, Gant, and Bowles, 1996) and

animals (Moss, 1997), and the annexin-mediated inhibition of callose synthase form the foundation for the hypothesis that annexins
may function in directing the subcellular location and modulating
activity of callose synthase via interaction with both the glucan
synthase and the cytoskeleton (Andrawis, Solomon, and Delmer,
1993). Two cotton cDNA clones, *AnnGh1* and *AnnGh2*, isolated
from a 21 dpa fiber library, encode annexins that are 67 percent
identical at the amino acid level (Potikha and Delmer, 1997). Within each of the four highly conserved repeats of seventy amino acids
is an endonexin fold, a conserved motif believed to bind both Ca^{2+}
and phospholipids (Moss, 1997). The genes encoding AnnGh1 and
AnnGh2 are differentially and developmentally expressed in cotton
fibers. AnnGh1 and AnnGh2 transcipts, which are highly abundant
during fiber elongation, gradually decline as the fibers enter the
stage of active secondary cell wall synthesis $\geq$ 21 dpa. AnnGh1
mRNA remains approximately eight- to tenfold more abundant than
AnnGh2 messages throughout fiber development (Potikha and
Delmer, 1997), indicating that AnnGh2 is a minor annexin species
and may perform a related or entirely different functional role. It
has not, as yet, been determined if the cDNA clones encode any of
the three annexin isoforms identified in late-developing cotton fibers. However, the high level of gene expression during fiber elongation is enough to suggest that AnnGh1 and AnnGh2 perform expansion-specific tasks in addition to, or apart from, the proposed role in
regulating callose synthase. In light of the functions currently assigned to the annexin family, and the voluminous number of vesicles transporting cargo to specific subcellular destinations during
rapid fiber expansion, the involvement of AnnGh1 and/or AnnGh2
in vesicular trafficking is a very promising possibility as well.

Fiber Maturation

Virtually nothing is known about the maturation process in developing fibers at the molecular and cellular levels, primarily because
the thick secondary cell wall restricts access to the inner workings of
the fiber. At present, it is assumed that the final stage of maturation
entails rapid death of the fibers, resulting from desiccation of the
fibers upon boll opening. The possibility that developmentally regulated processes, such as programmed cell death, are involved in fiber

maturation in any way is merely speculative. However, the fiber-enriched FbL2A protein may function in protecting cellular structure during fiber desiccation, although this hypothesis has not been tested (Rinehart, Petersen, and John, 1996). Based on transcriptional activity of the *FbL2A* gene promoter fused to reporter genes in transgenic plants, gene expression of *FbL2A* is initially detected during the period of active secondary cell wall synthesis. *FbL2A* expression increases rapidly to peak levels around 35 dpa, coincident with the maximum deposition of cellulose. Yet, structural features of the FbL2A protein are more compatible with a potential protective role during fiber dehydration. FbL2A is a hydrophilic secretory protein of 43.4 kD that is enriched in lysine and glutamic acid amino acids. Similar to other plant proteins, including dehydrins, the major portion of the polypeptide consists of repeats motifs, although each of the four FbL2A direct tandem repeats, comprised of fifty-five residues, is considerably longer in length. For the time being, where FbL2A is localized within the fiber cell, and how critical this protein is to both the structure and quality of mature cotton fibers, remains unknown.

CONCLUSIONS

The genetic manipulation of cotton fiber properties using molecular strategies relies on the identification and isolation of genes that control fiber development and/or directly affect a particular structural property of fibers. In fact, biotechnology has, in all likelihood, been the primary driving force for the notable upswing in the isolation of cotton genes. During the past decade, the prospect for genetic engineering of cotton fibers has soared in proportion to recent technological advances in the field of biotechnology. In 1996, the first two transgenic cotton cultivars, engineered for insect and herbicide resistance, were released for commercial production in the United States. Clearly, the production of transgenic cotton exhibiting increased yields, enhanced fiber quality, and/or novel fiber properties holds significant promise to have widespread impact on both U.S. and global economies. Although small gains in yield and fiber quality continue to be made by conventional breeding programs, genetic improvement of these agronomic traits is beginning to plateau as a result of an increasingly narrow germplasm base for selection. To counteract this current trend, the use of genetic engineering of cotton fibers will become increasingly commonplace as a means of bolstering breeding efforts.

In many cases, the functions of cotton genes in fiber development are as yet unknown; this is especially true of fiber-specific genes. No doubt fiber-specific genes will be important to bioengineering efforts in the long term; however, their impact will probably be negligible in the short term.

Another point worthy of mention is that many developmental studies are limited solely to RNA and/or protein analysis, which may bias our interpretation of a gene's importance to fiber development as a consequence. Moreover, some danger is associated with studying the expression of a single gene out of context. To fully grasp the functional role of a gene in fiber development, it is becoming increasingly important in expression work (1) to include a battery of genes to keep variability between labs to a minimum and (2) to conduct exhaustive analysis at as many levels as possible to thoroughly explore subtle aspects of the regulation of fiber genes and their impact on fiber development. For instance, gene products that are regulated almost exclusively at the posttranslational level will prove to be especially recalcitrant to genetic engineering efforts, regardless of their importance to fiber development.

Molecular studies have served as the cornerstone for elucidating the dynamics of developmental processes that determine the size and shape of mature cotton fibers and, hence, fiber properties that are of paramount importance to textile manufacturers. Understanding the molecular aspects of fiber development is essential to the strategic design of biotechnological approaches intended to enhance fiber quality. Based on the combined efforts of the cotton research community, major strides have been made in dissecting each of the major developmental stages into discrete substages that involve differential gene expression, as well as the coordinate regulation of subcellular events at the molecular level. For instance, fiber expansion entails the spatial and temporal regulation of gene expression in relation to coordinated cellular processes, including maintenance of cell turgor, extension of the cell wall in response to cell turgor, and cytoskeletal dynamics. Yet, the multitude of genes involved in each of these processes may be differentially regulated at the transcriptional, post-transcriptional, and posttranslational levels in response to similar or dissimilar signal cascades. Although we are only beginning to explore the molecular genetics of cotton fiber development in any detail, cotton researchers will be in a better position than ever before to successfully employ genetic engineering as we enter the new millennium, especially as this technology becomes further refined.

REFERENCES

Amor, Y., C.H. Haigler, S. Johnson, M. Wainscott, and D.P. Delmer (1995). A membrane-associated form of sucrose synthase and its potential role in synthesis of cellulose and callose. *Proceedings of the National Academy of Sciences USA* 2: 9353-9357.

Andrawis, A., M. Solomon, and D.P. Delmer (1993). Cotton fiber annexins: A potential role in the regulation of callose synthase. *The Plant Journal* 3: 763-772.

Beasley, C.P. (1979). Cellulose content in fibers of cottons which differ in their lengths and extent of fuzz. *Physiologia Plantarum* 45: 77-82.

Beasley, C.A. and I.P. Ting (1973). The effects of plant growth substances on *in vitro* fiber development from fertilized cotton ovules. *American Journal of Botany* 60: 130-139.

Beasley, C.A. and I.P. Ting (1974). The effects of plant growth substances on *in vitro* fiber development from unfertilized cotton ovules. *American Journal of Botany* 61: 188-194.

Calvert, C.M., S.J. Gant, and D.J. Bowles (1996). Tomato annexins p34 and p35 bind to F-actin and display nucleotide phosphodiesterase activity inhibited by phospholipid binding. *The Plant Cell* 8: 333-342.

Carpita, N.C. and D.M. Gibeaut (1993). Structural models of primary cell walls in flowering plants: Consistency of molecular structure with the physical properties of the walls during growth. *The Plant Journal* 3: 1-30.

Cosgrove, D.J. and D.M. Durachko (1994). Autolysis and extension of isolated walls from growing cucumber hypocotyls. *Journal of Experimental Botany* 45: 1711-1719.

Darvill, A., C. Augur, C. Bergmann, R.W. Carlson, J.-J. Cheong, S. Eberhard, M. G. Hahn, V.-M. Ló, V. Marfà, B. Meyer, et al. (1992). Oligosaccharins—Oligosaccharides that regulate growth, development and defence responses in plants. *Glycobiology* 2: 181-198.

Delmer, D.P., J.R. Pear, A. Andrawis, and D.M. Stalker (1995). Genes encoding small GTP-binding proteins analogous to mammalian Rac are preferentially expressed in developing cotton fibers. *Molecular and General Genetics* 48: 43-51.

Diekmann, D., A. Abo, C. Johnston, A.W. Segal, and A.W. Hall (1994). Interaction of Rac with p67phox and regulation of the phagocytic NADPH oxidase activity. *Science* 265: 531-533.

Dirk, L.M.A. and T.A. Wilkins (in press). Current status of cotton molecular biology. In *Cotton Physiology* Volume II, Eds. J. McD. Stewart, D. Oosterhuis, and J. Heitholt. Memphis, TN: The Cotton Foundation.

Dixon, D.C., R.W. Seagull, and B.A. Triplett (1994). Changes in the accumulation of α- and ß-tubulin isotypes during cotton fiber development. *Plant Physiology* 105: 1347-1353.

Ferguson, D.L., R.B. Turley and R.H. Kloth (1997). Identification of a δ-TIP cDNA clone and determination of related A and D genome subfamilies in *Gossypium* species. *Plant Molecular Biology* 34: 111-118.

Fry, S.C., R.C. Smith, P.R. Hetherington, and I. Potter (1992). Endotransglycosylation of xyloglucan: A role in wall yielding? *Current Topics in Plant Biochemistry and Physiology* 11: 42-62.

Fry, S.C., R.C. Smith, K.F. Renwick, D.J. Martin, S.K. Hodge, and K.J. Matthews (1992). Xyloglucan endotransglycosylase, a new wall-loosening enzyme activity from plants. *The Biochemical Journal* 282: 821-828.

Hasenfratz, M.-P., C.-L. Tsou, and T.A. Wilkins (1995). Expression of two related vacuolar H^+-ATPase 16-kilodalton proteolipid genes is differentially regulated in a tissue-specific manner. *Plant Physiology* 108: 1395-1404.

Hayashi, T., Y.-S. Wong, and G.A. Maclachlan (1984). Pea xyloglucan and cellulose. II. Hydrolysis by pea endo-β-1,4-glucanases. *Plant Physiology* 75: 605-610.

Hsieh, Y.-L., E. Honic, and M.M. Hartzell (1995). A developmental study of single fiber strength: Greenhouse grown SJ-2 Acala cotton. *Textile Research Journal* 65: 101-112.

Hülskamp, M., S. Misera, and G. Jürgens (1994). Genetic dissection of trichome cell development in *Arabidopsis*. *Cell* 76: 555-566.

Iraki, N.M., R.A. Bressan, P.M. Hasegawa, and N.C. Carpita (1989). Alteration of the physical and chemical structure of the primary cell wall of growth-limited plant cells adapted to osmostic stress. *Plant Physiology* 91: 39-47.

John, M.E. (1995). Characterization of a cotton (*Gossypium hirsutum* L.) fiber mRNA (Fb-B6). *Plant Physiology* 107: 1477-1478.

John, M.E. (1996). Structural characterization of genes corresponding to cotton fiber mRNA, E6: Reduced E6 protein in transgenic plants by antisense gene. *Plant Molecular Biology* 30: 297-306.

John, M.E. and L.J. Crow (1992). Gene expression in cotton (*Gossypium hirsutum* L.) fiber: Cloning of the mRNAs. *Proceedings of the National Academy of Sciences USA* 89: 5769-5773.

John, M.E. and G. Keller (1995). Characterization of mRNA for a proline-rich protein of cotton fiber. *Plant Physiology* 108: 669-676.

Kaldenhoff, R., A. Kölling, J. Meyers, U. Karmann, G. Ruppel, and G. Richter (1995). The blue light-responsive *AthH2* gene of *Arabidopsis thaliana* is primarily expressed in expanding as well as in differentiating cells and encodes a putative channel protein of the plasmalemma. *The Plant Journal* 7: 87-95.

Kloth, R.H. (1989). Changes in the levels of tubulin subunits during development of cotton (*Gossypium hirsutum*) fiber. *Physiologia Plantarum* 76: 37-41.

Kohel, R.J., J.E. Quisenberry, and C.R. Benedict (1974). Fiber elongation and dry weight changes in mutant lines of cotton. *Crop Science* 14: 471-474.

Ludevid, D., H. Höfte, E. Himelblau, and M.J. Crispeels (1992). The expression pattern of the tonoplast intrinsic protein δ-TIP in *Arabidopsis thaliana* is correlated with cell enlargement. *Plant Physiology* 100: 1633-1639.

Ludueña, R.F. (1993). Are tubulin isotypes functionally significant? *Molecular Biology of the Cell* 4: 445-457.

Ma, D.-P., H. Tan, Y. Si, R.G. Creech, and J.N. Jenkins (1995). Differential expression of a lipid transfer protein gene in cotton fiber. *Biochimica et Biophysica Acta* 1257: 81-84.

Ma, D.-P., H.-C. Liu, H. Tan, R.G. Creech, J.N. Jenkins, and Y.-F. Chang (1997). Cloning and characterization of a cotton lipid transfer protein gene specifically expressed in fiber cells. *Biochimica et Biophysica Acta* 1344: 111-114.

Marx-Figini, M. (1966). Comparison of the biosynthesis of cellulose *in vitro* and *in vivo* in cotton bolls. *Nature* 210: 754-755.

Maurel, C., R.T. Kado, J. Guern, and M.J. Crispeels (1995). Phosphorylation regulates the water channel activity of the seed-specific aquaporin α-TIP. *The European Molecular Biology Organization Journal* 14: 3028-3035.

McQueen-Mason, S. and D.J. Cosgrove (1994). Disruption of hydrogen bonding between plant cell wall polymers by proteins that induce wall extension. *Proceedings of the National Academy of Sciences USA* 91: 6574-6578.

McQueen-Mason, S., S.C. Fry, D.M. Durachko, and D.J. Cosgrove (1993). The relationship between xyloglucan endotransglycosylase and *in vitro* cell wall extension in cucumber hypocotyls. *Planta* 190: 327-331.

Meinert, M.C. and D.P. Delmer (1977). Changes in biochemical composition of the cell wall of the cotton fiber during development. *Plant Physiology* 59: 1088-1097.

Moss, S.E. (1997). Annexins. *Trends in Cell Biology* 7: 87-89.

Nakamura, S., H. Mori, F. Sakai, and T. Hayashi (1995). Cloning and sequencing of a cDNA for poplar endo-1,4-β-glucanase. *Plant and Cell Physiology* 36: 1229-1235.

Nayyar, H., K. Kaur, A.S. Basra, and C.P. Malik (1989). Hormonal regulation of cotton fibre elongation in *Gossypium hirsutum* L. *in vitro* and *in vivo*. *Biochemie und Physiologie der Pflanzen* 185: 415-421.

Nayyar, H., K. Kaur, C.P. Malik, and A.S. Basra (1989). Regulation of differential fibre development in cotton by endogenous plant growth regulators. *Proceedings of Indian National Science Academy* B55: 463-468.

Nolte, K.D., D.L. Hendrix, J.W. Radin, and K.E. Koch (1995). Sucrose synthase localization during initiation of seed development and trichome differentiation in cotton ovules. *Plant Physiology* 109: 1285-1293.

Oppenheimer, D.G., P.L. Herman, S. Sivakumaran, J. Esch, and M.D. Marks (1991). A *MYB* gene required for leaf trichome differentiation in *Arabidopsis* is expressed in stipules. *Cell* 67: 483-493.

Pear, J.R., Y. Kawagoe, W.E. Schreckengost, D.P. Delmer, and D.M. Stalker (1996). Higher plants contain homologs of the bacterial *celA* genes encoding the catalytic subunit of cellulose synthase. *Proceedings of the National Academy of Sciences USA* 93: 12637-12642.

Potikha, T.S. and D.P. Delmer (1997). cDNA clones for annexin AnnGh1 and AnnGh2 from *Gossypium hirsutum* (cotton). *Plant Physiology* 113: 305.

Pyee, J. and P.E. Kolattukudy (1995). The gene for the major cuticular wax-associated protein and three homologous genes from broccoli (*Brassica oleracea*) and their expression patterns. *The Plant Journal* 7: 49-59.

Ramsey, J.C. and J.D. Berlin (1976). Ultrastructure of early stages of cotton fiber differentiation. *Botanical Gazette* 137: 11-19.

Rinehart, J.A., M.W. Petersen, and M.E. John (1996). Tissue-specific and developmental regulation of cotton gene *FbL2A*: Demonstration of promoter activity in transgenic plants. *Plant Physiology* 112: 1331-1341.

Robinson, D.G., H. Sieber, W. Kammerloher, and A.R. Schäffner (1996). PIP1 aquaporins are concentrated in plasmalemmasomes of *Arabidopsis thaliana* mesophyll. *Plant Physiology* 111: 645-649.

Sadava, D., F. Walker, and M.J. Chrispeels (1973). Hydroxy-proline-rich cell wall protein (extensin): Biosynthesis and accumulation in growing pea epicotyls. *Developmental Biology* 30: 42-48.

Seagull, R.W. (1990). The effects of microtubule and microfilament disrupting agents on cytoskeletal arrays and wall deposition in developing cotton fibers. *Protoplasma* 159: 44-59.

Seagull, R.W. (1992). A quantitative electron microscopic study of changes in microtubule arrays and wall microfibril orientation during *in vitro* cotton fiber development. *Journal of Cell Science* 101: 561-577.

Shedletzky, E., M. Shmuel, D.P. Delmer, and D.T.A. Lamport (1990). Adaptation and growth of tomato cells on the herbicide 2,6-dichlorobenzonitrile leads to production of unique cell walls virtually lacking a cellulose-xyloglucan network. *Plant Physiology* 94: 980-987.

Shimizu, Y., S. Aotsuka, O. Hasegawa, T. Kawada, T. Sakuno, F. Sakai, and T. Hayashi (1997). Changes in levels of mRNAs for cell wall-related enzymes in growing cotton fiber cells. *Plant and Cell Physiology* 38: 375-378.

Shin, D.H., J.Y. Lee, K.Y. Hwang, K.K. Kim, and S.W. Suh (1995). High resolution crystal structure of the non-specific lipid-transfer protein from maize seedlings. *Structure* 3: 189-199.

Showalter, A.M. (1993). Structure and function of the plant cell wall proteins. *The Plant Cell* 5: 9-23.

Smart, L.B., F. Vojdani, M. Maeshima, and T.A. Wilkins (1998). Genes involved in osmoregulation during turgor-driven cell expansion of developing cotton fibers are differentially regulated. *Plant Physiology* 116: 1539-1549.

Song, P. and R.D. Allen (1997). Identification of a cotton fiber-specific acyl carrier protein cDNA by differential display. *Biochimica et Biophysica Acta* 1351: 305-312.

Stewart, J. McD. (1986). Integrated events in the flower and fruit. In *Cotton Physiology*, Eds. J.R. Mauney and J. McD. Stewart. Memphis, TN: The Cotton Foundation, pp. 261-300.

Sussman, M.R. (1994). Molecular analysis of proteins in the plant plasma membrane. *Annual Review of Plant Physiology and Plant Molecular Biology* 45: 211-234.

Thoma, S., U. Hecht, A. Kippers, J. Botella, S. DeVries, and C. Somerville (1994). Tissue-specific expression of a gene encoding a cell wall-localized lipid transfer protein from *Arabidopsis. Plant Physiology* 105: 35-45.

Tiwari, S.C. and T.A. Wilkins (1995). Cotton (*Gossypium hirsutum* L.) seed trichomes expand via a diffuse growing mechanism. *Canadian Journal of Botany* 73: 746-757.

Vojdani, F., W. Kim, and T.A. Wilkins (1997). Phosphoenolpyruvate carboxylase cDNAs from developing cotton (*Gossypium hirsutum*) fibers (accession nos. AF008939 and AF008940) (PGR 97-135). *Plant Physiology* 115: 315.

Wan, C.Y. and T.A. Wilkins (1994). Isolation of multiple cDNAs encoding the vacuolar H$^+$-ATPase subunit B from developing cotton (*Gossypium hirsutum* L.) ovules. *Plant Physiology* 106: 394-395.

Wilkins, T.A. (1993). Vacuolar H$^+$-ATPase 69-kilodalton catalytic subunit cDNA from developing cotton (*Gossypium hirsutum* L.) ovules. *Plant Physiology* 102: 679-680.

Wilkins, T.A. (1996). Bioengineering fiber quality: Molecular determinants of fiber length and strength. *Proceedings of the Beltwide Cotton Conferences*, pp. 1679-1680.

Wilkins, T.A. and S.C. Tiwari (1994). Cotton fiber morphogenesis: A subcellular odyssey. *Proceedings of the Biochemistry of Cotton Workshop*, pp. 89-94.

Wilkins, T.A. and S.C. Tiwari (1995). Biogenesis of distinct vacuole-types during cell differentiation of seed trichomes (Abstract). Keystone Symposia on Molecular and Cellular Biology, Taos, NM.

Wilkins, T.A. , C.-Y. Wan, and C.-C. Lu (1994). Ancient origin of the vacuolar H$^+$-ATPase 69-kilodalton catalytic subunit superfamily. *Theoretical and Applied Genetics* 90: 514-524.

Wu, S.-C., J.M. Blumer, A.G. Darvill, and P. Albersheim (1996). Characterization of an endo-ß-1,4-glucanase gene induced by auxin in elongating pea epicotyls. *Plant Physiology* 110: 163-170.

Ye, Z.-H. and J.E. Varner (1991). Tissue-specific expression of cell wall proteins in developing soybean tissues. *The Plant Cell* 3: 23-37.

Chapter 10

Genetic Engineering Strategies for Cotton Fiber Modification

Maliyakal E. John

INTRODUCTION

The combined total world consumption of synthetic and natural fibers was approximately 15 billion pounds in 1996 (see Table 10.1). Man-made fibers (9,950 million pounds) outperform natural fibers in the marketplace due to continuous improvements in manufacturing and marketing. Examples of innovations are microdenier fibers, superior blending capabilities, and aggressive pricing (The Freedonia Group, 1992). Due to the economic dependence of many countries on a continued strong cotton market, it is necessary to focus on reducing the cost of production, increasing yield and quality, and diversifying the product spectrum of fiber. Another important objective for cotton research is to enable farming practices, as well as the processing of fiber, to become more environmentally friendly. Recombinant DNA technol-

I acknowledge the expert technical help of Greg Keller, Jennifer Rinehart, Lori Spatola, and Michael Petersen. I am thankful to Dr. B. Chowdhury, MATECH Associates, for conducting the thermal studies and useful discussions. I am grateful to Cheryl Scadlock, Kim McEvilly, and Andrea Kersten for editorial assistance. Agreement for the use of *A. eutrophus* genes was obtained from Metabolix, Inc., Cambridge, MA. Antibodies to PHA synthase and reductase were obtained from S. Pagette of Monsanto (MO). Transmission electron microscopy analysis was conducted at University of Wisconsin, Madison. IAA and cytokinins in fiber were measured by Drs. Xiaoyne Li and Clifford LaMotte of Iowa State University. This study was partially funded by the U.S. Department of Commerce, National Institute of Technology, Advanced Technology Program Grant #70NANB5H1061.

271

TABLE 10.1. World Fiber Consumption (× Million Pounds)

Application	Cotton	Wool	Synthetic
Apparel	3,175	140	2,460
Home Furnishing	1,300	15	935
Floor Covering	20	25	3,055
Industrial	100	10	1,115
Medical and Other	215	—	2,385
TOTAL	4,810	190	9,950

Source: The Freedonia Group (1992).

ogy is one of the many new sciences being used to achieve these objectives. The marketing of insect-resistant Bt cotton and the development of herbicide-resistant varieties are examples of new technologies in the cotton sector (Jenkins, McCarty, and Wofford, 1995; Conner, 1996; Stalker et al., 1996).

Early application of recombinant DNA technology was directed toward producing insect- and herbicide-resistant cotton. More recently, the importance of improving fiber traits for superior product performance and new applications has also been recognized (John, 1994a; John and Keller, 1996). As evident from Table 10.1, the apparel sector is a major market for fiber. Thus, fiber traits such as strength, length, fineness, dye binding, wrinkle resistance, and shrinkage resistance are targets for improvement. Fiber strength and length have been increased by plant breeding (Culp, 1992; Meredith, 1992). However, this technology is not applicable to traits that are not found in other cotton cultivars. This limitation can be overcome through genetic engineering that enables the transfer of genes from any source into cotton. The crucial challenge in fiber modifications is the identification of useful genes. The particle bombardment method of gene transfer allows one to insert one or many genes into cotton and to test their usefulness in a timely fashion. This chapter provides an overview of this method and a brief description of various lines of investigations to illustrate two broad strategies for fiber modifications: (1) modifications of fiber proteins or growth regulators and (2) metabolic pathway engineering.

GENE TRANSFER INTO COTTON

Two well-documented methodologies exist for cotton transformation, *Agrobacterium* infection and particle bombardment. *Agrobacterium tumefaciens* infects a wound site in cotton and transfers a plasmid (Ti) segment into cells. The Ti plasmid is engineered to carry antibiotic selection and other desirable transgenes. Antibiotic selection enables one to isolate cells that have integrated the T-vector into their genome. These cells are then regenerated by tissue culture into whole plants (Barton et al., 1983; Fraley, Rogers, and Horsch, 1986; Umbeck et al., 1987; Trolinder and Goodin, 1987). The *Agrobacterium*-mediated transformation has proven to be reliable and has been used for a number of single gene trait transfers in cotton, including insect resistance by *Bacillus thuringiensis* (Bt) and genes for herbicide resistance for 2,4-D(2,4-dichlorophenoxyacetic acid) and bromoxynil (Perlak et al., 1990; Bayley et al., 1992; Stalker et al., 1996). Since *Agrobacterium*-mediated transformation is limited to only cotton cultivars that can be regenerated by tissue culture, only Coker varieties are now routinely transformed by *Agrobacterium*. At the present time, commercialization of transgenic cotton involves the introduction of genes into Coker lines and then backcrossing them into elite varieties. Several years of breeding are required to select suitable lines for marketing.

Methods based on the physical introduction of DNA into growing cotton tissue can circumvent the drawbacks of the *Agrobacterium*-mediated method. Particle bombardment relies on shooting metal microparticles coated with DNA into plant cells. The propulsion force can be electric discharge, pressurized gas, or gun powder explosion (Klein et al., 1987; McCabe et al., 1988; Sanford et al., 1991). The microparticles are carefully targeted toward the growing tips, where the meristematic cells take up and incorporate the DNA into their genome. The particle bombardment-mediated cotton transformation of elite cotton cultivars has been successful (McCabe and Martinell, 1993; John and Crow, 1992; John and Keller, 1996). The meristems are bombarded with gold particles (1 to 3 μm [micrometers] in diameter) containing one or several plasmids. A marker gene, β-glucuronidase (*GUS*), is included in one of the plasmids. The meristems are allowed to grow for 4 to 6 weeks; the leaves are then tested for GUS by histochemical staining, and the unstained parts are pruned. Nodes or axillary buds subtending GUS and expressing leaves are allowed to grow. The process is repeated until a transgenic plant is obtained (McCabe and Martinell, 1993).

The particle bombardment of cotton results in two types of transformants, epidermal and germ line. Selective pruning and the forcing of the axillary buds cause a few transformed cells from the L1, L2, or L3 layers to gain ascendancy and populate an entire tissue layer of its origin (McCabe and Martinell, 1993). If the transformed layer is responsible for germ line, the transgenes will be passed on to the subsequent progeny of the plant. Those plants are referred to as germ line transformants. Germ line transformants are detected by histochemical staining for GUS in the vascular tissues and pollen. If only the L1 layer (epidermis) of cells is transformed, then the transgene is not passed on to the plant's progeny. These transformants are propagated vegetatively. Since cotton fiber originates from ovule epidermal cells, the epidermal transformants are useful in studying transgene expression in fibers. Table 10.2 shows typical efficiencies for particle bombardment-mediated cotton transformation. As indicated, a minimum of 1000 seed axes is necessary to successfully generate two or three transformants. Mature transgenic fibers from particle bombardment-mediated transformants can be obtained in about 6 months. This is in contrast to the *Agrobacterium* method,

TABLE 10.2. Particle Bombardment-Mediated Cotton Transformation

| Gene | Cultivar | # Explants | Epidermal | | Germ Line | | Transform. |
			# Plants	Efficiency %	# Plants	Efficiency %	Efficiency %
Anti-E6	DP50	4,797	4	0.08	3	0.06	0.15
Anti-H6	DP50	5,713	7	0.12	0	0	0.12
Anti-FbL2A	DP50	4,993	6	0.12	4	0.08	0.20
IPT	DP50	12,367	16	0.13	3	0.02	0.15
TMO	DP50	350	2	0.57	0	0	0.57
IAH	DP50	250	0	0	1	0.4	0.4
TMO & IAH	DP50	9,902	8	0.08	1	0.01	0.09
TOTAL		38,372	43	0.11	12	0.03	0.14

Note: Transgenic plants were identified by GUS screening. TMO = iaaM; IAH = iaaH.

which requires 11 to 12 months. The faster cycle enables researchers to test larger numbers of potential genes for fiber modifications.

The transformation frequency of particle bombardment is low. However, this technique inserts genes directly into elite varieties of cotton germplasm. For example, genes were inserted into DP50 and 90, Coker 312, Pima S6 and S7, Sea Island, Acala-Prema, GC 356, and SG125 and 1001 (McCabe and Martinell, 1993; John and Keller, 1996; M.E. John, unpublished results). Particle bombardment allows biotechnologists to introduce new transgenic cotton lines into markets several years sooner than previously possible.

GENETIC STRATEGIES FOR FIBER MODIFICATION

Manipulation of Cotton Genes

Thousands of genes arc expressed in fiber cells (John and Crow, 1992). The majority of these genes are also expressed in other tissues, such as the ovule, leaves, petals, and roots. Several hundred mRNAs are predominantly expressed in fiber, and many are found exclusively in fiber. Assuming that these fiber-specific mRNAs may have critical functions in the development and architecture of fiber cells, one may be able to achieve fiber modifications through the modulation of their expression. Thus, a positive correlation between a given gene and a fiber trait is valuable. However, it is to be kept in mind that many proteins may be involved in a complex trait such as strength. My laboratory identified several fiber-specific mRNAs through differential library screening. Selected ones are listed in Table 10.3. Extensive structural characterization of three mRNAs, E6, H6, and FbL2A, were undertaken (John and Crow, 1992; John and Keller, 1995; Rinehart, Petersen, and John, 1996). All three mRNAs were found to be specifically expressed in fiber and to be developmentally regulated. We also attempted to assess the functions of these mRNAs in fiber development and fiber traits through antisense technology. Antisense genes transcribe RNAs that are complementary to the given mRNAs and, therefore, can hybridize to form double-stranded RNAs inhibiting expression. Attempts have been made to reduce or eliminate the expression of endogenous genes by antisense genes and to evaluate their functions (Cannon et al., 1990;

TABLE 10.3. mRNAs Predominantly Expressed in Cotton Fiber

cDNA	Insert Size bp	mRNA Size (bases)	Tissue Specificity
E6[1]	983	1,100	Fiber
H6[2]	913	950	Fiber
C12	659	1,100	Fiber
B8	690	1,100	Fiber
A11	989	1,100, 900	Fiber
D7	498	500	Fiber
C2	668	1,100	Fiber (petal)
C12	609	1,100	Fiber (petal; pollen)
C1	432	450	Fiber (petal; pollen)
A8	320	1,000	Fiber (petal; pollen)
A9	399	750	Fiber (leaf; petal; pollen)
D4	455	500	Fiber (petal)
B6[3]	1,100	1,200	Fiber (leaf)
A12	868	900	Fiber (leaf)
E9	1,283	1,300	Fiber (petal)
FbL2A[4]	974	1,300	Fiber

Note: The construction and screening of cDNA library are described in John and Crow, 1992. mRNA sizes are approximate and were estimated from Northern blots. All the inserts were sequenced. Tissue specificity was determined by Northern blots. Tissues that showed weak expression are shown in parentheses.
[1] John and Crow (1992).
[2] John and Keller (1995).
[3] John (1995).
[4] Rinehart, Petersen, and John (1996).
All other cDNAs are described in U.S. patent 5,521,078.

Pnueli et al., 1994; Paul et al., 1995). A similar strategy was attempted to decipher the functions of three cotton genes, the E6, H6, and FbL2A. Measurement of the fiber traits of transformants with reduced E6, H6, and FbL2A expression may reveal their roles in fiber development. The antisense E6, H6, and FbL2A genes were engineered and linked to either the 35S, E6, or FbL2A promoters and introduced into DP50 cotton through particle bombardment. Several transformants were identified for each antisense construct through GUS screening. The transgenic fibers from antisense E6 transformants were tested for E6 expression; an example is shown in Figure 10.1. Four transformants were shown to have reduced E6 mRNA

FIGURE 10.1. Down-Regulation of E6 Gene in Transgenic Cotton

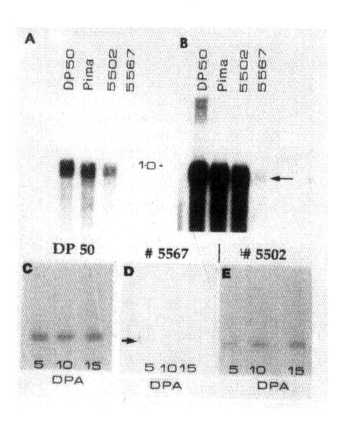

Note: The total fiber RNA (20 μg each) from control (DP50 and Pima) and transgenic (#5502 and #5567) cotton was size fractionated on denaturing formaldehyde agar gels and blotted to nitrocellulose. The blot was hybridized with [^{32}P]-labeled E6 cDNA and washed under stringent conditions (0.1 × SSC, 52°C). Autoradiography was done at 70°C. Molecular weight standards were used to estimate the sizes of the RNA. (A) Blot exposed for 12 h. (B) Same blot overexposed (72 h). Arrow marks the E6 mRNA. (C) Western blot of the E6 proteins in developing DP50 fibers. DPA = days postanthesis. (D) Western blot of the E6 proteins from transgenic #5567. The arrow marks the position of E6. (E) Western blot for the E6 protein in #5502. Details of the experimental conditions are given in John (1996).

levels. Transgenic cotton #5567 expressed less than 2 percent of the normal E6 mRNA (John, 1996). This result was confirmed by Western blot assay, using the E6 antibody (see Figure 10.1). Similarly, transformants for antisense H6 and antisense FbL2A were tested. Transformant #5701 showed a 90 percent reduction in H6 mRNA and protein, while plants #1066-2 and #1066-4 showed 98 percent lower FbL2A mRNA levels (see Figure 10.2; FbL2A, not shown). When fiber traits were measured, the strength, length, and micronaire of various transformants, including #5567, were similar to that of control DP50 (see Table 10.4). These results suggest that E6, H6, and FbL2A are not critical to normal fiber development or traits. Alterna-

FIGURE 10.2. Down-Regulation of H6 Gene in Transgenic Cotton

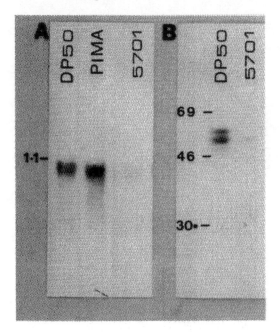

Note: (A) Northern blot containing 20 μg each of control (DP50 and Pima) and transgenic (#5701) fiber RNA was hybridized to H6 cDNA. (B) Western blot of the fiber proteins. The experimental conditions are described in John and Keller (1995).

TABLE 10.4. Fiber Properties of Transgenic Cotton

Gene Construct and Transformant	Gene Effect	Fiber Traits*		
		Strength gm/tex	Length (inches)	Micronaire
DP50 Control	—	19.3 ± 0.83	1.16 ± 0.05	4.1 ± 0.4
Anti-E6 #5567	E6 reduced by 98 percent	17.7	1.10	4.2
Anti-FbL2A #1066-4	FbL2A reduced by 98 percent	20.6	1.20	3.2
Anti-FbL2A #1066-2	FbL2A reduced by 98 percent	21.6	1.20	3.6
iaaM/iaaH #5492	2- to 8-fold higher IAA in fiber	17.7	1.15	3.7
iaaM/iaaH #5453	2-fold increase IAA in 20-day fiber	17.7	1.18	3.2
iaaM/iaaH #5458	2-fold higher IAA in 15-day fiber	18.5	1.09	3.8

*All measurements were done by Stelometer.

tively, very low levels of these proteins are sufficient to maintain normal fiber development and properties (John, 1996). Detailed chemical compositions of these transgenic fibers and the control may yet reveal differences.

It is possible that manipulating levels of other fiber-specific mRNAs may influence fiber properties. However, identification of such mRNAs through antisense genes remains a time-consuming and risky proposition.

MANIPULATING HORMONE LEVELS IN DEVELOPING FIBERS

Measurement of the hormone levels in developing fibers, as well as external applications of various hormones in situ and in vitro, have allowed assessment of the influence of gibberellins, auxins (indoleacetic acid, IAA), cytokinins, abscisic acid, and ethylene in fiber development (reviewed in Basra and Malik, 1984; Kosmidou-Dimitropoulou, 1986). Thus, IAA has been implicated in fiber initiation and elongation. Higher levels of IAA in relation to gibberellins were also postulated to favor fiber elongation and smaller nucleoli (Beasley and Ting, 1973; Popova et al., 1979).

IAA is one of several phytohormones that collectively regulate growth and differentiation of plant cells. Shoot gravitropism, maintenance of apical dominance, and differentiation of vascular tissue were shown to be dependent on IAA (Tamas, 1987; Aloni, 1988). Two genes, *iaaM* and *iaaH*, involved in the biosynthesis of IAA from *Agrobacterium tumefaciens* T-DNA, have been characterized (Thomashow, Reeves, and Thomashow, 1984; Van Onckelen et al., 1986). The *iaaM* gene encodes a tryptophan monooxygenase that converts tryptophan to indole-3-acetamide (IAM). This is the first step in the two-step IAA biosynthesis pathway. IAM is further hydrolyzed to IAA by indoleacetamide hydrolase (*iaaH*). A number of investigators have shown that plants transformed with *iaaM* and *iaaH* genes expressed high levels of IAA (Klee et al., 1984; Sitbon et al., 1991). To establish the levels of free IAA in different cultivars that differ in fiber traits, we measured the IAA levels in DP50, Pima S6, and Sea Island. Both Pima S6 and Sea Island had superior strength, length, and micronaire than the DP50. The free IAA was measured by solvent partitioning and high-performance liquid chromatography (HPLC), followed by gas chromatography-mass spectrometry (GC-MS) using selected-ion monitoring (Li et al., 1992). Isotope dilution was used to correct for incomplete recovery. The results are shown in Figure 10.3. Sea Island and Pima fibers showed higher levels of IAA in five-day and older fibers compared to DP50. We investigated whether increasing the levels of IAA during DP50 fiber development would have any effect on its fiber properties by introducing the *A. tumefaciens* genes *iaaM* and *iaaH* into DP50. To avoid any growth abnormalities, the *iaaH* and *iaaM* genes were linked to the fiber-specific E6 promoter (John, 1994b). Several transformants were identified by GUS screening. All of the transformants showed normal morphology and growth. This was expected, as the E6 promoter is active only in fiber cells. Figure 10.3 (B) shows the free IAA content of transgenic fibers of transformant #5492 during development. The #5492 fibers contained two- to eightfold higher levels of IAA than DP50. However, fiber length, strength, and micronaire showed no difference when compared to DP50 fibers (see Table 10.4).

Naithani, Rama Rao, and Singh (1982) measured the IAA levels in developing fibers from three Indian cotton cultivars. No positive relationship was found between IAA and the rate of elongation. Similarly, Nayyar and colleagues (1989) found no relationship between IAA levels and the extent of fiber elongation in *Gossypium arboreum*.

FIGURE 10.3. The Free IAA Content in Developing Fibers

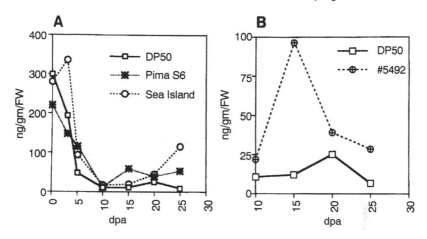

Source: John, M.E. (1994b). Progress in genetic engineering of cotton for fiber modifications. In *Proceedings of the World Cotton Research Conference I,* Eds. G. A. Constable and N. W. Forrester. Brisbane, Australia, CSIRO, pp. 292-296.

Note: (A) The free IAA content of fibers of various ages was measured after purification by solvent partitioning and HPLC as described by Li et al., 1992. Samples 0, 3, and 5 dpa are a mixture of ovule and fiber tissues. The remaining samples are fibers only. Measurements were repeated two to four times for each time point and the average value is shown. (B) The free IAA content of transgenic cotton #5492 was measured and compared to the DP50 control.

These results are in agreement with our results that showed no discernible effect of increased free IAA content in transgenic cotton DP50.

A second set of experiments was conducted by introducing another hormone gene, isopentenyl transferase (*ipt*), into cotton. The *A. tumefaciens ipt* gene is involved in the biosynthesis of cytokinin and catalyzes the condensation of AMP (adenosine monophosphate) and isopentenyl pyrophosphate to form isopentenyl AMP. AMP is the precursor for all other cytokinins (Akiyoshi et al., 1984). Overexpression of the *ipt* gene in transgenic plants produced frequent incited shoots, suppression of root formation, increased growth of axillary buds, and short, intensely green plants (Klee, Horsch, and Rogers, 1987; Smigocki, 1991; Smigocki et al., 1993). The *ipt* gene was linked to the constitutive promoter 35S and introduced into DP50 plants. Transformants were selected based on GUS expression.

The transformants exhibited a range of phenotypes that included bushy, short plants with large numbers of axillary branches. The leaves were dark green in color. Some of the transformants were severely stunted and did not produce flowers. We did not measure cytokinin levels in these plants; we only made observations regarding their phenotypes. Transformants that produced flowers and bolls exhibited no change in fiber traits (M. E. John, unpublished data).

From these experiments, we conclude that the overexpression of auxins and cytokinins has no influence on fiber elongation, strength, or micronaire.

Synthesis of New Biomaterials in Fiber

A third strategy being considered for fiber modification is to synthesize a new biopolymer in cotton (John and Keller, 1996). Depending upon the biopolymer selected and its level, the new compound may impart desirable traits to a fiber. As a model system to test this concept, we selected a polyhydroxyalkanoate (PHA), poly-D(-)-3-hydroxybutyrate (PHB) for synthesis in fiber. PHAs are biodegradable, thermoplastic compounds with similar chemical and physical properties as polypropylene (Steinbuchel, 1991; Muller and Seebach, 1993). PHAs are produced by many bacteria as inclusion bodies and serve as carbon sources and electron sinks (Steinbuchel et al., 1992). *Alcaligenes eutrophus* produces PHB from acetyl CoA through a three enzyme pathway (see Figure 10.4). Two molecules of acetyl CoA are joined by β-ketothiolase (*phaA*) to form acetoacetyl CoA, which is then reduced by acetoacetyl CoA reductase (*phaB*) to R-(-)-3-hydroxybutyryl CoA. This activated monomer is then polymerized by PHA synthase (*phaC*) to form PHB. PHB has been produced in transgenic plants, with the objective of large-scale isolation (Poirier et al., 1992).

Metabolic pathway engineering for the synthesis of PHB in cotton requires only the integration of functional *phaB* and *phaC* genes, since β-ketothiolase, which is involved in the synthesis of mevalonate, is ubiquitous in plants (Poirier et al., 1992; John and Keller, 1996). Reduction in growth and seed production was observed in transgenic *Arabidopsis* as a result of *phaB* overexpression in leaves and other tissues (Poirier et al., 1992). The production of PHB in the chloroplasts of higher plants has also been accom-

FIGURE 10.4. Enzymatic Pathway for Biosynthesis of PHB from Acetyl CoA

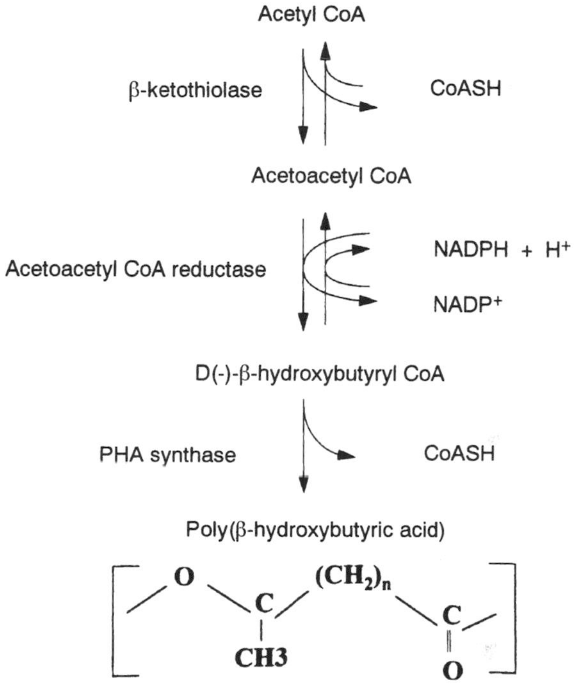

plished by linking the *phaA*, *phaB*, and *phaC* genes to plastid transit signal peptide sequences (Nawrath, Poirier, and Somerville, 1994).

Polymerase chain reaction was used to amplify the coding region of the *phaB* and *phaC* genes from the genomic DNA of *Alcaligenes eutrophus*. To avoid any detrimental effects of PHB production during plant growth, one of the genes, *phaB*, was linked to cotton promoter E6 or FbL2A, both of which are predominantly active in fiber. The genes were linked to poly(A) addition signal (280 bp [base pairs]) of *Agrobacterium* nopaline synthase gene at the 3′ ends. The *phaC* gene was linked to the cauliflower mosaic virus 35S promoter (John and Crow, 1992; Rinehart, Petersen, and John, 1996). A 35S promoter-linked *GUS* gene was added to these plasmids to generate cotton expression vectors (John and Keller, 1996). A second set of genes (phaA, phaB, and phaC/35S) was constructed by linking pea chloroplast transit signal sequences (Nawrath, Poirier, and Somerville, 1994). The chimeric genes were introduced into cotton by particle bombardment as described by McCabe and Martinell (1993).

A total of 105 R0 cotton plants was scored as transgenic, based on GUS expression in leaves. Seventy-six of these were epidermal transformants, while the remaining were germ line transformants. The R0 epidermal transformants and the R1 progeny of germ line transformants were grown and examined for transgene expression. Reductase activities varied between a low of 0.006 to a high of 0.52 μmol/min/mg (micromoles per minute per milligram) protein in fiber. PHA synthase activities also showed large variations among transformants. All transgenic cotton that expressed phaB and phaC showed normal growth and morphology. The presence of reductase and PHA synthase proteins in transgenic fibers was confirmed by Western blot analysis (John and Keller, 1996).

Comparison of reductase levels in developing fibers of plants each carrying different promoter phbB genes showed that both 35S and E6 promoters are active during early fiber development, whereas the FbL2A promoter directs reductase gene activity in late fiber development (Rinehart, Petersen, and John, 1996). The FbL2A promoter also appeared to be stronger than the 35S and E6 promoters in fiber. The optimal levels of reductase and PHA synthase activities, as well as their time of expression in fiber, are likely to be important for maximum synthesis of PHB. From the limited number of plants studied, it appears that the high levels of reductase and PHA synthase during early fiber development are conducive for higher levels of PHB synthesis. Transgenic plants, which contained the FbL2A-phbB gene (active in later fiber development), showed moderately high levels of acetoacetyl CoA reductase and PHA synthase activities, but did not result in high levels of PHB (see Table 10.5). It is possible that decreased levels of acetyl CoA during late fiber development may be a contributing factor for reduced PHB synthesis (Rinehart, Petersen, and John, 1996).

A number of analytical tests, such as epifluorescence microscopy, transmission electron microscopy, GC-MS, and HPLC, were conducted to detect and identify PHB in transgenic cotton (John and Keller, 1996). Representative examples of transformants and the levels of PHB are shown in Table 10.5. Electron micrographs of leaf showed

TABLE 10.5. PHB-Producing Transgenic Cotton

Transformant	Promoter Gene	Targeting Sequence	PHB (µg/gm)	
			Leaf	Fiber
#7148 Epidermal	*E6-pha B* *35S-pha C*	—	0	3,440
#8801 Epidermal	*FbL2-pha B* *35S-pha C*	—	0	423
#6888-7 Epidermal	*E6-pha B* *35S-pha C*	—	0	30
#14216 Epidermal	*35S-pha A* *35S-pha B* *35S-pha C*	Pea Chloroplast	2,350	489
#12334 Epidermal	*E6-pha A* *E6-pha B* *35S-pha C*	Pea Chloroplast	550	437
#12347 Epidermal	*35S-pha A* *35S-pha B* *35S-pha C*	Pea Chloroplast	7,455	0

Note: PHB was quantitated by HPLC analysis of crotonic acid. Yield is given in µg/gm dry weight. #7148, #8801, and #6888-7 were described in John and Keller (1996).

electron lucent granules that were in the range of 0.15 to 0.3 µm in diameter (see Figure 10.5, A and B). The granules were observed in the chloroplasts of transformants containing the 35S promoter and chloroplast transit peptide signal sequence-linked genes. Similar granules were not found in the control leaf (see Figure 10.5[B]). Transgenic fibers also showed electron lucent granules (see Figure 10.6).

Further identification of the chemical nature of these granules comes from HPLC, GC, and mass spectral studies. PHB was converted by acid hydrolysis to crotonic acid and detected by HPLC. Similarly, PHB was treated with ethanol-chloroform-hydrochloric acid mixtures to convert it into ethyl ester derivatives that were then separated by GC for mass spectral analysis. These studies confirmed that the compound found in the transgenic fiber extract has similar mass fragmentation patterns as that of the bacterial PHB. Moreover, these two patterns matched with that of the reference compound, ethyl ester hydroxybutyrate. Thus, GC-MS data provide evidence that transgenic plants produce PHB in fibers (John and Keller, 1996).

FIGURE 10.5. PHB in Transgenic Cotton Leaf

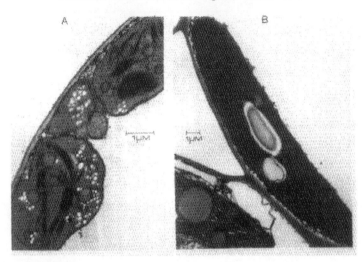

Note: (A) The cotton leaves were fixed with 1 percent paraformaldehyde/2 percent glutaraldehyde in 0.05 M phosphate buffer (pH 7.2) for 2 h at room temperature before embedding in Spurrs epoxy resin. The sections were placed on nickel (300 mesh) grids and stained with 2 percent uranyl acetate followed by Reynolds lead citrate. Transmission electron (JEOL 100CXII) micrographs of sections of transgenic A (#12347) and control DP50 (B) are shown. The electron lucent granules are visible within the chloroplasts of the #12347 sample. Plant #12347 contained *phaA*, *phaB*, and *phaC* genes linked to chloroplast transit peptide signal and 35S promoter.

Thus, the bioplastic genes, when introduced into the cotton genome, resulted in the production of a new biopolymer in the fiber lumen. The temporal pattern of PHB production was determined by the promoters that drive the bioplastic genes. The chemical and physical properties of PHB in cotton fibers were very similar to those of bacterial PHB.

To assess whether any of the physical and chemical properties of fibers containing PHB were different from those of the control, a number of studies were carried out. Results from the thermal property evaluations are described in the following material.

The thermal properties of transgenic fibers were compared with those of the control by differential scanning calorimetry (DSC) and thermogravimetric (TSG) analyses. DSC measures qualitative and quantitative heat and temperature transitions by measuring the heat

FIGURE 10.6. PHB in Transgenic Fiber

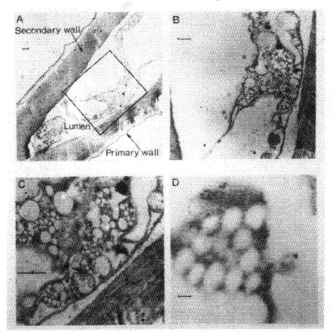

Note: Thirty-five-day-old fiber samples of transgenic cotton #7148 were treated as described for Figure 10.5. The thick secondary wall is visible in A and B. At higher magnifications (C and D), electron lucent granules are visible. The bioplastic genes (*phaB* and *phaC*) in transformant #7148 were not linked to plastid targeting signal sequences, and, therefore, are present in the cytoplasm.

flow rate through the sample. The DSC measurements of one of the transgenic fiber samples, #7148, when compared to DP50, indicated that the onset of decomposition of #7148 fiber was advanced. The PHB content of #7148 was 3440 μg/g (micrograms per gram) (see Table 10.5). The total heat uptake for #7148 (690.7 J/g [joules per gram]) was 11.6 percent higher than for DP50 (619.0 J/g). The heat uptake measurements were repeated with DP50 and #7148 samples. The heat uptake for three independent samples of DP50 was 618.8, 620.1, and 623.5 J/g each, whereas the values of 695.3 and 692.0 J/g were obtained from repeat measurements of #7148 samples. Other transgenic fibers containing varying amounts of PHB also showed positive changes in heat uptake, and the results are summarized in Table 10.6. Thus, the heat uptake capacity appears to be related to the amount of PHB present in fiber. The precise relationship between thermal properties and the amounts of PHB in fiber can be established as fibers with greater PHB content are developed. The transgenic fibers (#7148) were spun into yarn by miniature spinning (Starlab VY-5 direct sliver-to-yarn spinning frame) and knitted into cloth by a knitting machine. Unbleached and undyed fabric was then subjected to thermal property measurements by DSC, along with the control fabric (DP50). The DSC measurements showed a heat uptake of 695.4 J/g for #7148 and 617.8 J/g for DP50 fabric, respectively (John and Keller, 1996).

The relative heat transmission capacities of #7148 and DP50 fibers were determined by thermal conductivity (TC) measurements. The TC of #7148 fibers (0.264 W/m-K) was 6.7 percent lower than for DP50 (0.283W/m-K), indicating slower cooling of the material (John and Keller, 1996). Thus, the DP50 fibers have faster heat dissipation properties.

The heat retention of samples was determined by specific heat measurements at two temperatures (36°C and 60°C). Sample #7148 showed a 8.6 percent higher heat retention than the DP50 sample at 36°C, while the difference was 44.5 percent higher for the transgenic fiber at 60°C. Thus, these results agree with the TC measurements and confirm that sample #7148 has higher heat capacity (John and Keller, 1996).

TABLE 10.6. Heat Uptake of Transgenic Fibers Containing PHB

# Sample	PHB µg/g Fiber	DSC Heat Uptake J/g	% Difference with DP50 Control
DP50 Control	—	619.0	0.0
#7148[1]	3,440	690.7	11.6
#8801[1]	423	642.3	3.8
#6888-7[1]	30	627.5	1.4
#12334[2]	437	642.5	3.8

[1] John and Keller (1996).
[2] Chowdhury and John (unpublished results).

CONCLUSIONS

The fiber modification strategy based on the modification of the expression of the cotton genes E6, H6, and FbL2A was not successful. The transgenic plants maintained normal fiber development, strength, length, and micronaire. It is possible that minute levels of these proteins are sufficient to maintain the proper development and architecture of the fiber. Alternatively, changes in chemical composition may have occurred, but were not identified in the present study. Further chemical analysis of these fibers is under way. The overexpression of the plant growth regulators IAA and cytokinins also did not affect the gross fiber traits. Thus, these genes appear to have no utility in fiber modifications. On the other hand, production of a new polymer, polyhydroxybutyrate, enhanced fiber properties, as shown by the studies of thermal characteristics. The transgenic cotton fibers exhibited measurable changes in thermal properties that suggest enhanced insulation characteristics. The transgenic fibers conducted less heat, cooled down slower, and took up more heat than conventional cotton fibers. Modified fibers with superior insulating properties may have applications in winter wear or other textile uses for which enhanced insulating properties are advantageous. However, the changes in thermal properties are relatively small, as expected from the small amounts of PHB in fibers (0.34 percent fiber weight). It is likely that a severalfold increase in PHB synthesis is required for product applications. Nevertheless, the positive changes in fiber qualities demonstrated here are an indication of the potential of this technology.

REFERENCES

Akiyoshi, D.E., H. Klee, R.M. Amasino, E.W. Nester, and M.P. Gordon (1984). T-DNA of *Agrobacterium tumefaciens* encodes an enzyme of cytokinin biosynthesis. *Proceedings of the National Academy of Sciences USA* 81: 5994-5998.

Aloni, R. (1988). Vascular differentiation within the plant. In: *Vascular Differentiation and Plant Growth Regulators,* Ed. T.E. Timell. Berlin: Springer-Verlag, pp. 36-62.

Barton, K.A., A.N. Binns, A.J.M. Matzke, and M.D. Chilton (1983). Regeneration of intact tobacco plants containing full length copies of genetically engineered T-DNA and transmission of T-DNA to R1 progeny. *Cell* 32: 1033-1043.

Basra, A.S. and C.P. Malik (1984). Development of the cotton fiber. *International Review of Cytology* 89: 65-113.

Bayley, D., N. Trolinder, C. Ray, M. Morgan, J.E. Quisenberry, and D.W. Ow (1992). Engineering 2,4-D resistance into cotton. *Theoretical and Applied Genetics* 83: 645-662.

Beasley, C.A. and I.P. Ting (1973). The effects of plant growth substances on *in vitro* fiber development from fertilized cotton ovules. *American Journal of Botany* 60: 130-139.

Cannon, M., J. Platz, M. O'Leary, C. Sookdeo, and F. Cannon (1990). Organ-specific modulation of gene expression in transgenic plants using antisense RNA. *Plant Molecular Biology* 15: 39-47.

Conner, C. (1996). Genetically altering cotton: Has transgenic cotton found a niche in Mississippi? *Cotton Farming* 40: 8-12.

Culp, T.W. (1992). Simultaneous improvement of lint yield and fiber quality in Upland cotton. In *Proceedings from Cotton Fiber Cellulose: Structure, Function and Utilization Conference*, Eds. C.R. Benedict and G.M. Jividen. Memphis, TN: National Cotton Council, pp. 247-288.

Fraley, R.T., S.G. Rogers, and R.B. Horsch (1986). Genetic transformation in higher plants. *Critical Reviews in Plant Sciences* 4: 1-46.

The Freedonia Group (1992). *Textile Fibers: Industry Study #420*. Cleveland, OH: The Freedonia Group, Inc., pp. 54-78.

Jenkins, J.N., J.C. McCarty, and T. Wofford (1995). Bt cotton—A new era in cotton production. In *Proceedings of the Beltwide Cotton Conferences*, Volume 1. Memphis, TN: National Cotton Council, pp. 171-173.

John, M.E. (1994a). Re-engineering cotton fibre. *Chemistry and Industry*, 17 (September 5): 676-679.

John, M.E. (1994b). Progress in genetic engineering of cotton for fiber modifications. In *Proceedings of the World Cotton Research Conference I*, Eds. G.A. Constable and N.W. Forrester. Brisbane, Australia: CSIRO, pp. 292-296.

John, M.E. (1995). Characterization of a cotton (*Gossypium hirsutum* L.) fiber-mRNA (Fb-B6). *Plant Physiology* 107: 1478.

John, M.E. (1996). Structural characterization of genes corresponding to cotton fiber mRNA E6: Reduced E6 protein in transgenic plants by antisense gene. *Plant Molecular Biology* 30: 297-306.

John, M.E. and L.J. Crow (1992). Gene expression in cotton (*Gossypium hirsutum* L.) fiber: Cloning of the mRNAs. In *Proceedings of the National Academy of Sciences USA* 89: 5769-5773.

John, M.E. and G. Keller (1995). Characterization of mRNA for a proline-rich protein of cotton fibers. *Plant Physiology* 108: 669-676.

John, M.E. and G. Keller (1996). Metabolic pathway engineering in cotton: Biosynthesis of polyhydroxybutyrate in fiber cells. *Proceedings of the National Academy of Sciences USA* 93: 12768-12773.

Klee, H., R. Horsch, and S. Rogers (1987). *Agrobacterium*-mediated plant transformation and its further applications to plant biology. *Annual Review of Plant Physiology* 38: 467-486.

Klee, H., A. Montoya, F. Horodyski, C. Lichtenstein, D. Garfinkel, S. Fuller, C. Flores, J. Peschon, E. Nester, and M. Gordon (1984). Nucleotide sequence of the *tms* genes of the pTiA6NC octopine Ti plasmid: Two gene products involved in plant tumorigenesis. *Proceedings of the National Academy of Sciences USA* 81: 1728-1732.

Klein, T.M., E.D. Wolf, R. Wu, and J.C. Sanford (1987). High-velocity microprojectiles for delivering nucleic acids into living cells. *Nature* 327: 70-73.

Kosmidou-Dimitropoulou, K. (1986). Hormonal influence in fiber development. In *Cotton Physiology*, Eds. J.R. Mauney and J.M. Stewart. Memphis, TN: The Cotton Foundation, pp. 361-373.

Li, X., C.E. La Motte, C.R. Stewart, N.P. Cloud, S. Wear-Bagnall, and C.Z. Jiang (1992). Determination of IAA and ABA in the same plant sample by a widely applicable method using GC-MS with selected-ion monitoring. *Journal of Plant Growth Regulation* 11: 55-65.

McCabe, D.E. and B.J. Martinell (1993). Transformation of elite cotton cultivars via particle bombardment of meristems. *Bio/Technology* 11: 596-598.

McCabe, D.E., W.F. Swain, B.J. Martinell, and P. Christou (1988). Stable transformation of soybean (*Glycine max*) by particle acceleration. *Bio/Technology* 6: 923-926.

Meredith, W.R. (1992). Improving fiber strength through genetics and breeding. In *Proceedings from Cotton Fiber Cellulose: Structure, Function and Utilization Conference*. Memphis, TN: National Cotton Council, pp. 289-302.

Muller, H.M. and D. Seebach (1993). Poly(hydroxyalkanoates), a fifth class of physiologically important organic biopolymers. *Angewandte Chemie International Edition* 32: 477-502.

Naithani, S.C., N. Rama Rao, and Y.D. Singh (1982). Physiological and biochemical changes associated with cotton fiber development. *Physiologia Plantarum* 54: 225-229.

Nawrath, C., Y. Poirier, and C. Somerville (1994). Targeting of the polyhydroxybutyrate biosynthetic pathway to the plastids of *Arabidopsis thaliana* results in high levels of polymer accumulation. *Proceedings of the National Academy of Sciences USA* 91: 12760-12764.

Nayyar, H., K. Kaur, A.S. Basra, and C.P. Malik (1989). Hormonal regulation of cotton fiber elongation in *Gossypium arboreum* L. *in vitro* and *in vivo*. *Biochemie und Physiologie der Pflanzen* 185: 415-421.

Paul, M.J., J.S. Knight, D. Habash, M.A.J. Parry, D.W. Lawlor, S.A. Barnes, A. Loynes, and J.C. Gray (1995). Reduction in phosphoribulokinase activity by antisense RNA in transgenic tobacco: Effect on CO_2 assimilation and growth in low irradiance. *Plant Journal* 7: 535-542.

Perlak, F.J., R.W. Deaton, T.A. Armstrong, R.L. Fuchs, S.R. Sims, J.T. Greenplate, and D.A. Fischhoff (1990). Insect resistant cotton plants. *Bio/Technology* 8: 939-943.

Pnueli, L., D. Hareven, S.D. Rounsley, M.F. Yanofsky, and E. Lifschitz (1994). Isolation of the tomato *Agamous* gene *TAG1* and analysis of its homeotic role in transgenic plants. *Plant Cell* 6: 163-173.

Poirier, Y., D.E. Dennis, K. Klomparens, and C. Somerville (1992). Polyhydroxybutyrate, a biodegradable thermoplastic, produced in transgenic plants. *Science* 256: 520-523.

Popova, P.Y., F.N. Nuritdinova, A.I. Imamaliev, and K.D. Madzhitova (1979). Differences in phenol-compound content and character of fiber formation in wild and cultivated cotton. *Sel' skokhozyaistvennykh Nauk im.V. I. Lenina* #8: 22-25.

Rinehart, J.A., M.W. Petersen, and M.E. John (1996). Tissue-specific and developmental regulation of cotton gene FbL2A. *Plant Physiology* 112: 1331-1341.

Sanford, J.C., M.J. Devit, J.A. Russell, F.D. Smith, P.R. Harpending, M.R. Roy, and S.A. Johnston (1991). An improved helium-driven biolistic device. *Technique* 3: 3-16.

Sitbon, F., B. Sundberg, O. Olsson, and G. Sandberg (1991). Free and conjugated indoleacetic acid IAA contents in transgenic tobacco plants expressing the *iaaM* and *iaaH* IAA biosynthesis genes from *Agrobacterium tumefaciens. Plant Physiology* 95: 480-485.

Smigocki, A.C. (1991). Cytokinin content and tissue distribution in plants transformed by a reconstructed isopentenyl transferase gene. *Plant Molecular Biology* 16: 105-115.

Smigocki, A., J.W. Neal, I. McCanna, and L. Douglass (1993). Cytokinin-mediated insect resistance in *Nicotiana* plants transformed with the *ipt* gene. *Plant Molecular Biology* 23: 325-335.

Stalker, D.M., J.A. Kiser, G. Baldwin, B. Coulombe, and C.M. Houck (1996). Cotton weed control using the BXN system. In *Herbicide-Resistant Crops: Agricultural, Environmental, Economic, Regulatory, and Technical Aspects*, Ed. S.O. Duke. New York: Lewis Publishers, pp. 93-105.

Steinbuchel, A. (1991). Polyhydroxyalkanoic acids. In *Biomaterials: Novel Materials from Biological Sources*, Ed. D. Byrom. New York: Stockton Press, pp. 124-213.

Steinbuchel, A., E. Hustede, M. Liebergesell, U. Pieper, A. Timm, and H. Valentin (1992). Molecular basis for biosynthesis and accumulation of polyhydroxyalkanoic acids in bacteria. *Federation of European Microbiological Society Reviews* 9: 217-230.

Tamas, I.A. (1987). Hormonal regulation of apical dominance. In *Plant Hormones and Their Role in Plant Growth and Development*, Ed. P.J. Davies. Boston, MA: Kluwer Academic, pp. 349-410.

Thomashow, M.F., S. Reeves, and M.F. Thomashow (1984). Crown gall oncogenisis: Evidence that a T-DNA gene from the *Agrobacterium* Ti plasmid pTiA6 encodes an enzyme that catalyzes synthesis of indoleacetic acid. *Proceedings of the National Academy of Sciences USA* 81: 5071-5075.

Trolinder, N.L. and J.R. Goodin (1987). Somatic embryogenesis and plant regeneration in cotton (*Gossypium hirsutum* L.). *Plant Cell Reports* 6: 231-234.

Umbeck, P., G. Johnson, K. Barton, and W. Swain (1987). Genetically transformed cotton (*Gossypium hirsutum* L.) plants. *Bio/Technology* 5: 263-266.

Van Onckelen, H., E. Prinsen, D. Inze, P. Rudelsheim, M. Van Lijsebettens, A. Follin, J. Schell, M. Van Montagu, and J. Greef (1986). *Agrobacterium*-tumefaciens plasmid T DNA gene 1 codes for tryptophan 2-monooxygenase activity in tobacco crown gall cells. *Federation of European Biochemical Societies Letters* 198: 357-360.

Chapter 11

Postharvest Management of Fiber Quality

W. Stanley Anthony

OVERVIEW OF COTTON GINNING

Cotton gins are responsible for converting cotton into marketable commodities, such as bales of lint, cottonseed, "motes," compost, and so forth. In a growing number of instances, they are vertically integrated enterprises that include components ranging from farm supplies to warehousing.

Cotton possesses its highest fiber quality and best potential for spinning when it is on the stalk (Anthony, 1994). Qualities of the cotton in the bale depend on many factors, including variety, weather conditions during production, cultural and harvesting practices, moisture and trash content, and ginning processes. The principal functions of the cotton gin are to separate lint from seed and to maintain fiber quality, but the gin must also be equipped to remove foreign matter that would significantly reduce the value of the ginned lint. A ginner must have two objectives: (1) to produce lint of satisfactory quality for the grower's market and (2) to gin the cotton with minimum reduction in fiber spinning quality so that the cotton will meet the demands of its ultimate users—the spinner and the consumer. Accordingly, quality preservation during ginning re-

Author's note: Mention of a trade name, proprietary product, or specific equipment does not constitute a guarantee or warranty by the U.S. Department of Agriculture and does not imply approval of a product to the exclusion of others that may be suitable.

quires the proper selection and operation of each machine that is included in a ginning system.

Recommendations for the sequence and amount of gin machinery to dry and clean spindle-harvested cotton are based on research by the Cotton Ginning Laboratories of the U.S. Department of Agriculture, Agricultural Research Service, and by cooperating organizations. The recommendations are as follows: drier, cylinder cleaner, stick machine, drier, cylinder cleaner, and extractor-feeder and saw-type gin stand, followed by one or two stages of saw-type lint cleaning (see Figure 11.1). These recommendations were designed to achieve satisfactory bale value and to preserve the inherent quality of cotton. The recommendations consider marketing system premiums and discounts as well as the cleaning efficiency and fiber damage resulting from various gin machines. Some variation from these recommendations is necessary for special harvesting conditions.

Foreign matter levels in seed cotton usually range from 5 to 10 percent before gin processing, but levels of 12 to 14 percent do occur. The foreign matter level dictates the cleaning level needed.

FIGURE 11.1. Typical Cotton Ginning Sequence for Mechanically Harvested Cotton

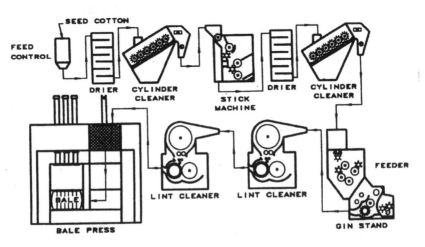

Obviously, any machinery that is not necessary for a particular lot of cotton should be bypassed. Driers, seed cotton cleaners and extractors, and lint cleaners should be provided with bypasses to allow the cotton to skip these machines when extra-clean, dry cotton is brought to the gin.

Extensive use of module storage during the last decade has increased the amount of relatively dry cotton that enters the ginning system. In addition, special market conditions also influence the amount of equipment needed. Machine-picked Upland cotton varieties from across the Cotton Belt have similar cleaning characteristics. Some varieties, typically those which have large quantities of trichrome (hair) on plant surfaces, are difficult to clean. One additional stage of cleaning may be required for hairy leaf cottons. Driers should be adjusted to supply the gin stand with lint having a moisture content of 6 to 7 percent. Research has shown that cotton at this moisture level is more able to withstand the stresses of ginning without breaking.

The quality of ginned lint is directly related to the quality of the cotton before ginning. High grades will result from cotton that comes from clean fields harvested by properly adjusted machines in good condition. Lower grades will result from cotton that comes from grassy, weedy fields in which poor harvesting practices are used.

When gin machinery is used in the recommended sequence, 75 to 85 percent of the foreign matter is usually removed from cotton. Unfortunately, this machinery also removes small quantities of good quality cotton in the process of removing foreign matter, so the quantity of marketable cotton is reduced during cleaning. Cleaning cotton is therefore a compromise between foreign matter level and fiber loss and damage.

Mechanical handling and drying may modify the natural quality characteristics of cotton. At best, a ginner can only preserve the quality characteristics inherent in the cotton when it enters the gin.

EFFECTS OF GIN MACHINERY ON COTTON QUALITY

Cotton quality is affected by every production step, including selecting the variety, harvesting, and ginning. Certain quality characteristics are highly influenced by genetics, whereas others are determined mainly by environmental conditions or by harvesting and ginning practices. Problems during any step of production or processing can cause irreversible damage to fiber quality and reduce profits for the producer as well as the textile manufacturer. Fiber quality is highest the day a cotton boll opens. Weathering, mechanical harvesting, handling, ginning, and manufacturing can diminish the natural quality.

Many factors can indicate the overall quality of cotton fiber. The most important ones include strength, fiber length, short-fiber content (fibers shorter than one-half inch), length uniformity, maturity, fineness, trash content, color, seed coat fragment and nep content, and stickiness (Mayfield et al., 1994).

The ginning process can significantly affect fiber length, uniformity, and the content of seed coat fragments, trash, short fibers, and neps. The two ginning practices that have the most impact on quality are (1) the regulation of fiber moisture during ginning and cleaning and (2) the degree of lint cleaning used.

Moisture Content

The recommended lint moisture range for ginning is 6 to 7 percent. Gin cleaners remove more trash at low moisture, but not without more fiber damage. Higher fiber moisture preserves fiber length but results in ginning problems and poor cleaning (see Figure 11.2).

Lowering fiber moisture decreases individual fiber strength, causing more fibers to break during ginning. For each 1 percent reduction in fiber moisture content below 5 percent, the number of short fibers is increased by almost 1 percent. The effects of ginning cotton below 5 percent moisture are decreased yarn strength and yarn appearance and increased short fibers in the card sliver. Also, overheating may cause increased fiber breakage from the mechanical action of cleaning in the gin and textile mill.

If drying is increased to improve trash removal, yarn quality is reduced. Although yarn appearance improves with drying, up to a point, because of increased foreign matter removal, the effect of increased short-fiber content outweighs the benefits of foreign matter removal.

FIGURE 11.2. Moisture-Ginning Cleaning Compromise for Cotton

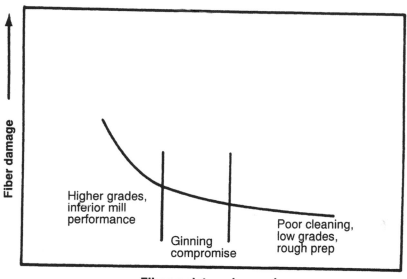

Gin Cleaning

Choosing the degree of gin cleaning is a compromise between fiber trash content and fiber quality. Lint cleaners are much more effective in reducing the lint trash content than are seed cotton cleaners, but lint cleaners can also damage fiber quality and reduce bale weight (turnout) by discarding some good fiber with the waste. Cleaning does little to change the true color of the fiber, but combing the fibers and removing trash changes the perceived color. Lint cleaning can sometimes blend fiber so that fewer bales are classified as spotted or light spotted. Ginning does not affect fineness and maturity. Each mechanical or pneumatic device used during cleaning and ginning increases the nep content, but lint cleaners have the most pronounced influence (Mangialardi, 1985). The number of seed coat fragments in ginned lint is affected by the seed condition and ginning action. Lint cleaners decrease the size, but not the number, of fragments (Anthony et al., 1988). Yarn strength, yarn appearance, and spinning-end breakage are three important spinning quality elements. All are affected by length uniformity and, therefore, by the proportion of short or broken fibers. These three elements are usually preserved best when cotton is ginned with minimum drying and cleaning machinery.

Seed cotton cylinder cleaners decrease the lint trash content, but they slightly decrease the yarn strength. The yarn appearance is improved by cylinder cleaners, but using more than fourteen cylinders in a gin can cause quality problems.

Balancing Fiber Quality and Cleaning

Whether it is done in a gin or in a textile mill, cleaning generally lowers most of the important fiber quality characteristics, other than the trash content, and reduces the amount of usable fiber (turnout). Ginners must compromise between trash removal and fiber damage when choosing their cleaning machinery. For machine-picked cotton, ginners should use twelve to fourteen cylinders of seed cotton cleaning along with a stick machine and one or two lint cleaners, depending on seed cotton trash content and color potential. For stripped cotton, a second stick machine and an air line cleaner or cleaning separator should be included. To deliver the absolute highest quality products, growers and ginners must take care during production, harvesting, ginning, and textile manufacturing to avoid practices that may diminish fiber quality.

SEED COTTON STORAGE AND HANDLING

Seed cotton is stored in trailers or modules. Trailers were historically used to store seed cotton. Trailers (typically 8 ft [feet] wide and 24 to 36 ft long) are an efficient, inexpensive method of delivering cotton to the gin, but they are expensive if used for seed cotton storage. The cost of owning numerous trailers limited the amount of storage, and harvesting was frequently delayed when the trailers were full. Modules increased in use for storing harvested cotton in the United States after 1972, and more than half of the U.S. cotton crop was stored in modules in 1992. Cotton modules are freestanding stacks of cotton that are produced by dumping harvested material into a form known as a module builder (Lalor, Willcutt, and Curley, 1994). A module builder is equipped with a mechanism that compacts the harvested material to a density of about 12 lb/ft^3 (pounds per cubic foot), thus giving the stack integrity to be free-standing after the module builder is removed. Specially built, self-loading trucks and trailers are used to transport the intact modules. Modules provide reservoirs in which harvested cotton may be stored until ginning, permitting the harvesting operation to proceed independent of ginning. Having a supply of cotton leads to a more predictable, manageable, and economical gin operation. Module builders, in two sizes (24 ft or 32 ft long), became commercially available in 1972. Stripper-harvester models are 2 ft higher than spindle-picker models. In general, modules should not weigh over 21,000 lb (pounds).

Quality Changes During Storage

Several variables affect seed and fiber quality during seed cotton storage. Moisture content is the most important. Other variables include length of storage, amount of high-moisture foreign matter, variation in moisture content throughout the stored mass, initial temperature of the seed cotton, temperature of the seed cotton during storage, weather factors during storage (temperature, relative humidity, rainfall), and protection of the cotton from rain and wet ground. Some light spotting occurs in seed cotton stored at a moisture level above 13 percent. A comparison of grades between cotton stored in modules and in trailers shows no significant differences below 13 percent moisture. Moisture content causes yellowness to increase sharply at levels above 13 to 14 percent, especially when the storage period exceeded 45 days. For long storage periods, moisture should be below about 12 percent.

Yellowing is accelerated at high temperatures. Both temperature rise and maximum temperature are important. Temperature rise is probably more related to the heat generated by biological activity than to heat gained from the environment. Seed cotton moisture of 12 percent or less will allow safe, long-term storage, assuming that production, harvesting, and storage guidelines are followed. Higher seed cotton moisture can be tolerated for short storage periods. The rate of lint yellowing, however, begins to increase sharply at moisture levels above 13 percent and can increase even after the module temperature drops.

Seed Quality

When seed cotton is stored, the length of the storage period is important in preserving seed quality and should be based on the moisture content of the seed cotton. Seed quality is sacrificed (germination is reduced and free fatty acid content and aflatoxin level are increased) if the relationship between moisture content and storage length is not understood. Seed cotton moisture should not exceed 10 percent for module storage when the seed will be saved for planting. Oil quality, on the other hand, appears to be less sensitive, so seed cotton saved for use as oil can be preserved at 12 percent moisture content during storage.

SEED COTTON UNLOADING SYSTEMS

The essential requirements of seed cotton unloading systems are to remove seed cotton from the transport vehicle that delivers it to the gin site and to feed cotton into the gin at a constant and uniform rate. Auxiliary functions are to remove rocks, metal, or other hazardous material and to remove green bolls and some sand and dirt. The functions of the unloading system vary somewhat with the harvest method and with the transport and storage method used between field and gin. The main storage methods used for harvested seed cotton in the U.S. cotton industry involve trailers or modules. Two types of modern seed cotton unloading systems are associated with trailer or module storage: (1) pneumatic suction through swinging telescopes that remove cotton directly from the trailer or module and (2) module disperser systems that break up the module mechanically and deposit the seed cotton onto a conveyor that delivers it to a fixed suction pickup point (Laird et al., 1994).

MOISTURE CONTROL

The moisture content of seed cotton is very important in the ginning process. Seed cotton having too high a moisture content will not clean or gin properly and will not easily separate into single locks; rather, it will form wads that may choke and damage gin machinery or entirely stop the ginning process. Cotton with too low a moisture content may stick to metal surfaces as a result of static electricity generated on the fibers and cause machinery to choke and stop. Fiber dried to very low moisture content becomes brittle and will be damaged by the mechanical process required for cleaning and ginning. When pressing and baling such low-moisture cotton, it is often difficult to achieve the desired bale weight and density without adding moisture. Drying cotton at high temperatures may damage the cotton fiber. There is an optimum fiber moisture content for each process in the gin, but the overall gin-moisture relationship (see Figure 11.2) is a compromise between cleaning and quality preservation (Hughs, Mangialardi, and Jackson, 1994).

Both constituents of seed cotton—fiber and seed—are hygroscopic, but at different levels. Dry cotton placed in damp air will gain moisture, and wet cotton placed in dry air will lose moisture. For every combination of ambient air temperature and relative humidity, there are corresponding equilibrium moisture contents for the seed cotton, fiber, and seed. For example, if seed cotton is placed in air at 50 percent relative humidity and 70°F (Fahrenheit), the fibers will tend to reach a moisture content (wet basis) of approximately 6 percent; the seed will tend to reach a moisture content of about 9 percent; and the composite mass will approach a moisture content of 8 percent. The equilibrium moisture content at a given relative humidity is also a function of the temperature and barometric pressure.

Moisture occurs not only in fibers and seed (hygroscopic moisture) but also, sometimes, on their exterior surfaces (surface moisture). The moisture contents (both hygroscopic and surface) of the fiber, seed, and trash of the seed cotton are influenced by weather, method of harvest, and storage time. Seed cotton that is damp or wet from rain or dew may have excessive surface moisture, whereas seed cotton exposed to moist air will have a high hygroscopic moisture content.

The effects of atmospheric conditions, particularly relative humidity, must be considered when harvesting seed cotton. As discussed previously, ambient conditions at the time of harvest influence the moisture content of the harvested seed cotton. The effect of relative humidity on cotton moisture is relatively simple, useful, and easily understood for ambient conditions but is less useful or meaningful as air is heated in a gin drying system. As the air and seed cotton move through a drier, the air temperature will drop (1) because heat is lost, (2) because heat is used to increase the temperature of the cotton, and (3) because moisture is vaporized from the cotton. Fiber temperature will increase until it is about the same as that of the surrounding air.

Most of the moisture removed during the short drying time in commercial gin driers comes from the fibers rather than from the seed and trash. The seed constitutes about 60 percent of the weight of spindle-harvested seed cotton. Moisture results based on oven drying of seed cotton do not necessarily indicate the ginning condition of the fibers. The moisture content of the seed is considerably less important from a ginning standpoint than the moisture content of the fibers, unless the seeds are so wet that they are soft or mushy. For satisfactory ginning, seed moisture content should not exceed 12 percent.

Cotton should be dried at the lowest temperature that will allow satisfactory gin operation. Laboratory tests have shown that fibers will scorch at 450 to 500°F, ignite at 450°F, and flash at 550 to 600°F. In no case should the temperature in any portion of the drying system exceed 350°F.

Drying Systems

Drying systems across the Cotton Belt include reel type, tower (see Figure 11.3), tower hybrid, and towerless systems. Modern drying systems can seriously overdry cotton and must be used properly to avoid reducing cotton quality. Longer-term drying at low temperatures is much less harmful than rapid drying at high temperatures. The air velocities and volumes necessary for proper operation of seed cotton driers vary considerably, depending on the type of drier and the rate of ginning. Air volumes used vary from 20 ft^3/lb of seed cotton in tower drying systems to 40 to 50 ft^3/lb in towerless systems. In the tower type, the air velocity must be sufficient to keep the cotton moving 4,000 to 5,000 ft/min (feet per minute) in the piping and 1,500 to 2,000 ft/min in the driers. The cotton-handling capacity of tower driers may range from 8 to 30 bales/hr (hour), depending

FIGURE 11.3. Typical Tower Drier Used in a Cotton Gin

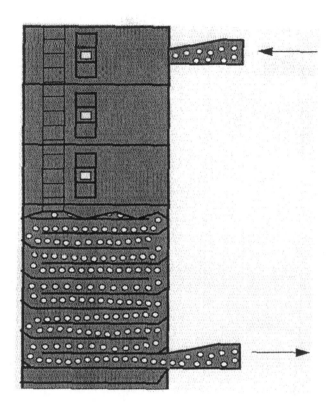

upon drier width, drier shelf spacing, condition of the cotton, and quantity of air moving through the system.

In pneumatic conveying systems, the cotton and drying air move at different speeds. Damp, heavy, dense cottons move much slower than the air. As the cotton is opened and fluffed by moisture removal and mechanical action, it becomes more easily airborne, and its final velocity more nearly approaches that of the conveying air. Cotton is normally retained in a tower drier 7 to 10 sec (seconds). The typical gin drying system consists of a heater, one or more fans, piping, tower drier, and seed cotton cleaner, which serves the dual purpose of separating the warm moist air from the cotton and cleaning the cotton.

SEED COTTON CLEANING AND EXTRACTING

The term "seed cotton cleaning" is often used interchangeably by the ginning industry when referring to the process performed by either the total cleaning and extracting system of a gin or by specific types of machines within the system. In its more restrictive sense, "cleaning" refers to the use of various types of cylinder cleaners designed primarily for removal of dirt and small pieces of leaves, bracts, and other vegetative matter. "Extracting," on the other hand, refers strictly to those processes designed to remove large trash, such as burs and sticks, from the seed cotton. Bur machines, stick machines, extractor-feeders, and combination bur and stick machines are examples of extracting-type machinery.

The cleaning and extracting systems in a modern gin serve a dual purpose. First, large trash components, such as burs, limbs, and branches, must be extracted from the seed cotton so that the gin stand will operate at peak efficiency and without excessive down-time. Second, seed cotton cleaning is often necessary to obtain optimum grades and market values, especially when ginning high-trash-content cotton. Also, cleaners and extractors help open the seed cotton for more effective drying, which is usually done concurrently with cleaning. The amount of cleaning and extracting machinery required to satisfactorily clean cotton varies with the trash content of the seed cotton, which depends, in large measure, on the method of harvest.

Virtually all cotton produced in the United States today is harvested mechanically by either pickers (74 percent) or strippers (25 percent). A small amount (1 percent) of machine-scrapped cotton is recovered from the ground each year by specially designed salvage machines after the regular harvest. The trash contents of seed cotton vary widely as a result of the different harvesting methods employed and the year-to-year variations in the weather during the cropping season. Although these variations in trash level appear to be very wide when viewed from a Beltwide perspective, variations within a given ginning community are usually not nearly as great. Most gins process either picked or stripped cotton and are usually equipped with only the amount and type of cleaning and extracting machinery required for the most severe conditions expected in their trade area

(Baker, Anthony, and Sutton, 1994). For less severe conditions, part of the system should be bypassed to prevent excessive weight losses and to reduce the possibility of overprocessing the cotton. Seed cotton cleaning should be restricted to that which is necessary to ensure smooth, trouble-free ginning and that which is needed to obtain optimum bale values.

Cylinder Cleaners

Cylinder cleaners are used for removing finely divided particles and for opening and preparing the seed cotton for the drying and extraction processes (Mayfield et al., 1983). The cylinder cleaner (see Figure 11.4) consists of a series of spiked cylinders, usually four to seven in number, that agitate and convey the seed cotton across cleaning surfaces containing small openings or slots. The cleaning surfaces may be either concave screen or grid rod sections or serrated disks, such as those found in impact cleaners. Foreign matter that is dislodged from the seed cotton by the action of the cylinders falls through the screen, grid rod, or disk openings for collection and disposal. A typical screen is made of two-mesh woven galvanized wire cloth. Screen-type cylinder cleaners have been largely replaced by the more durable grid-type cylinder cleaners, but screens continue to be popular in some areas for the last inclined cleaner in the cleaning sequence. Cylinder cleaners can be further classified with respect to how they are used in the gin. In this respect, they are either air line, air-fed, or gravity-fed cleaners (Garner and Baker, 1977).

Extractors

Bur Machine

The bur machine was developed in the 1920s in response to hand snapping and mechanical stripping of cotton in Texas. At one time, the bur machine was the most popular bulk extractor in use by the ginning industry, especially in stripper-harvesting areas. Because of its limited operating capacity and its low stick removal efficiency, the bur machine has been largely replaced in most modern cotton

FIGURE 11.4. Six-Cylinder, Air-Fed Inclined Cylinder Cleaner Equipped with Screen Sections

Source: Courtesy of Continental Eagle Corporation, Prattville, Alabama.

gins by other more efficient extractors. However, the bur machine continues to be used in many older, low-capacity cotton gins and is important from a historical viewpoint. The bur machine, more than any other machine, made it possible for a gin to satisfactorily handle hand-snapped and machine-stripped cotton (Pendleton and Moore, 1967).

The bur machine is based on a dislodging or stripping principle. Seed cotton is presented to a large-diameter saw cylinder by a kicker-conveyer equipped with special flippers. Seed cotton adheres to the

saw cylinder and is carried past a flighted stripper-roller that dis-
lodges burs and sticks from the cotton on the surface of the cylinder.
The dislodged material finds its way through the incoming stream of
seed cotton to the kicker-conveyer, which moves the material to one
end of the machine. At this point, the material falls onto a spiked
conveyer and is moved back along the entire length of the saw
cylinder so that seed cotton can be reclaimed from the dislodged
material. Fine particles and dirt sift through a screened trough to an
auger located underneath the spiked conveyer.

Stick Machine

Stick machines (see Figure 11.5) utilize the sling-off action of
high-speed saw cylinders to extract burs and sticks from seed cotton
by centrifugal force (Franks and Shaw, 1959). Seed cotton is fed onto
the primary sling-off saw cylinder and wiped onto the sawteeth by
one or more stationary brushes. Foreign matter and some seed cotton
are slung off the saw cylinders by centrifugal force 25 to 50 times the
force of gravity. Grid bars are strategically located about the periph-
ery of the saw cylinder to help control the loss of seed cotton and to
aid in the extraction process. However, some loss of seed cotton is
inevitable to obtain satisfactory cleaning. Additional saw cylinders
are used to reclaim the seed cotton extracted with the burs and sticks.
Reclaimer saw cylinders resemble the primary saw but usually operate at slower speeds and are equipped with more grid bars.

Extractor-Feeders

Gin stand feeders with extracting capabilities have been used since
the early 1900s (Bennett, 1962). Early extractor-feeders were based
on the stripping or dislodging principle utilized by bur machines.
However, after development of the stick machine, most manufacturers abandoned the stripping principle in favor of the stick machine's
more efficient sling-off feature. Also, the introduction of high-capacity gin stands in the late 1950s forced manufacturers to simplify and
streamline feeders to achieve desired capacities. This was accomplished by adopting the sling-off feature for extractor-feeders. In
addition, many modern extractor-feeders enhanced fine-trash remov-

FIGURE 11.5. Gravity-Fed Three-Saw Stick Machine

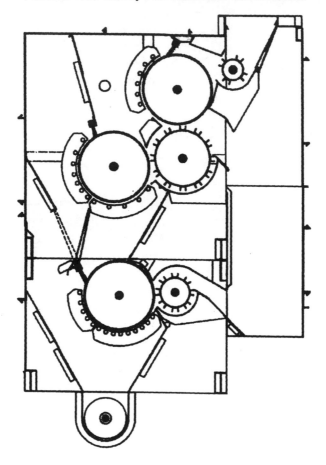

Source: Courtesy of Consolidated Cotton Gin Company, Inc., Lubbock, Texas.

al by also employing cleaning cylinders similar to those found in an inclined cleaner.

The primary function of a modern high-capacity extractor-feeder (see Figure 11.6) is to feed seed cotton to the gin stand uniformly and at controllable rates, with extracting and cleaning as a secondary function. The feed rate of seed cotton is controlled by the speed of two star-shaped feed rollers located at the top of the feeder directly

FIGURE 11.6. Extractor-Feeder (Top Machine) Equipped with Cleaning and Extracting Cylinders Feeding a Saw Gin Stand (Lower Machine)

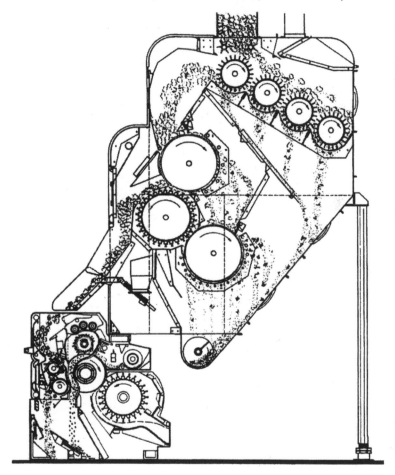

Source: Courtesy of Lummus Corporation, Savannah, Georgia.

under the distributor hopper. Many of the systems are designed to maintain constant seed-roll densities. This is usually accomplished by regulating the speed of the feed rolls in response to feedback control signals from the gin stand. The signals are based on monitoring the power consumption of the electric motor driving the gin stand, measuring displacements of the cove board in the seed-roll box, or monitoring the pressure required to drive the hydraulically powered seed-roll agitator.

Cleaning and Extracting Efficiency

The efficiency of a cleaner or extractor depends on many factors, including machine design; cotton moisture level; processing rates; adjustments, speed, and condition of the machine; the amount and nature of trash in the cotton; distribution of cotton across the machine; and the cotton variety. The following discussion on cleaning efficiency is of a general nature and is included only as a point of reference. Although the efficiency values for various machines quoted in this section are all based on actual test data, one should also recognize that these values depend, to some extent, on the specific conditions under which the tests were conducted. This information should, however, give the reader an idea of the general efficiency level of each machine and illustrate major differences among various types of cleaners and extractors.

The total trash removal efficiency of cylinder cleaners is generally low. However, they are not usually used alone but rather in combination with other machines. Studies using both machine-picked and machine-stripped cottons have shown that the total trash removal efficiency of a six-cylinder inclined cleaner with grid rods generally ranges from 10 to 40 percent (Cocke, 1972; Read, 1972; Baker, Boving, and Laird, 1982; Anthony, 1990). These efficiencies, however, were based on the test cotton's total trash content, including burs and sticks, which were not removed to any great extent by the inclined cleaners. The performance of an inclined cleaner is much more impressive when efficiency figures are based entirely on fine-trash levels before and after cleaning. Fine-trash removal efficiencies as high as 50 to 55 percent have been reported for both grid-rod and screen-type inclined cleaners when processing stripped and picked cotton (Laird, Baker, and Childers, 1984; Anthony, 1990).

The cleaning efficiencies of stick machines vary widely, depending on the condition of the seed cotton and on machine design variables. For machine-stripped cotton, a modern commercial stick machine can be expected to remove about 65 percent of the burs, 50 percent of the sticks, and 10 to 35 percent of the fine trash (Baker, 1971; Baker, Boving, and Laird, 1982). The total cleaning efficiency for stripped cotton is normally in the 60 to 65 percent range for the latest models. The efficiency for machine-picked cotton is highly variable because of wide variations in the amount of burs and sticks present in this type of cotton. The total cleaning efficiency can range from about 20 percent for cleanly picked seed cotton to as high as 50 percent for picked cotton containing significant amounts of burs and sticks (Read, 1972; Anthony, 1990).

Extractor-feeders are efficient cleaners. Seed cotton is usually well dispersed when it enters an extractor-feeder, and the feed rate through this machine is often lower than the feed rate of other seed cotton cleaning machinery. Studies, wherein all seed cotton cleaners prior to the extractor-feeder were bypassed, have indicated that the extractor-feeder removes 70 percent of the hulls, 15 percent of the motes, and 40 percent of the remaining trash components and has an overall cleaning efficiency of about 40 percent for machine-picked cotton (Anthony, 1974).

Cleaning efficiencies for sequences of four seed cotton machines consisting of a cylinder cleaner, a stick machine, a second cylinder cleaner, and an extractor-feeder range from 40 to 80 percent, depending on the factors previously discussed. The amount of each type of trash in cotton also varies substantially. Spindle-harvested cotton normally contains 75 to 150 lb of foreign material per bale of seed cotton. Burs and motes represent the majority of the trash. Each type of seed cotton cleaner is designed to remove different types of trash, and any calculation of machine efficiency is predicated on the type of trash involved.

GIN STANDS

The gin stand (see Figure 11.6) is the heart of the ginning system (Columbus et al., 1994). The capacity of the system and the quality and potential spinning performance of the lint depends on the operating condition and adjustment of the gin stand (Wright and Moore, 1977). A number of adjustments are common to nearly all gin stands, and attention to these details will contribute to better operation of the gin and to preservation of lint quality (Anthony, 1985). It is very important that the saw speed be very close to the recommended speed, since as slight a variation as 20 to 25 rpm (revolutions per minute) from the manufacturer's recommendations can make an appreciable difference in performance. Gin manufacturers generally have made their gin stands safe to operate. The gin stand should always be stopped and the power turned off and locked out before making adjustments to the settings or attempting maintenance work.

Ginning Effects on Quality

The high-capacity and super-high-capacity gin stands now on the market are the result of years of research on, and experience with, millions of bales of cotton. They will give good service as long as they are properly adjusted, kept in good condition, and operated at or below design capacity. If gin stands are overloaded, the quality of the cotton may be reduced. Griffin (1977, 1979) showed that short-fiber content increased as the ginning rate was increased above the manufacturer's recommendation. He also found that short-fiber content increased as saw speed increased. Mangialardi, Bargeron, and Rayburn (1988) found that increasing the ginning rate resulted in increases in Uster yarn imperfections. Seed damage can also result from increasing the ginning rate, especially when the seeds are dry. Watson and Helmer (1964) found that variations in ginning rate and seed moisture could cause seed damage ranging from 2 to 8 percent in gin stands. Thus, it is paramount to maintain the gin stand in good mechanical condition, to gin at recommended moisture levels, and to not exceed the capacity of the gin stand or other components of the system.

LINT CLEANING

Lint cleaners were developed specifically for removing leaf particles, motes, grass, and bark that remain in cotton after seed cotton cleaning, extracting, and ginning. They were developed and improved in conjunction with the transition from manual to mechanized harvesting of cotton during the 1950s. Virtually all gins in the United States have lint-cleaning facilities, and over four-fifths of the gins have two or more stages of lint cleaning (Mangialardi et al., 1994). The lint cleaners now being marketed are of two general types, flow-through air type and controlled-batt saw type.

Flow-Through Air Lint Cleaner

The flow-through air lint cleaner, commercially known as the Air Jet/Super Jet, Centrifugal Cleaner, or Super Mote Lint Cleaner, has no saws, brushes, or moving parts. It is usually installed immediately behind the gin stand. Loose lint from the gin stand is blown through a duct within the chamber of the air lint cleaner. Air and cotton moving through the duct change direction abruptly as they pass across a narrow trash-ejection slot. Foreign matter that is heavier than the cotton fibers and not too tightly held by fibers is ejected through the slot by inertial force. The amount of trash taken out by the air jet is controlled by opening and closing this cleaning slot. Fans pull the air and lint to suction condensers, where the air and cotton are separated. A critical factor in the operation of air cleaners is that a vacuum of 2 to 2.5 inches of water must be maintained in the duct on the discharge side of the cleaner. In most cases, the gin stand brush or air blast system provides the pressure to accelerate the air, lint, and trash to a high velocity as it enters the trash-ejection slot. In some cases, particularly in roller gin installations, boost air is added to maintain an air velocity of 10,000 to 12,000 ft/min at the cleaning nozzle.

Flow-through air lint cleaners are less effective in improving the grade of cotton than saw lint cleaners because these air lint cleaners do not comb the fibers. However, air lint cleaners do remove less weight from the bale. Fiber length, fiber strength, and neps are unaffected by the air lint cleaner (Griffin and McCaskill, 1957). This type of cleaner is also commonly used in roller gins processing extra-long-staple (ELS) cotton.

Controlled-Batt Saw Lint Cleaner

The controlled-batt saw cleaner (see Figure 11.7) is now the most common lint cleaner in the ginning industry. Lint from the gin stand

FIGURE 11.7. Unit Controlled-Batt Saw Lint Cleaner with Brush Doffing (Lummus Model 86 or 108 Lint Cleaner)

Source: Courtesy of Lummus Corporation, Savannah, Georgia.

or another lint cleaner is formed into a batt on a condenser screen drum. The batt is then fed through one or more sets of compression rollers, passed between a very closely fitted feed roller and feed plate or bar, and fed onto a saw cylinder. Each set of compression rollers rotates slightly faster than the preceding set and causes some thinning of the batt. The feed roller and plate grip the batt so that a combing action takes place as the saw teeth seize the fibers; the feed plate clears the saw by about one-sixteenth of an inch. The teeth of the saw cylinder convey the fibers to the discharge point. While the fibers are on the saw cylinder, which may be 12 to 24 inches in diameter, they are cleaned by a combination of centrifugal force, scrubbing action between saw cylinder and grid bars, and gravity assisted by an air current.

Multiple Lint Cleaning

Lint cleaners are referred to as either unit or battery (bulk) cleaners, depending on whether they process lint from one or more gin stands. The unit cleaner is located behind a gin stand and receives lint from only that stand. A battery cleaner receives lint from two or more gin stands. Two lint cleaners, either unit or battery, can be placed in series so that the same lint passes through both of them, resulting in what is variously called tandem, dual, or double lint cleaning. The number of stages of saw cleaning refers to the number of saws over which the fibers pass.

Feed Rate

Lint fed to the cleaning machinery at high rates will result in decreased cleaning efficiency and perhaps lower bale values. For efficient cleaning, feed rates should average about 1.0 bale/hr/ft of saw-cylinder length. Many gins handle up to 1.5 bales/hr/ft of saw with no noticeable problem; this rate corresponds to about 7 bales/hr for a 66-inch model lint cleaner and about 10 bales/hr for an 86-inch model cleaner (82-inch saw cylinder).

Some gins have two lint cleaners behind the gin stand and have valves that allow choosing modes of operation. The entire output of the gin stand can flow through first one lint cleaner and then the other, or the lint can be split so that half of the total output flows through each of the lint cleaners. If the lint is split, the speed of the condenser and feed works should be slowed; thus, the combing ratio of each lint cleaner will generally be doubled (Griffin, LaFerney, and Shanklin, 1970; Lummus Industries, Inc., 1984).

The saw cylinder is covered with toothed wire wound in a spiral from one end to the other. Usually there are eight spiral wraps of wire per inch of saw-cylinder length. There are normally 5 to 6 teeth per linear inch of wire, creating a cylinder population of about 45 teeth/in^2 (square inch).

In saw lint cleaners, the uniformity and thickness of the batt and the manner in which it is delivered to the saw teeth are important in the effective operation of the cleaner. These conditions are controlled, to some extent, by the machine design but, in some cases, may be modified by the ginner by regulating the ginning rate, combing ratio (the ratio of the rim speed of the saws to the rim speed of the fluted or final feed roller), or saw speed of the lint cleaner.

Increasing the combing ratio and saw speed usually increases the cleaning efficiency of the cleaner, but may adversely affect the spinning qualities of the cotton. Most lint cleaners normally operate within a combing ratio of about 16 to 28 and at saw-cylinder speeds of 800 to 1,200 rpm (saw-tip speed of 3,350 to 5,700 ft/min). These settings should not be changed without consulting the manufacturer. Higher than normal ratios and saw speeds will sometimes result in high fiber breakage and nep counts, while lower ratios and speeds considerably decrease cleaning efficiency (Mangialardi, 1970; Baker, 1978).

The number of grid bars in a modern lint cleaner may vary from four to eight, depending on the model used. Clearance gauges are used to set the grid bars with respect to the saw cylinder. The nose of the grid bar is set $1/32$ to $1/16$ inch from the saw. If the nose or leading edge of the grid bar is farther away from the saw than the body of the bar, a wedge will form and force sticks into the saw. The leading edges of the grid bars should have the same clearance all the way across the machine with reference to the saw. A keen edge should also be maintained on the nose of the grid bar. Worn bars will waste substantially more fiber than new, sharp bars, decreasing lint turnout (Baker and Brashears, 1988).

Cotton-ginning plants utilize grid-bar air wash to improve lint cleaner efficiency, pick up and remove waste, and reduce air pollution within the gin plant. Air movement across the grid-bar area should average 350 to 400 ft^3/min/ft of grid-bar length. Thus, the air movement would total about 2,700 ft^3/min for an 86-inch model lint cleaner (Lummus Industries, Inc., 1984).

Lint Cleaner Waste

The amount of material removed by lint cleaners varies with harvesting practices and grades of cotton being ginned. When multiple stages of saw lint cleaners are used, the first cleaner removes the most weight. Typical quantities of waste removed by one, two, and three stages of lint cleaning, respectively, are 22, 30, and 36 lb for spindle-picked cotton and 31, 41, and 45 lb for machine-stripped cotton (Baker, 1972; Mangialardi, 1972; Mangialardi, 1988.) Lint cleaner waste from relatively clean cottons contains a greater percentage of lint than that from trashier cottons. Nonlint content of the lint cleaner waste for one, two, and three stages of lint cleaning, respectively, averages about 75 percent, 70 percent, and 67 percent. This material is composed of motes, leaf particles, grass, stems, bark, bracts, and seed-coat fragments.

Bale Value

Lint cleaning generally improves the grade classification of the lint. However, the extent of grade improvement decreases with each successive cleaning. In addition, lint cleaners blend Light-Spotted cottons so that some of these pass into the White grades. Lint cleaners can also decrease the number of bales that are reduced in grade because of grass and bark content. However, they also reduce bale weights and may decrease staple length, thus affecting bale value. In some cases, the net effect of multiple-stage lint cleaning is loss in bale value.

When price spreads between grades are small, the grower can most often maximize bale value by using one saw lint cleaner on early season clean cottons and two stages of lint cleaning on late-season, trashier, or Light-Spotted cottons. This holds for both spindle-harvested and machine-stripped cottons (Looney et al., 1963; Baker, 1972; Mangialardi, 1972). Smooth-leaf variety cottons will generally require one less lint cleaner than hairy-leaf varieties for maximum market value of the bale (Mangialardi, 1988).

PACKAGING LINT COTTON

Bale packaging is the final step in processing cotton at the gin. The packaging system consists of a battery condenser, lint slide, lint feeder, tramper, bale press, and bale-tying mechanism (Anthony, Van Doorn, and Herber, 1994). This system may be supplemented with systems for bale conveying, weighing, and wrapping. The bale press consists of a frame, one or more hydraulic rams, and a hydraulic power system. Tying subsystems may be entirely manual, semi-automated, or fully automated.

Bale presses are described primarily by the density of the bale that they produce, such as low density (flat or modified flat) or universal density (gin or compress). Other descriptions include up-packing, down-packing, fixed box, and doorless. Regardless of description, they all package lint cotton so that it can be handled in trade channels and at the textile mills (Anthony and McCaskill, 1977).

Tramper

The purpose of the tramper (see Figure 11.8) is to pack the lint into the press box. Mechanical and hydraulic trampers are available. Regardless of type, care should be taken to prevent contamination of lint beneath the tramper by hydraulic fluid or grease from the tramper mechanism. Motors that have 10 to 15 hp (horsepower) and are equipped with a fail-safe brake are usually used on mechanical trampers. Motors used on hydraulic trampers vary from 25 to 75 hp.

A rate of about 10 tramper strokes/min is recommended for gins with capacities of 12 to 15 bales/hr. For higher capacity gin plants, both the tramper stroke length and speed are increased to increase the size of the charges and the number of charges per minute. The lint feeder and tramper should have a capacity greater than the gin plant capacity to be able to accommodate the extra lint accumulated in the lint slide during turning of the press boxes.

Bale Press

There are four types of gin presses; each type is named according to the bale it produces—flat, modified flat (bales to be sent for

FIGURE 11.8. Universal Density Press (Lummus Doorless) with a Tramper on the Left and the Compression Ram Beneath the Floor on the Right.

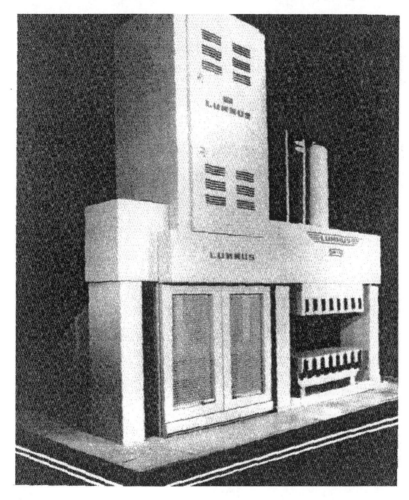

Source: Courtesy of Lummus Corporation, Savannah, Georgia.

recompression to become compress universal density bales), gin standard, and gin universal. One type of universal density cotton bale press is shown in Figure 11.8. Flat-bale presses are being phased out in favor of modified flat-bale presses because bales produced by the latter can be converted to universal bales (55 by

26 to 28 by 20 to 21 inches) without applying side pressure to the bale. Bales from a flat-bale press (27 by 54 inches) can easily be converted to modified-flat-press bales (24 by 54 inches) (National Cotton Council, 1972). The long-range goal is to have only one type of bale—gin universal density. About 96 percent of the bales produced at gins in the United States in 1995 were gin universal or gin standard density.

Bale Ties

After the bale is compressed to a given density or press platen separation, ties are applied around the circumference of the bale to restrain the lint within prescribed dimensions. Bale ties are normally either wire or flat, cold-rolled steel bands, or wire, and are placed at intervals along the length of the bale. Usually, six, eight, or nine ties per bale are used. The weakest point of a bale tie is the connection. To increase the holding capacity of the tie, connections should be positioned near the top or bottom of the crown of the bale. The tie force is considerably less at that point, and the connection is protected because it tends to recess inside the fiber. Flat and modified-flat bales are normally packaged with six 10-gauge steel, interlocking, reusable wire ties having a minimum connected strength of 1,850 lb. Gin standard-density bales are normally packaged with eight steel bands having a connected strength of 2,700 lb or with wire having a connected strength of 3,040 lb. Gin universal bales are normally packaged with eight flat, cold-rolled steel bands having a connected strength of 3,400 lb or with wire having a connected strength of 3,040 lbs. If the wire connections are placed on the top of the bale, connected strength requirements are 2,100 lb. Other configurations are approved and used in limited quantities. Information on currently approved bale-tie materials can be obtained from the National Cotton Council, Memphis, Tennessee.

Bale Covering

Bales should be fully covered, and all bale-covering material should be clean, in sound condition, and of sufficient strength to adequately protect the cotton. The material must not have salt or other corrosive additives and must not contain sisal or other hard fibers or any other material that will contaminate or adversely affect cotton. Recommended bale coverings are published annually by the National Cotton Council, Joint Cotton Industry Bale Packaging Committee, Memphis, Tennessee, a publication that should be consulted for current guidance. Since the adoption of net weight trading, the weight of bale-packaging materials has declined. Tare weights for bagging and ties are published annually by the Commodity Credit Corporation of Washington, DC.

Moisture Change in Cotton Bales

Cotton is usually dried, cleaned, ginned, and packaged at a moisture content well below its eventual equilibrium moisture content in storage. The moisture content and other physical responses of cotton to environmental conditions vary, depending primarily upon the surrounding atmosphere (Hearle and Peters, 1960; Griffin, 1974).

When the humidity of the storage environment is higher than the humidity within the bale, the bale will gain moisture until it equilibrates with the environment. As the bale gains or loses moisture during storage, the weight of the bale changes. The rate of moisture gain is significantly influenced by the density of the cotton bale. Modified-flat bales have a lower density ($14 \ lb/ft^3$) than universal-density bales ($28 \ lb/ft^3$) and gain moisture more rapidly (Anthony, 1982). The rate of moisture gain is also influenced by the type of bale covering. The rates of moisture gain for bales covered with polypropylene, burlap, and extrusion-coated polypropylene are similar. Bales packaged at universal density and covered with a relatively impermeable material, such as polyethylene, will take more than a year to equilibrate with the environment. Bales of cotton ginned at less than 5 percent lint moisture, covered with polypropylene, and stored for over 60 days at high humidities (75 to 80 percent) will absorb about 10 to 15 lb of water, which will increase the lint moisture to 7 or 8 percent.

PNEUMATIC AND MECHANICAL HANDLING SYSTEMS

Cotton gins use the movement of air to propel seed cotton, cottonseed, trash, and lint through conveying pipes much like the wind blows leaves, dirt, and other materials in nature. In gins, pneumatic conveying systems are very important because they are the principal means by which gins move materials from one processing stage to another throughout the entire ginning plant (McCaskill, Baker, and Stedronsky, 1977). When the conveying air is heated or humidified, the pneumatic conveyor becomes a drying or moisture-adding system.

Cotton gins utilize enormous quantities of air for pneumatic conveying. It is not uncommon for a gin to use 150,000 ft^3 or more of air per minute in its various conveying processes (Baker et al., 1994). Since air weighs 0.075 lb/ft^3 at standard conditions, a gin can easily handle 11,250 lb air/min or 675,000 lb/hr. This weight of air per hour usually exceeds the total weight of seed cotton handled per hour by severalfold. Thus, it is understandable why these air systems consume over half of the total power required in a modern cotton gin. It is also obvious why it is so important to maximize the efficiency of pneumatic systems. Efficient pneumatic systems not only lessen the gin's energy costs but also promote smooth, trouble-free ginning with a minimum of downtime.

ABATEMENT OF AIR POLLUTION AND DISPOSAL OF GIN WASTE

Air Pollution

The goal of air pollution control is to minimize deterioration of air resources so that the public can breathe the best-quality air possible. Typically, a construction permit must be obtained from the state air pollution control agency prior to initiation of gin construction. In addition, an operating permit must be approved by this agency prior to operating, and this permit must be kept current (Parnell, Columbus, and Mayfield, 1994). Construction permits are also needed before modifying existing facilities if the modifications may increase emissions. State air pollution agencies usually have authority to administer penalties and fines to violators.

In some states, EPA standards titled "Particulate Emission Factors for Cotton Gins with Controls" (Environmental Protection Agency, 1985), are used for permitting gins. According to EPA emission factors, a gin with controls processing 10 bales/hr should emit no more than 22.4 lb/hr of total particulate, with the major emissions being from the unloading fan (3.2 lb/hr) and from the condenser above the first stage of lint cleaning (8.1 lb/hr). Other information regarding emission factors is available in EPA standards (1975, 1978); Kirk, Wright, and Read (1979); National Enforcement Investigations Center and EPA Region IX (1978); and Parnell and Baker (1973).

Gins processing picked and stripped cotton utilize 7,000 and 8,000 ft^3/min of air/bale/hr rated capacity, respectively. For example, a gin rated at 10 bale/hr of stripped cotton uses approximately 80,000 ft^3/min. Approximately 40 percent of this total (32,000 ft^3/min) is associated with axial-flow fan (condenser) exhausts, which require a relatively large rate of flow to obtain uniform batts of lint for lint cleaners and for the battery condenser at the lint slide above the press. The remaining 60 percent of the air used for pneumatic conveying in a gin is attributed to centrifugal or high-pressure fan exhausts.

Cotton Gin Trash

The amount of seed cotton needed to produce one 480 lb bale of lint is about 1,500 lb for picked cotton and 2,260 lb for stripped cotton. The trash and dust in a bale ranges from 75 to 150 lbs for picked cotton and 700 to 1,000 lbs for stripped cotton. In a typical year, cotton gins in the United States processing spindle-picked cotton will handle 500,000 to 1 million tons of cotton gin trash. Those processing stripped cotton will manage 1 to 1.5 million tons of trash.

Common disposal methods for cotton gin trash include the following: (1) incineration, (2) composting, (3) using it for cattle feed, and (4) direct application to land using spreader trucks. Caution should be used when feeding gin trash to cattle, since pesticide residues may be present in the trash. Cotton gin trash from a crop treated with arsenic acid should never be fed to cattle. Incineration is not allowed in most states and will likely be even less acceptable in the future. The composting of gin trash offers the potential to reduce the negative attributes of "raw" gin trash. If this material is composted properly, the live weed seeds and live disease organisms should be minimized, and the trash volume should be reduced 40 percent. The resulting compost may have a market value.

The most common method of cotton gin trash disposal is direct application to land using spreader trucks. Each ginner using this technology spends approximately $10/ton of trash disposed. This cost is dependent upon the distance the trucks must travel to get to the disposal site. It is becoming difficult in some areas for ginners to acquire sites for trash disposal. At $10/ton to spread trash on the land, the cotton ginning industry would spend $15 to $25 million each year for solid waste disposal. The gin trash, however, does return nutrients to the soil.

COTTONSEED HANDLING AND STORAGE

For every bale of cotton ginned, about 800 lb of seed must be handled and placed in either a temporary or long-term storage facility. Traditionally, gins were only equipped with overhead-type storage houses used to temporarily hold accumulated seed until it could be taken to a cottonseed oil mill. Recently, many gins have built on-site long-term storage facilities that can be filled directly from the gin. Normally 5 to 7 million tons of cottonseed are produced annually in the United States (Willcutt and Mayfield, 1994). About 60 percent is processed into oil, meal, and hulls, 38 percent is fed to livestock, and the remaining 2 percent is used for planting.

Cottonseed quality is determined by the degree of weathering that the seed cotton received in the field, conditions of seed cotton storage, mechanical damage during harvesting and ginning, conditions during storage after ginning, and contamination. Cottonseed stored at excessive moisture will likely develop a rancid flavor and contain an elevated level of free fatty acid. Seed quality tests that determine seed value and suitability for specific uses are available from public and private labs. Oil mill grade reports reflect a composite grade computed from the percentage of oil, ammonia, foreign matter, moisture, and free fatty acid.

Contaminants may either reduce seed value or render the seed completely useless. Contamination may occur during the growing season, harvesting and handling of seed cotton, ginning process, or subsequent handling of seed en route to the final point of use.

Contaminants may be excessive residues from pesticides and defoliants applied to the crop or bulky items such as plant debris, rocks, scrap metal, hardware, plastic, rubber, or pieces of clothing.

Whole, fuzzy cottonseed has some unique characteristics that make it impossible to handle with a tube grain auger. Thus, common grain bins and grain-handling facilities are unsatisfactory for cottonseed storage. Research in 1996 suggested that cottonseed can be coated with a starch material and the handling characteristics greatly improved. Cottonseed has a density of about 20 lb/ft^3 and 1,800 to 2,400 cottonseed are required per pound.

Dry, cool seed and sufficient aeration are the keys to minimizing storage losses and quality deterioration. At ginning, cotton may have a seed moisture content as low as 6 percent or as high as 18 percent, but the majority is within the 10 to 15 percent range. With proper aeration, seed having a moisture content below 19 percent can be successfully stored. However, cottonseed for planting should not be stored with a moisture content above 12 percent.

Aeration

Long-term cottonseed storage facilities must be equipped with an aeration system. Most aeration systems are designed so that the air flows downward through the cottonseed to prevent tunneling and to help minimize moisture accumulation in the top layers of seed. If the airflow is not downward, the top layer will become moist when warm, moist air moves upward into the cooler top layers of seed. Downward airflow also counteracts any natural convectional air movement. The temperature and odor of the exhaust air from the fan can give an indication of cottonseed condition.

A safe airflow rate for aerating cottonseed in flat storage is 10 ft^3/min/ton. At this rate, cottonseed with as much as 15 percent moisture can be safely stored. Drier seed and well-managed storage facilities do not require such a fast airflow rate. Well-designed installations may require an aeration rate of only 5 ft^3/min/ton.

Managing Aeration Systems

Good judgment should be used in selecting cottonseed to store. Cottonseed should not be stored if it is from cotton picked in the early morning after a heavy dew or picked soon after a rain or snow. Seed from cotton that may have gotten wet on a trailer or module should also not be stored. The cottonseed moisture content should not exceed 12 percent if the seed will be stored at temperatures above 60°F; however, the moisture content can be as high as 15 percent if the seed temperature is reduced to 50 to 60°F during storage. Being hygroscopic, cottonseed will absorb moisture from, or give up moisture to, its surrounding air. Therefore, aeration fans should not be operated during high humidity periods or during rain or fog. Ideally, cottonseed in storage should be cooled to 50 to 60°F by selecting cool, dry days to run the fans.

GINNING RECOMMENDATIONS FOR PROCESSING MACHINE-STRIPPED COTTON

Because machine-stripped cotton contains 6 to 10 times as much foreign matter as machine-picked cotton, ginning systems in stripper areas have to be more elaborate than those in picker areas (Baker, 1994). A preferred option is to use an extractor mounted directly on the stripper to remove some of the foreign material before ginning. Otherwise, additional extraction equipment is required to handle large amounts of burs and sticks. Unless removed, burs and sticks will seriously lower gin stand performance and result in unacceptably high trash contents for cottonseed and ginned lint. Provisions also have to be made for removing green bolls and sand from stripped cotton early in the seed cotton-cleaning process.

Even though the cleaning requirements of stripped cottons vary from year to year and, to some extent, from variety to variety, the following array of gin machinery is near optimum for most conditions: green boll separator, air line cleaner, drier, cylinder cleaner, combination bur and stick extractor, drier, cylinder cleaner, stick machine, extractor-feeder, and saw-type gin stand, followed by two stages of saw-type lint cleaning. The machinery recommendations are general in that they are appropriate for most gins handling stripper-harvested cotton at typical conditions. Under such conditions, the recommended machinery arrangement will produce satisfactory lint grades and near-maximum bale values for most cottons.

However, modification of the recommendations may be necessary in some production areas to meet special needs or unusual growing conditions. Cotton containing excessive amounts of foreign matter, particularly stick material, or hard-to-clean varieties of cotton may require more cleaning than can be obtained with the basic machinery arrangement. An extra stick machine or extra cylinder cleaner in the seed cotton-cleaning sequence will usually overcome special problems.

On the other hand, very clean cotton can often be satisfactorily cleaned with less machinery. For this reason, seed cotton extractors and lint cleaners should be provided with bypasses to allow for selection of less cleaning machinery. Generally, ginners should select the minimum amount of machinery required to maximize bale value for the producer. In this way, they can minimize ginning costs and, at the same time, do a better job of preserving the inherent qualities of the cotton fiber.

ROLLER GINNING

Roller-type gins provided the first mechanically aided means of separating lint from seed (Bennett, 1960). The Churka gin consisted of two hard rollers that ran together at the same surface speed, pinching the fiber from the seed and producing about 2 lbs of lint per day (Gillum et al., 1994). During the 1700s, a series of developments using the roller principle followed, but ginning rates remained low. In 1840, Fones McCarthy invented a more efficient roller gin that consisted of a leather ginning roller, a stationary knife held tightly against the roller, and a reciprocating knife that pulled the seed from the lint as the lint was held by the roller and stationary knife. Although the McCarthy gin was a major improvement over the Churka-type gin, machine vibration due to the reciprocating knife, along with attendant gin stand maintenance problems, prohibited high ginning rates.

In the late 1950s and early 1960s, a rotary-knife roller gin was developed by the U.S. Department of Agriculture, Agricultural Research Service's Southwestern Cotton Ginning Research Laboratory, gin manufacturers, and private ginneries. This gin is currently the only roller-type gin used in the United States. The roller and stationary knife of this gin are similar to those of the McCarthy gin, but a rotary knife is used instead of a reciprocating knife. The rotary knife vibrates less and is more efficient than the reciprocating knife, which wasted time during each backstroke. The rotary-knife gin allows faster ginning rates than the McCarthy gin allowed. During the 1989-1990 season, there were forty-nine roller-ginning plants in the United States—twenty-eight in Arizona, eight in Texas, six each in California and New Mexico, and one in Mississippi (Hughs and Gillum, 1991).

Since only 6 percent (Supima Association of America, 1989; U.S. Department of Agriculture, 1989) of the world cotton production is ELS cotton, a specialty market exists for this type of cotton, and textile mills pay a premium for it. Pima cotton, the only ELS cotton grown in the United States, successfully competes with other ELS cottons worldwide. Pima cotton belongs to the species *Gossypium barbadense* (Niles and Feaster, 1984). Pima cotton fiber is long ($1^7/_{16}$-inch staple length), strong (35 g/tex strength), fine (3.8 micronaire) and is used in the finer apparel.

THE CLASSIFICATION OF COTTON

Historically, the U.S. Department of Agriculture (USDA) has classed (graded) about 97 percent of the U.S. cotton production for growers each year (Moore, 1994). Since 1981, growers have paid for the service. The classification system has undergone significant change, moving from heavy reliance on the human senses to the utilization of precision instruments that analyze more quality factors with greater accuracy. Currently, some USDA quality determinations are still made by classers, but the majority are determined by High-Volume Instruments, commonly referred to as HVI.

Classer Determinations—Upland Cotton

Color Grade

There are twenty-five color grades. Fifteen of these grades are represented in physical form by samples prepared and maintained by the USDA. The remaining standards for color are descriptive. The range of each color grade having a physical standard is represented by six samples placed adjacent to one another in a standards box. For practical considerations, the color and leaf grade standards are contained in the same box. For instance, the standards box containing the Strict Low Middling color grade also contains the size and amount of leaf that would be described as leaf grade 4. Each descriptive standard provides a description for cotton that lies above, below, or between certain physical standards.

Color grades fall into five different color groups as follows: White, Light Spotted, Spotted, Tinged, and Yellow Stained. White color grades are grouped into the following: Good Middling, Strict Middling, Middling, Strict Low Middling, Low Middling, Strict Good Ordinary, and Good Ordinary. The color is affected, to great extent, by weather and length of exposure to weather conditions after the bolls open. It may also be affected by varietal characteristics and by harvesting and ginning practices. When Upland cotton opens normally, it has a bright, white color. Abnormal color indicates a deterioration in quality. Continued exposure to weather and the action of microorganisms can cause the white cotton to lose its brightness and to become darker in color. When plant growth is stopped prematurely by frost, drought, or other weather conditions, the cotton may have a yellow color that varies in intensity. Cotton may also become discolored or spotted by insects or fungi.

Leaf Grade

Leaf grade describes the amount of leaf (plant parts) content in the cotton. There are seven leaf grades, and all are represented by physical standards. In the spinning industry, leaf material is viewed as waste, creating an additional cost factor associated with its removal. Leaf content is affected by different harvesting methods, harvesting conditions, and ginning procedures. The amount of leaf material remaining in the lint after ginning depends on the amount present in the seed cotton before ginning and on the type and the amount of cleaning and drying equipment used during ginning. Even with the most careful harvesting and ginning methods, a small amount of leaf will remain in the cotton lint.

Preparation

Preparation is a measure of the degree of roughness or smoothness of the ginned lint cotton. As a general rule, smooth cotton has less spinning waste and produces a smoother and more uniform yarn than rough cotton. Various methods of harvesting, handling, and ginning can produce readily apparent differences in preparation. Because of improvements in equipment and practices, abnormal preparation now occurs in less than one-half of 1 percent of the crop during harvesting and ginning. Abnormal preparation is noted in the remarks of the classification data.

Extraneous Matter

In a cotton sample, extraneous matter is any substance that is not cotton fiber or leaf material and that is not discernible in the official cotton standards. Examples of extraneous matter are bark, grass, spindle twists, dust, and oil. The classer determines whether the amount of extraneous matter present in the sample is sufficient enough to be noted.

HVI Determinations—Upland Cotton

Fiber Length

Fiber length is a varietal characteristic. The length decreases when temperatures during the early stages of fiber development exceed the optimum for the variety or when moisture is limited. Fiber length is measured by passing a small tuft of parallel fibers through a sensing point of the HVI system. The tuft is formed by grasping the fibers with a clamp and paralleling them by combing and brushing. Reported length is the average or mean length of the longest one-half of the fibers (upper half mean length). Results are reported to the nearest one-hundredth of an inch.

Length Uniformity

Length uniformity is the ratio of the average or mean length of the fibers to their upper half mean length and is expressed as a percentage. The same tuft of fiber used for length measurement is used to determine length uniformity. If all fibers in a sample are the same length, the length uniformity index would be 100.

Fiber Strength

Fiber strength is determined by the HVI system on the same tuft of fiber used for the length measurement. The following is a general description of fibers of different strengths in g/tex: very strong (30 and over), strong (26 to 29), intermediate (24 to 25), weak (21 to 23), very weak (20 and below). Fiber strength is a varietal characteristic and is less influenced by adverse growing conditions than are length and micronaire.

Micronaire

The term micronaire refers to an airflow measurement that indicates fiber fineness and maturity. Micronaire is determined by passing air compressed to a standard volume through a cotton specimen of standard weight and standard volume. The volume of airflow through the specimen is expressed as the micronaire, which may be referred to as the mike reading, or simply mike. Optimum micronaire, which is defined as 3.5 to 4.9, is dependent upon many things, including the variety of the cotton and the relative importance of strength and appearance in a yarn or fabric. Different cotton varieties vary in micronaire at full maturity.

Fiber fineness is a varietal characteristic that is also affected by growing conditions in the latter stages of fiber development. Favorable growing conditions result in fully mature fibers and high mike readings. Unfavorable conditions, such as lack of moisture, early freeze, or any other conditions that interrupt plant processes, will result in immature fibers and low mike readings.

Color

Presently, the USDA classification system includes a determination of color by the classer and also by HVI. The HVI color determinations are in terms of grayness (measured as Rd) and yellowness (measured as +b). Grayness (or percent reflectance) indicates the lightness or darkness of the sample. Cotton Rd values are usually within the 48 to 82 range. Yellowness indicates the amount of yellow coloration in the sample and is usually within the 5.0 to 17.0 range. Normally, opened cotton will have an Rd of 70 or higher and a +b of 9.0 or lower. The various combinations of grayness and yellowness can be converted into color values, plotting the Rd and +b values on an official color chart.

Leaf

In addition to the classer's leaf grade, USDA provides an instrument determination of the trash content in the sample. The trash content measured by the HVI system is determined by scanning the sample surface and recording the particles present. Results are reported as the percentage of the sample surface covered by nonlint particles. The maximum is less than 5.0 percent. The following lists the average trash meter readings for the different grades of Upland cotton: Strict Middling (0.1), Middling (0.2), Strict Low Middling (0.4), Low Middling (0.7), Strict Good Ordinary (1.1), and Good Ordinary (1.5).

American Pima Grades

American Pima grade standards are also represented in physical form. There are six American Pima grades, numbered 1 through 6. American Pima and Upland grade standards differ. American Pima cotton has a deeper yellow color than Upland cotton. The leaf content of American Pima standards is unique to this cotton and does not match that of Upland standards. The preparation is very different from the preparation for Upland standards, since American Pima cotton is normally ginned on roller gins and is more stringy and lumpy. Upland cotton is usually cleaned with saw-type lint cleaners that produce a smooth, blended, combed sample. Roller-ginned cotton is usually cleaned with an air or cylinder-type cleaner and is rough in appearance.

CONCLUSION

Cotton gins convert seed cotton into useable products such as cotton fiber, seed, and plant parts. Many specialized machines are used to clean and dry the cotton as well as to remove the fiber from the seed and to package the fiber. Cotton gins are crucial to the production of sufficient quantities of high-quality fiber to meet worldwide needs.

REFERENCES

Anthony, W.S. (1974). Development and evaluation of a small-scale cotton ginning system. *U.S. Department of Agriculture, Agricultural Research Service, ARS-S-36.* Washington, DC: U.S. Government Printing Office, 9 pp.

Anthony, W.S. (1982). Effect of bale covering and density on moisture gain of cotton bales. *Cotton Ginners' Journal and Yearbook* 50(1): 44-48.

Anthony, W.S. (1985). The effect of gin stands on cotton fiber and seed. *Cotton Gin and Oil Mill Press* 86(16): 14-18.

Anthony, W.S. (1990). Performance characteristics of cotton ginning machinery. *Transactions of the ASAE* 33(4): 1089-1098.

Anthony, W.S. (1994). Overview of the ginning process. In *Cotton Ginners Handbook, U.S. Department of Agriculture, Agricultural Handbook 260,* Eds. W.S. Anthony and W.D. Mayfield. Washington, DC: U.S. Government Printing Office, pp. 43-46.

Anthony, W.S. and O.L. McCaskill (1977). Packaging of lint cotton. In *Cotton Ginners Handbook, U.S. Department of Agriculture, Agricultural Handbook 503.* Washington, DC: U.S. Government Printing Office, pp. 43-52.

Anthony, W.S., W.R. Meredith, Jr., J.R. Williford, and G.J. Mangialardi Jr. (1988). Seedcoat fragments in ginned lint: The effect of varieties, harvesting and ginning practices. *Textile Research Journal* 58(2): 111-116.

Anthony, W.S., D.W. Van Doorn, and D. Herber (1994). Packaging lint cotton. In *Cotton Ginners Handbook, U.S. Department of Agriculture, Agricultural Handbook 260,* Eds. W.S. Anthony and W.D. Mayfield. Washington, DC: U.S. Government Printing Office, pp. 119-142.

Baker, R.V. (1971). Comparative performances of a stick machine and a bur machine on machine-stripped cotton. *U.S. Department of Agriculture, Technical Bulletin 1437.* Washington, DC: U.S. Government Printing Office, 16 pp.

Baker, R.V. (1972). Effects of lint cleaning on machine-stripped cotton: A progress report. *Cotton Gin and Oil Mill Press* 73(21): 6-7, 22.

Baker, R.V. (1978). Performance characteristics of saw-type lint cleaners. *Transactions of the ASAE* 21(6): 1081-1087, 1091.

Baker, R.V. (1994). Ginning recommendations for processing machine-stripped cotton. In *Cotton Ginners Handbook, U.S. Department of Agriculture, Agricultural Handbook 260,* Eds. W.S. Anthony and W.D. Mayfield. Washington, DC: U.S. Government Printing Office, pp. 242-243.

Baker, R.V., W.S. Anthony, and R.M. Sutton (1994). Seed cotton cleaning and extracting. In *Cotton Ginners Handbook, U.S. Department of Agriculture, Agricultural Handbook 260,* Eds. W.S. Anthony and W.D. Mayfield. Washington, DC: U.S. Government Printing Office, pp. 69-90.

Baker, R.V., P.A. Boving, and J.W. Laird (1982). Effects of processing rate on the performance of seed cotton cleaning equipment. *Transactions of the ASAE* 25: 187-192.

Baker, R.V. and A.D. Brashears (1988). Influence of grids on lint cleaner performance. ASAE Paper No. 88-1621. St. Joseph, MI: American Society of Agricultural Engineers, 12 pp.

Baker, R.V., E.P. Columbus, R.C. Eckley, and B.J. Stanley (1994). Pneumatic and mechanical handling systems. In *Cotton Ginners Handbook, U.S. Department of Agriculture, Agricultural Handbook 260,* Eds. W.S. Anthony and W.D. Mayfield. Washington, DC: U.S. Government Printing Office, pp. 143-171.

Bennett, C.A. (1960). Roller Cotton Ginning Developments. Dallas, TX: Texas Cotton Ginners' Association, 90 pp.

Bennett, C.A. (1962). Cotton Ginning Systems and Auxiliary Developments. Dallas, TX: Texas Cotton Ginners' Association, 101 pp.

Cocke, J.B. (1972). Effect of processing rates and speeds of cylinder-type cleaners on ginning performance and cotton quality. *U.S. Department of Agriculture, Agricultural Research Service, ARS 42-199.* Washington, DC: U.S. Government Printing Office, 9 pp.

Columbus, E.P., D.W. Van Doorn, B.M. Norman, and R.M. Sutton (1994). Gin stands. In *Cotton Ginners Handbook, U.S. Department of Agriculture, Agricultural Handbook 260,* Eds. W.S. Anthony and W.D. Mayfield. Washington, DC: U.S. Government Printing Office, pp. 90-102.

Environmental Protection Agency (1975). Source assessment document No. 27, cotton gins. *Environmental Protection Agency EPA-600/2-78-004a,* 66 pp.

Environmental Protection Agency (1978). Emission test report, Westside Farmer's Cooperative Gin #5, Tranquility, CA. *Environmental Protection Agency PN3370-2-D,* 25 pp.

Environmental Protection Agency (1985). Compilation of air pollutant emission factors. Volume 1, stationary and area sources, Fourth Edition. *Environmental Protection Agency AP-42,* pp. 6.3-1-6.3-6.

Franks, G.N. and C.S. Shaw (1959). Stick remover for cotton gins. *U.S. Department of Agriculture, Marketing Research Report 852,* 30 pp.

Garner, W.E. and R.V. Baker (1977). Cleaning and extracting. In *Cotton Ginners Handbook, U.S. Department of Agriculture, Agricultural Handbook 503.* Washington, DC: U.S. Government Printing Office, pp. 18-29.

Gillum, Marvis N., D.W. Van Doorn, B.M. Norman, and Charles Owen (1994). Roller Ginning. In *Cotton Ginners Handbook, U.S. Department of Agriculture, Agricultural Handbook 260,* Eds. W.S. Anthony and W.D. Mayfield. Washington, DC: U.S. Government Printing Office, pp. 244-258.

Griffin, A.C. Jr. (1974). The equilibrium moisture content of newly harvested cotton fibers. *Transactions of the ASAE* 17(2): 327-328.

Griffin, A.C. Jr. (1977). Quality control with high capacity gin stands. *Cotton Ginners' Journal and Yearbook* 45(1): 25, 28-29.

Griffin, A.C. Jr. (1979). High-capacity ginning and fiber damage. *Textile Research Journal* 49(3): 123-126.

Griffin, A.C., P.E. LaFerney, and H.E. Shanklin (1970). Effects of lint-cleaner operating parameters on cotton quality. *U.S. Department of Agriculture, Marketing Research Report 864*, 15 pp.

Griffin, A.C. and O.L. McCaskill (1957). Tandem lint cleaning-air-saw cylinder combination. *Cotton Gin and Oil Mill Press* 58(6): 25, 52-53.

Hearle, J.W.S. and R.H. Peters (1960). *Moisture in Textiles*. New York: Interscience Publishers, Inc.

Hughs, S.E. and M.N. Gillum (1991). Quality effects of current roller-gin lint cleaning. *Transactions of the ASAE* 7(6): 673-676.

Hughs, S.E., G.J. Mangialardi Jr., and S.G. Jackson (1994). Moisture control. In *Cotton Ginners Handbook, U.S. Department of Agriculture, Agricultural Handbook 260*, Eds. W.S. Anthony and W.D. Mayfield. Washington, DC: U.S. Government Printing Office, pp. 58-68.

Kirk, I.W., T.E. Wright, and K.H. Read (1979). Particulate emissions from commercial cotton ginning operations. *Transactions of the ASAE* 22: 894-898.

Laird, J.W., R.V. Baker, and R.E. Childers (1984). Screen and grid dimensions and feeding method effects on performance of a cylinder cleaner at the gin. *Proceedings of the Beltwide Cotton Production Research Conference*. Memphis, TN: National Cotton Council, pp. 159-160.

Laird, J.W., B.M. Norman, S. Stuller, and P. Bodovsky (1994). Seed cotton unloading systems. In *Cotton Ginners Handbook, U.S. Department of Agriculture, Agricultural Handbook 260*, Eds. W.S. Anthony and W.D. Mayfield. Washington, DC: U.S. Government Printing Office, pp. 46-58.

Lalor, W.F., M.H. Willcutt, and R.G. Curley (1994). Seed cotton storage and handling. In *Cotton Ginners Handbook, U.S. Department of Agriculture, Agricultural Handbook 260*, Eds. W.S. Anthony and W.D. Mayfield. Washington, DC: U.S. Government Printing Office, pp. 16-26.

Looney, Z.M., L.D. LaPlue, C.A. Wilmot, W.E. Chapman Jr., and F.E. Newton (1963). Multiple lint cleaning at cotton gins: Effects on bale value, fiber properties, and spinning performance. *U.S. Department of Agriculture, Marketing Research Report 601*, 53 pp.

Lummus Industries, Inc. (1984). *Service Manual: Model 86 and 108 Lint Cleaners. No. 0801*. Columbus, GA: Lummus Industries, Inc. 26 pp.

Mangialardi, G.J. Jr. (1970). Saw-cylinder lint cleaning at cotton gins: Effects of saw speed and combing ratio on lint quality. *U.S. Department of Agriculture, Technical Bulletin 1418*, 73 pp.

Mangialardi, G.J. Jr. (1972). Multiple lint-cotton cleaning: Its effect on bale value, fiber quality, and waste composition. *U.S. Department of Agriculture, Technical Bulletin 1456*, 69 pp.

Mangialardi, G.J. Jr. (1985). An evaluation of nep information at the cotton gin. *Textile Research Journal* 55(12): 761-767.

Mangialardi, G.J. Jr. (1988). Multiple lint cleaning of Midsouth cottons for maximum market value. In *Annual Report-Fiscal Year 1988, U.S. Cotton Ginning Laboratory,* Stoneville, MS: U.S. Department of Agriculture, Agricultural Research, pp. 303-329.0987.

Mangialardi, G.J. Jr., R.V. Baker, D.W. Van Doorn, B.M. Norman, and R.M. Sutton (1994). Lint cleaning. In *Cotton Ginners Handbook, U.S. Department of Agriculture, Agricultural Handbook 260,* Eds. W.S. Anthony and W.D. Mayfield. Washington, DC: U.S. Government Printing Office, pp. 102-119.

Mangialardi, G.J. Jr., J.D. Bargeron III, and S.T. Rayburn Jr. (1988). Gin stand feed rate effects on cotton quality. *Transactions of the ASAE* 31: 1844-1850, 1894.

Mayfield, W.D., W.S. Anthony, R.V. Baker, and S.E. Hughs (1994). Effects of gin machinery on cotton quality. In *Cotton Ginners Handbook, U.S. Department of Agriculture, Agricultural Handbook 260,* Eds. W.S. Anthony and W.D. Mayfield. Washington, DC: U.S. Government Printing Office, pp. 237-240.

Mayfield, W.D., R.V. Baker, S.E. Hughs, and A.C. Griffin (1983). Ginning cotton to preserve quality. *U.S. Department of Agriculture, Extension Service, Program Aid 1323,* 11 pp.

McCaskill, O.L., R.V. Baker, and V.L. Stedronsky (1977). Fans and piping. In *Cotton Ginners Handbook, U.S. Department of Agriculture, Agricultural Handbook 503,* pp. 57-67.

Moore, J.F. (1994). The classification of cotton. In *Cotton Ginners Handbook, U.S. Department of Agriculture, Agricultural Handbook 260,* Eds. W.S. Anthony and W.D. Mayfield. Washington, DC: U.S. Government Printing Office, pp. 287-292.

National Cotton Council, Universal Bale Committee (1972). A Universal Bale for Universal Benefits. Memphis, TN: National Cotton Council, 7 pp.

National Enforcement Investigations Center and EPA Region IX (1978). Cotton gin emissions tests, Marana gin, producers cotton oil company. *National Enforcement Investigations Center and EPA Region IX EPA-330/2-78-008,* 44 pp.

Niles, G.A. and C.V. Feaster (1984). *Cotton, ASA Monograph 24.* Madison, WI: American Society of Agronomy, pp. 201-231.

Parnell, C.B. Jr., and R.V. Baker (1973). Particulate emissions of a cotton gin in the Texas stripper area. *U.S. Department of Agriculture, Production Research Report 149.* Washington, DC: U.S. Government Printing Office, 18 pp.

Parnell, C.B. Jr., E.P. Columbus, and W.D. Mayfield (1994). Abatement of air pollution and disposal of gin waste. In *Cotton Ginners Handbook, U.S. Department of Agriculture, Agricultural Handbook 260,* Eds. W.S. Anthony and W.D. Mayfield. Washington, DC: U.S. Government Printing Office, pp. 172-194.

Pendleton, A.M. and V.P. Moore (1967). Ginning cotton to preserve quality. *U.S. Department of Agriculture, Extension Service, ESC-560.* Washington, DC: U.S. Government Printing Office, 19 pp.

Read, K.H. (1972). Cylinder cleaner speed influences cleaner efficiency. *Cotton Gin and Oil Mill Press* 73(24): 6-7.

Supima Association of America (1989). *Supima Newsletter.* Phoenix, AZ: Supima Association of America, June 26, 4 pp.

U.S. Department of Agriculture (1989). *Agricultural Statistics.* Washington, DC: U.S. Government Printing Office, p. 64.

Watson, H. and J.D. Helmer (1964). Cottonseed quality as affected by the ginning process—A progress report. *U.S. Department of Agriculture, Agricultural Research Service, ARS 42-107,* 8 pp.

Willcutt, M.H. and W.D. Mayfield (1994). Cottonseed handling and storage. In *Cotton Ginners Handbook, U.S. Department of Agriculture, Agricultural Handbook 260,* Eds. W.S. Anthony and W.D. Mayfield. Washington, DC: U.S. Government Printing Office, pp. 195-214.

Wright, T.E. and V.P. Moore (1977). Gin stands. In *Cotton Ginners Handbook, U.S. Department of Agriculture, Agricultural Handbook 503.* Washington, DC: U.S. Government Printing Office, pp. 29-35.

Chapter 12

Fiber-to-Fabric Engineering: Optimization of Cotton Fiber Quality

Yehia E. El Mogahzy

INTRODUCTION

The General Principle of the Fiber-to-Fabric Conversion System

The cotton fiber-to-fabric conversion system consists of several operations that must work in a coordinated and integrated fashion to achieve two main objectives: (1) good processing performance of material and (2) high-quality levels of end product at the lowest cost possible. As can be seen in Figure 12.1, two subsystems are used to convert fibers into fabrics: (1) the fiber-to-yarn conversion system and (2) the yarn-to-fabric conversion system. Although the two subsystems are fully integrated to produce the desired fabric, each subsystem is considered an independent operational phase with respect to preparation and processing criteria.

In the fiber-to-yarn conversion system, fibers are converted into yarn through a series of processes that, in principle, may be divided into two primary stages (see Figure 12.2): (1) spinning preparation and (2) spinning. In the first stage, a tightly packed mass of staple fibers (fiber bale) is gradually opened, cleaned, mixed, and finally drafted or attenuated to form a fiber strand that is ready to be spun. In the second stage, the fiber strand is further attenuated to obtain the desirable yarn thickness, and some form of fiber-to-fiber cohesion is introduced to provide the necessary strength to the yarn.

FIGURE 12.1. The Fiber-to-Fabric Conversion System

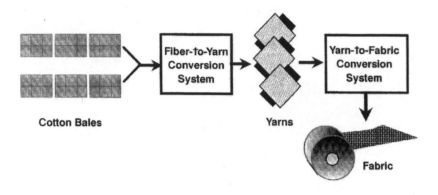

FIGURE 12.2. The Fiber-to-Yarn Conversion System

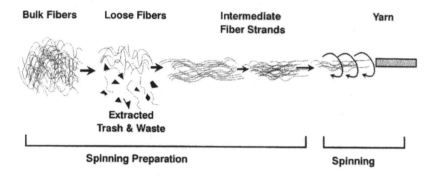

The principle of the yarn-to-fabric conversion system depends on the type of fabric to be produced (woven or knitted). In the case of woven fabric, two sets of yarns must be interlaced to produce the fabric (see Figure 12.3): warp yarns and filling (or weft) yarns. Warp yarns require preparation prior to weaving to enhance their surface integrity by adding a special chemical film (the sizing operation) and to form a yarn beam suitable for mounting on the weaving machine (the loom). Filling yarns, on the other hand, are fed directly to the weaving machine.

FIGURE 12.3. The Yarn-to-Fabric Conversion System

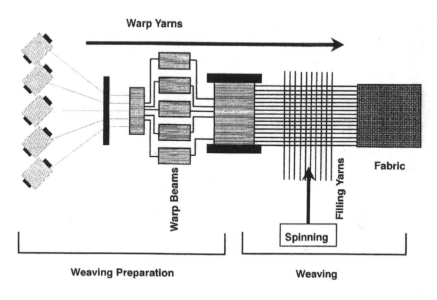

Fiber-to-Fabric Engineering: Basic Concepts

The concept of fiber-to-fabric engineering (FFE) is based on integrating the two systems described previously so that a fabric can be designed in view of the relationships between the desirable levels of fabric characteristics and the corresponding fiber and yarn characteristics. The key objective of an FFE program is to produce an optimum quality of fabric, that is, the desirable fabric quality level at the lowest cost possible.

The integration process relies mainly on the available information or the database associated with raw material, processing parameters, and end products. Fortunately, such databases are available even in small textile mills. The question is whether the data are scattered or trackable. Traditionally, data collected from the different zones of the fiber-to-fabric conversion system have provided independent and isolated clusters of information that are often difficult to integrate or relate to one another. In an FFE program, data should be trackable and organized to allow for maximum utilization of the information provided by the numbers.

An FFE program should consist of three basic elements that collectively assist in the production of an optimum quality product. These elements include the following:

1. Optimization of cotton fiber quality through proper bale management and fiber selection
2. Optimization of yarn quality through the proper setting of different machines in the fiber-to-yarn conversion system with respect to fiber/machine interaction and fiber/yarn relationships.
3. Optimization of fabric quality through the proper setting of different machines in the yarn-to-fabric conversion system with respect to yarn/machine interaction and fiber/yarn/fabric relationships (see Figure 12.4)

Among the previous three basic elements, the first element is undoubtedly the most critical one. The importance of this element is attributed to the following reasons:

FIGURE 12.4. Fiber-to-Fabric Engineering: Basic Elements

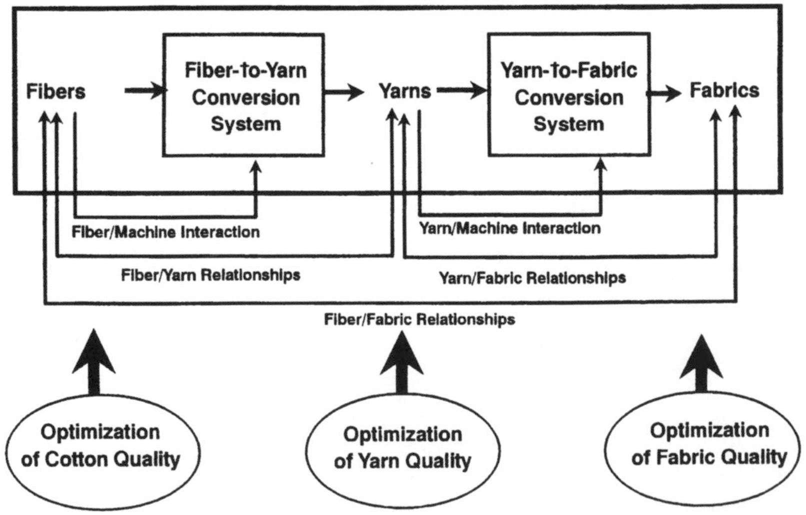

1. Optimization of cotton fiber quality will greatly assist in optimizing yarn and fabric quality.
2. Proper bale management and fiber selection will result in feeding a uniform cotton mix to the textile process and help to prevent several problems that may be encountered later in the process, when the option of prevention is replaced by the more expensive option of correction.
3. Many yarn and fabric quality problems are attributed to improper fiber selection. These include poor strength, excessive failure of yarn during spinning, and poor uniformity of yarn and fabric.

Accordingly, my focus in this chapter will be on the first element, that is, the optimization of cotton fiber quality through proper bale management and fiber selection. Specific procedures of this element are as follows:

1. Developing a systematic approach for determining the technological value of cotton fibers
2. Establishing reliable procedures of cotton bale management and fiber selection
3 Developing reliable relationships between the cotton cost profiles and corresponding profiles of quality and manufacturing cost

DETERMINING THE TECHNOLOGICAL VALUE OF COTTON

Different Views of Cotton Value

In a typical marketing system, cotton prices may be affected by several factors, including laws of supply and demand, regional factors, fiber attributes, and possible chaotic changes from one crop to another (e.g., disease, pests, and weather). In any situation, however, fiber attributes represent the primary factor in determining the premiums and discounts associated with the value of a certain cotton. Traditionally, these attributes have consisted of grade, staple length, micronaire, and extraneous matter (any substance other than fiber or leaf).

In recent years, the issue of what fiber attributes should constitute the market value of cotton has been reexamined. New ideas (El Mogahzy, Broughton, and Lynch, 1990; Deussen and Farber, 1994) have been presented in view of the revolutionary developments in spinning and weaving technologies and in the powerful fiber information systems, such as the High-Volume Instrument (HVI) and the Advanced Fiber Information System (AFIS). These efforts revealed that the market value of cotton has yet to reflect its technological worth.

Market analysis (Chen and Ethridge, 1996; Hudson and Ethridge, 1996) performed to determine the current trends of cotton purchasing in the United States indicated that traditional cotton attributes, particularly leaf grade and color grade, still dominate the cotton market. In addition, textile mills have not been paying strength premiums for high-strength cotton, nor discounting for low-strength cotton. More ironically, the analysis revealed that the pricing structures of cotton at the user's end of the market appear to be substantially different between the western and south central regions for all fiber attributes. These regional effects indicate that historical repetition, art, and experience are still greatly implemented among textile manufacturers in cotton purchasing.

The establishment of a market value of cotton that is truly representative of the actual technological worth of cotton faces three main challenges. The first challenge arises from the substantial differences in views of what constitutes fiber quality as expressed by different organizations involved in the cotton industry. The second challenge involves the impact of the current market structure on the value of cotton. The third challenge is the lack of a systematic model to scientifically evaluate the value of cotton. This section deals specifically with the third challenge. Because of the strong interrelationship among the three challenges, I will briefly discuss the impacts of the other two challenges.

From a cotton producer's viewpoint, the primary attribute that determines the value of cotton is the yield per acre. This means that any breeding improvement of a particular fiber quality parameter, to satisfy technological needs, will have to be achieved without impairing the yield per acre. According to Meredith (1995), with the exception of micronaire, the association between the yield per acre and most fiber properties is generally weak. The micronaire reading reflects both the maturity and the lint percentage and, thus, has a correlation of about 0.70 with the yield per acre. The author also indicated that in any breeding situation, when one feature (such as fiber quality) receives an increase in breeding priorities, progress in other features (such as yield) declines. According to Deussen and Farber (1994), certain cotton varieties with superior quality traits, but somewhat lower yields, fall victim to the fact that the current marketing system does not compensate for reduced yield with a premium on desirable fiber properties.

Another fiber attribute that is highly emphasized by the cotton producer is fiber appearance and cleanliness. This emphasis is primarily driven by the significant discount points associated with cottons of poor appearance (reflected in leaf and color grades) and high levels of extraneous matter. Accordingly, cotton undergoes extensive cleaning during ginning (two to three passes of lint cleaning). Since trash content is a nonfibrous material, it is worthless to the textile manufacturer. In addition, it has adverse effects on processing performance and end-product quality.

Although cotton users (textile manufacturers) fully understand the impact of trash on the value of cotton, they disagree with the extent of cotton cleaning during ginning. Studies in this regard (El Mogahzy and Kearny, 1992; Selker, 1992; Leifeld, 1994) indicated that excessive cleaning in the gin can result in fiber damage, seed-coat fragments, and fine trash. None of these quality problems is accounted for in the cotton's current market value. The studies also indicated that there should be a coordinated strategy between gin and mill cleaning. In support of this idea, Liefeld (1994) recommended a distribution of cotton cleaning whereby ginning would result in about 4 percent trash content in the bale, and the mill cleaners would reduce the levels in the rest from 4 percent to 0.4 percent and produce cards with levels from 0.4 percent to 0.04 percent. The essence of this recommendation is to achieve equal levels of cleaning efficiency in the three areas (i.e., 90 percent) and to provide gradual cleaning.

In addition to the attributes discussed previously, extraneous matter has assumed a solid place in the market criteria, particularly in recent years. As indicated earlier, any substance in cotton other than fiber or leaf is considered extraneous matter. The amount of extraneous matter in cotton is determined by the cotton classer. Two levels are usually reported: heavy (level 1) and light (level 2). These two levels are used to characterize the extent of preparation; the amount of bark, grass, seed-coat fragments, and oil; and spindle-twist intensity. Although the method used to characterize extraneous matter is still subjective, consideration of this type of attribute in market evaluation is certainly critical due to the obvious consequences of the presence of this matter on processing performance.

The previous discussion indicates that cotton producers are mainly motivated by incentives. As long as higher yield and excessively cleaned cottons provide overwhelming incentives over particular fiber traits, the cotton producer will be reluctant to breed higher-quality cottons at the expense of these motivational factors.

The cotton user (the textile manufacturer) views fiber quality from a different standpoint than the producer. In the marketplace, the primary interest of the cotton user is value discounts. In other words, cotton buyers are often motivated by the lowest possible price of cotton.

Traditionally, purchasing discount cottons has been based on the assumption that these cottons can be mixed with high-quality cottons to reach a desirable average quality level. The economical and technological feasibility of this assumption is quite clear, particularly when inherent fiber characteristics such as length, fineness, and strength are considered. In the case of induced attributes, such as short-fiber content, fine trash, and seed-coat fragments, the desirable average target should always be zero; this target value cannot be achieved by mixing poor- and high-quality cottons. Only the impact of these induced attributes on manufacturing cost (spinning ends down, filling stops, and yarn contaminants) should be considered in purchasing cottons.

As we approach a new century, the current views of both cotton producers and cotton users will inevitably change. The new era will be based solely on information technology; fiber-testing techniques will provide expert information, and machines will be operated by computers and artificial intelligence, with accurate fiber information playing a more critical role. These trends are already obvious in today's modern technology, and they will continue to develop.

Obviously, new developments bring about new constraints on fiber quality. In the textile process, numerous examples illustrate these constraints. For instance, the production of medium to fine yarns on rotor spinning requires levels of fiber fineness, strength, and elongation that are superior to the traditional levels; air-jet spinning, although superior in producing 100 percent polyester yarns of a wide range of yarn count, is still incapable of producing 100 percent cotton yarns because of the limited fiber length and the level of fine trash that is intolerable by this type of spinning. The performance quality levels of end products such as denim, wrinkle-free cotton textiles, and undershirts are directly associated with the quality of the fibers from which these products are made.

In the following sections, I discuss a systematic approach by which the cotton value can be determined with respect to the technological impact of cotton on the textile process. This approach is based on developing a technological value model of cotton fibers.

The Technological Value Model: Structural Elements

The main structural elements of the proposed technological value model are the model input variables, the relative association analysis, the model anchored parameters, and the technological premium/ discount index (TPDI). These elements, illustrated in Figure 12.5, will be discussed in the following sections. It should be pointed out that each of these elements actually represents a familiar task that a textile mill may undertake on a regular or occasional basis, using objective or subjective methods. It is the integration of these tasks, however, that provides the basis for developing the technological value model.

Model Input Variables

The model input variables represent all the important parameters that will be used for determining the technological value of fibers. These parameters may vary, depending on the type of process and the type of end product considered in the analysis. Typically, inpu variables will include fiber properties, yarn properties, processing parameters, and fabric properties. Examples of these variables are given in Tables 12.1 through 12.4. Also included in these tables are typical values for these variables.

FIGURE 12.5. Different Phases of Building the Cotton Technological Value Model

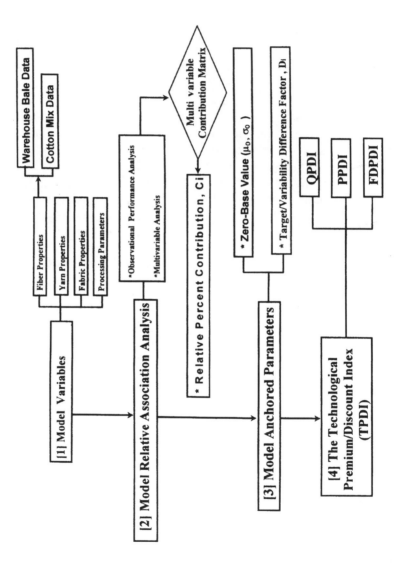

TABLE 12.1. Typical Values of Cotton Fiber Properties

Fiber Property	Plausible Range
HVI Parameters:	
Micronaire (Mic)	2.5-6.0
Fiber Length (FL, cm)	2.03-3.30
Fiber Bundle Strength (FS, g/tex)	18.0-40.0
Fiber Bundle Elongation (FE, %)	4.0-15.0
Length Uniformity (LU, %)	70.0-85.0
Color Reflectance (Rd)	60.0-85.0
Color Yellowness (+b)	5.0-16.0
Trash Area (TA)	0.1-1.0
AFIS Parameters:	
Maturity Ratio	0.5-1.2
Neps/g	50.0-800.0
Dust/g	200.0-2,000.0

TABLE 12.2. Ranges of Cotton Fiber Properties in Relation to Particular Selected Fabrics

Fabric Type	Fiber Length (UHM) inch	Fiber Strength (g/tex)	Micronaire	Maturity Ratio
Denim (open end)	0.92-1.15	24-32	3.0-5.0	0.80-0.90
Toweling	0.93-1.10	20-32	3.5-4.9	0.80-0.90
Print Cloth	1.06-1.18	23-32	3.5-4.9	0.85-0.95
Sheeting	1.06-1.14	22-32	3.8-4.6	0.85-0.95
Twills	1.00-1.09	22-32	3.5-4.9	0.85-0.95
Corduroy	1.06-1.16	22-32	3.8-5.5	0.90-1.00
Flannel	0.93-1.06	22-32	3.5-4.9	0.85-0.95
Poplins	1.09-1.13	22-32	3.5-4.9	0.85-0.95
Knits (18-28 cut)	1.09-1.16	22-32	3.5-4.9	0.85-1.00
Shirting	1.12-1.18	28-32	3.7-4.4	0.90-1.10
Nonwoven	0.95-1.10	23-32	4.9-5.5	0.85-1.00

Source: U.S. Cotton Fiber Chart (1995).

TABLE 12.3. Examples of Important Yarn Properties and Corresponding Recommended Values

Yarn Characteristics	Warp or Weft Yarn	Knitting Yarn
Count Variation:		
CV_{ct}%,cut length 100 m	< 2.0%	< 1.8%
CV_{ct}%,cut length 10 m	---	< 2.5%
Yarn Strength (cN/tex)	>11	> 10
Yarn Elongation (%)	> 5	> 5
Yarn Strength Variation (CV_{st}%)	< 10	< 10
Variation in Breaking Elongation (CV_E%)	< 10	< 10
Weak Places per 100,000m [Points of <60% of the mean breaking force]	< 1	< 1
Paraffin waxing/surface friction value	-----	ideal around 0.15μ

Source: Uster Statistics and Uster Report (1993).

TABLE 12.4. Examples of Important Performance Parameters

Performance Characteristics	Typical Values and Criteria
Cleaning efficiency (%) of the opening and cleaning line	Low—0.0%-20% Average—20%-35% High—35%-50%
Cleaning efficiency (%) of carding	Low—40%-60% Average—60%-75% High—75%-90%
Trash content of card or drawn sliver	Very Clean—Up to 0.05% Clean—0.05%-0.1% [for rotor-spinning] Average—0.1%-0.15% Fair—0.15%-0.2% High—Above 0.2%
Nep count [neps/g] of carded sliver	50-200
Uniformity of intermediate strands [Uster C.V%] : Card sliver Breaker drawn sliver Finished drawn sliver Roving	2.7-5.8 2.8-8.7 2.5-8.0 5.2-12.0
Spinning ends down: Ring spinning Rotor spinning	< 50 End breaks/1,000 spindle hours < 50 End breaks/1,000 rotor hours

Relative Association Analysis

The textile end product is normally evaluated based on several characteristics that must endure collectively to meet its expected performance. These characteristics result from manipulation of many factors, including yarn properties, fabric structure, and finishing treatments. The same concept holds for yarns for which manipulation of factors such as fiber properties and yarn structure determines the expected performance of yarn during weaving or knitting. The extent to which these factors must be manipulated depends on technological and economical constraints. Traditionally, art and experience have played a major role in manipulating material-related factors to produce an end product of specified performance characteristics. This approach has already proven to be costly and unprofitable. Alternatively, scientific product-development techniques should be implemented. A critical effort in this regard is the establishment of reliable relationships describing the association between different process and quality variables.

The association analysis can be performed using several techniques, ranging from simple bivariable correlation analysis to multivariable analysis. In the bivariable correlation analysis, the extent of association between the two variables can be determined using the simple coefficient of correlation, r, which ranges from minus 1 to plus 1. A zero value of r means no correlation between variables, and a value of $\pm$ 1 indicates perfect correlation. The analysis itself is a straightforward one; however, its success depends on the database used.

The multivariable association analysis aims at developing an equation relating a number of independent variables to a dependent variable (El Mogahzy, 1988; Kleinbaum, Kupper, and Muller, 1988; El Mogahzy and Broughton, 1989; El Mogahzy, Broughton, and Lynch, 1990). Verification of the reliability of the equation is then made, and the values of the relative contribution of different variables in the model are estimated.

The output of the association analysis used in developing the technological value model may be called the relative contribution matrix. It is a simple arrangement by which the percent relative contributions of different fiber properties to processing or end product parameters can be identified. The general form of the relative contribution matrix is as follows:

	X_1	X_2	$\cdots$	X_m
Y_1	$W_{y1.x1}$	$W_{y1.x2}$	$\cdots$	W_{y1xm}
Y_2	$W_{y2.x1}$	$W_{y2.x2}$	$\cdots$	$W_{y2.xm}$
Y_3	$W_{y3.x1}$	$W_{y3.x2}$	$\cdots$	$W_{y3.xm}$
$\cdots$	$\cdots$	$\cdots$	$\cdots$	$\cdots$
Y_n	$W_{yn.x1}$	$W_{yn.x2}$	$\cdots$	$W_{yn.xm}$
C_i	C_1	C_2	$\cdots$	C_m

In this matrix, the variables Y represent the desirable process characteristics (e.g., yarn characteristics, fabric characteristics, or other processing performance parameters), the x variables represent the fiber properties, and the term $W_{yi.xj}$ represents the association index. The C_i term, the contribution index, represents the overall percent relative contribution of a fiber property to the process under examination.

Model Anchored Parameters

To formulate the technological value model, a number of anchored parameters are required. These parameters are the contribution index obtained from the relative association analysis, a zero-base value, and a target or variability difference factor. The latter two parameters are discussed next.

Zero-base value (μ_o, σ_o). In general, we may define the zero-base value of a fiber parameter as an intermediate point on the premium/discount scale of the fiber parameter. The choice of the zero-base value of any fiber parameter will mainly depend on the type of application in which the technological value model is used. When the model is used for market analysis, the current market basis of cotton may be used as the zero-base value. This allows direct comparison between the market premium/discount profiles and corresponding technological profiles. When the model is used to evaluate the processing performance of cotton, zero-base values should be selected on the basis of the desired levels of fiber properties used in the cotton mix. In this regard, we recommend the use of the mean (μ_o) and standard deviation (σ_o) of fiber properties in the cotton mix as the zero-base values.

The difference factor (D_i). The difference factor represents the departure of the actual average or variability measure of a fiber property from the zero-base value. Thus, two types of difference factors may be used: (1) target difference factor and (2) variability difference factor. The target difference factor ($D_{\mu i}$) is given by the following equation:

$$D_{\mu i} = K_T \; \frac{(\overline{X}_i - \mu_{oi})}{\sigma_{oi}}$$

where X_i = the actual mean value of the ith fiber parameter; μ_{oi} = the zero-base value of the ith fiber parameter; σ_{oi} = the zero-base standard deviation of the ith fiber parameter; and K_T is a scaling constant.

Note that the target difference factor is normalized with respect to the standard deviation to produce a nondimensional value. This normalization allows elimination of the effect of units when different fiber parameters are considered in the model.

The expression used previously implies that fiber parameter values below the zero-base will be associated with discounts, whereas those above will be associated with premiums (see Figure 12.6a). This approach works well for parameters that follow "the larger, the better" pattern (e.g., fiber length, fiber strength, fiber elongation, and color Rd). Parameters such as trash content, short-fiber content, and color yellowness will follow a "the smaller, the better" pattern. For these parameters, the previous expression is modified by simply using a negative sign. In this case, any value of fiber parameter exceeding the zero-base value will always be associated with discount (see Figure 12.6b). We should point out that for these types of properties, the ideal zero base should be zero, irrespective of the process type or the quality level required. However, cotton, being a natural fiber, is expected to exhibit values of these properties that are significantly greater than zero.

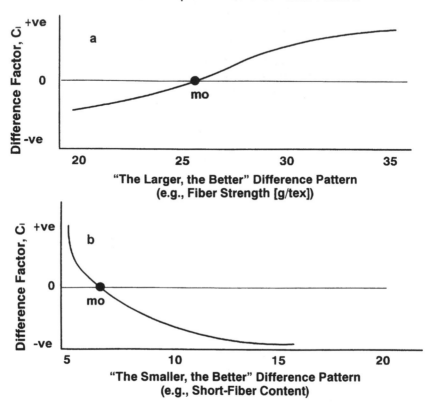

FIGURE 12.6. Examples of Difference Factor Patterns

In addition to the two types of fiber parameters discussed previously, another type is unique in its contribution to the processing performance and the quality of the end product. Examples of this type include micronaire and fiber friction. These two parameters provide optimum performance at some intermediate levels of their value ranges.

In the case of micronaire, too low values often indicate immaturity, and too high values indicate fiber coarseness. The adverse effects of these extreme levels on processing performance or end-product quality are well realized. Similarly, too low friction can be costly because of the need for more twist during spinning to reach the optimum yarn strength, and too high friction results in well-recognized problems during carding and drafting. The complex nature of the effects of this type of fiber parameter is extremely difficult to reveal by traditional association analysis. The main difficulty is finding cottons exhibiting extreme values of these parameters, while showing constant levels of other fiber parameters.

For parameters of this type, the difference factors should be modified to account for the nature of their contribution. Figure 12.7 illustrates one approach to this modification. In this case, two threshold zero-base values are assigned, $\mu_{o,min}$ and $\mu_{o,max}$. The difference factor for any parameter value falling around the minimum threshold zero-base value will follow a "the larger, the better" pattern, and that for any value falling around the maximum threshold zero-base value will follow a "the smaller, the better" pattern. The threshold values can be obtained from analysis of historical data. In Figure 12.7, the two threshold values of micronaire were obtained from extensive analysis of U.S. cotton crop data (1983 through 1990).

The variability difference factor ($D_{\sigma i}$) is given by the following equation:

$$D\sigma_i = 100\,K_v\,\left[\frac{\sigma_i}{X_t} - \frac{\sigma_{oi}}{\mu_{oi}}\right]$$

where σ_i = the actual standard deviation of the ith fiber parameter and K_v = scaling constant.

The previous equation can be rewritten using the familiar coefficient variation as follows:

$$D\sigma_i = K_v\,[\,C.V_i - C.V_{oi}\,]$$

where $C.V_i$ = the actual coefficient of variation of the ith fiber parameter and $C.V_{oi}$ = the zero-base coefficient of variation of the parameter.

Note that all variability difference factors will always follow a "the smaller, the better" pattern, irrespective of the process or the quality level desired.

FIGURE 12.7. Examples of Special Difference Factor Patterns (e.g., Fiber Micronaire)

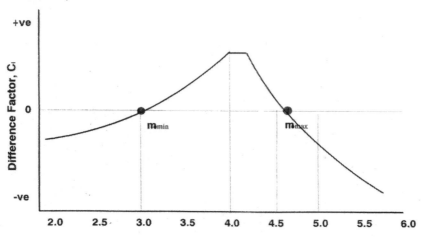

The Technological Premium/Discount Index

The technological premium/discount index (TPDI) represents the output of the technological value model. The general form of this index is as follows:

$$TPDI = \sum_{i=1}^{i=m} C_i D_i$$

where C_i = the percent relative contribution of fiber parameter i and D_i = the difference factor of fiber parameter i.

For target values, the TPDI will be as follows:

$$TPDI = \sum_{i=1}^{i=m} C_i D\mu_i = [\, C_{Mic}D_{\mu_{Mic}} + C_{FS}D_{\mu_{FS}} + C_{FL}D_{\mu_{FL}} + \dots \,]$$

For variability measures, the TPDI will be as follows:

$$TPDI = \sum_{i=1}^{i=m} C_i D_{\sigma i} = [\, C_{Mic}D_{\sigma_{Mic}} + C_{FS}D_{\sigma_{FS}} + C_{FL}D_{\sigma_{FL}} + \dots \,]$$

One of the important features of these TPDI expressions is the flexibility associated with using them in the technological value model. This flexibility is a result of the possibility of separating premium/discount components for different fiber parameters and of separating target premium/discount components from variability components. For example, a mill may decide that the most important fiber parameters to its process are micronaire and fiber strength. In this case, only TPDI components associated with these two parameters can be considered in the technological value model. Furthermore, a mill may decide that variability is only important for some fiber parameters, for example, micronaire and color. In this case, only TPDI variability components associated with these two parameters can be included in the model.

The linearly additive forms of the TPDI may somewhat represent oversimplification of the premium/discount contribution of each fiber factor to the TPDI. This is particularly true in the case of the variability index. However, alteration of this form to account for possible nonlinearity can easily be done, provided the relative association analysis has revealed such a nonlinearity.

The Premium/Discount Index (PDI) Trilobate System

To produce reliable premium/discount values, fiber attributes should be divided into two main classes. The first class represents the expected inherent fiber characteristics (e.g., length, fineness, strength, maturity, etc.). The second class represents attributes that should not exist under ideal fiber production conditions (e.g., trash content, short-fiber content, neps, and stickiness). These attributes will be called fiber defects (or contaminants). These defects do not inherently characterize a textile fiber, and they can be prevented or minimized through proper growing and ginning conditions. Although the current classing system admittedly considers these attributes as being defects or extraneous matter, only heavy trash and leaves are accounted for in the system.

When the first class of fiber attributes is under consideration, two distinct forms of contribution should be recognized: the contribution of fiber attributes to processing performance parameters and the contribution of fiber attributes to the quality of the end product (yarn or fabric). Examples of processing performance parameters include opening and cleaning waste, combing waste, spinning ends down, spinning potential, and filling stops. Examples of end product quality parameters include yarn strength, yarn evenness, and fabric strength. From the standpoint of process design, some fiber attributes can contribute to these two types of parameters in uniquely different manners, as discussed in the following material.

Processing performance mainly involves interaction between the fibers and the machine elements. The end-product quality parameter, on the other hand, involves fiber-to-fiber interaction. Accordingly, some fiber attributes contribute to processing performance in a uniquely different manner than to yarn quality. For example, it is well recognized that fine and long fibers are considered premium in relation to yarn quality. This is simply a result of the superior fiber-to-fiber interaction in the yarn that enhances both the integrity and the strength of the yarn. In relation to processing performance, very fine or very long fibers may result in excessive opening and carding waste and in high nep formation due to the high flexibility of these fibers. Similar arguments can be made for other fiber attributes, including friction and elongation.

In light of the previous discussion, we recommend three different premium/discount indexes that can collectively determine the technological value of cotton fibers. These are the processing performance premium/discount index (PPDI), the end product quality premium/discount index (QPDI), and the fiber defects premium/discount index (FDPDI). These three indexes form the PDI Trilobate System shown in Figure 12.8.

The PDI Trilobate provides an inclusive system that can assist in determining the technological value of cotton in view of the three major areas of fiber impact. Different companies may have different points of emphasis regarding the worth of cotton fibers in relation to their processes. The Trilobate System provides the necessary flexibility of valuing cottons depending on the company's emphasis and the level of quality needed. In addition, it integrates the various efforts of cost and quality optimization performed by a textile company into a systematic approach that can lead to more objective

FIGURE 12.8. The Technological Premium Discount Trilobate System

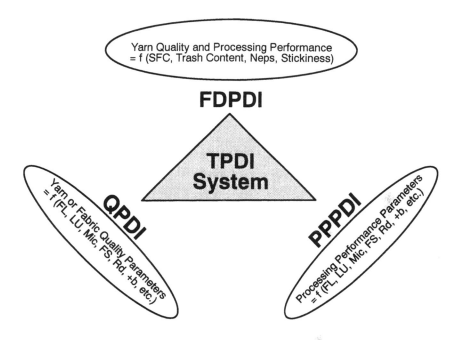

cotton-purchasing decisions and better utilization of cotton fibers than the traditional subjective approaches.

Applications of the Technological Model

The technological model discussed earlier can be used in several applications, such as the following:

1. Comparative analysis between the market value and the technological value of cotton
2. Making proper decisions regarding the selection of certain cottons suitable for producing particular end products
3. Establishing objective methods for determining the relative contribution of cotton cost to the total manufacturing cost

In the latter section of this chapter, I will discuss some of these applications.

COTTON FIBER SELECTION FOR BLENDING

The Importance of Proper Fiber Selection

The main technological challenge in any textile process is to convert the high variability in the characteristics of input fibers into a uniform end product. This critical task is mainly achieved in the blending process, provided that two basic requirements are met: (1) a consistent input fiber profile and (2) capable blending machinery. Uniform blending is the key factor to a stable process and a uniform product. Poor blending may result in the presence of clusters of cotton fibers that exhibit unfavorable characteristics such as excessive short-fiber content, clustered colored fibers, immature fibers, and stickiness. In end products, problems such as fabric barré, dimensional irregularity, and poor appearance can result from poor blending.

Over the years, development in blending techniques has been largely hindered by insufficient fiber information, resulting from a lack of capable and efficient testing methods. Accordingly, art and experience have been the primary tools. One of the common approaches has been massive blending, during which vast quantities of bales are mixed to reduce variability. Mixed cottons are then rebaled and fed to the opening line in a random order to further enhance the mixing effect. The traditional feed hoppers certainly accommodated this type of blending.

In today's technology, the rising cost of storage, labor, equipment, and raw material makes it impossible to perform the old blending approach. In a modern warehouse, a bale population is no longer static. Bales are supplied to the warehouse on an incremental basis, depending on the rate of bale consumption, warehouse space and arrangement, and other cost-related factors. Furthermore, utilization of different spinning techniques to meet various end-product demands has resulted in the use of several varieties and types of cotton that have to be grouped according to the type of equipment and desired end-product characteristics. These factors have led to a more complex warehouse in which proper bale management has become a necessity. Accordingly, advanced blending should be based on precise fiber selection from different groups in which fiber characteristics are categorized according to predetermined criteria.

In view of today's technology, cotton fiber selection should be considered as an integrated phase in the FFE program. In this regard, sources and causes of variability in cotton fiber characteristics should be understood (Earnest, 1993). The capability of advanced blending machinery should be evaluated (Leifeld, 1993). In marketing systems in which a gap between the technological value and the market value of cotton is likely to exist, fibers should be primarily selected on the basis of their technological merits (El Mogahzy, Broughton, and Lynch, 1990). Furthermore, economical constraints, including the cost of various components in the blend and bale inventory, should be considered in the optimization process (El Mogahzy, 1992, a,b).

Objectives of Fiber Selection

An "engineered-in" fiber selection approach should meet two main objectives: (1) to achieve a uniform profile of the characteristics of input fibers and corresponding end products and (2) to maintain the average values of output characteristics at their desired levels. The first objective can be met using proper algorithms of fiber selection for the given blending equipment. These algorithms should aim at picking cotton bales from the population warehouse according to prespecified criteria. The second objective can be met through the use of reliable models relating input to output characteristics. These models can act as a powerful tool for controlling yarn or fabric quality.

Cotton Fiber Selection: Basic Concepts

As indicated earlier, advanced fiber blending should aim at improving the uniformity of both input fiber and output yarn characteristics and maintain average values of output characteristics at their desired levels. Accordingly, fibers should be selected from the bale population in such a way that these objectives are being met on a mix-to-mix basis. From an economical viewpoint, a proper fiber selection strategy should result in better bale management, improved cotton bale acquisition, improved mill efficiency, and optimum utilization of cotton.

In general, a fiber selection program should involve the following basic steps:

1. Examination of the population(s) of fiber properties from which cotton bales are to be selected. This step requires testing samples from all cotton bales in the population using High-Volume Instrument (HVI) systems, generating data on fiber properties of all bales, and examination of the quality of data to detect outliers.
2. Developing frequency distributions of the fiber properties considered in the selection process.
3. Implementing reliable bale-picking schemes based on the distributions of fiber properties of bale population(s). This step requires knowledge of the size of the bale population, the size of the bale mix or "laydown," and average fiber characteristics.
4. Controlling average output characteristics by developing reliable fiber-yarn regression relationships. These relationships are then incorporated in the bale-picking scheme.
5. Verifying the effectiveness of the fiber selection program by monitoring the uniformity of bale laydown fiber characteristics and corresponding yarn characteristics.

In practice, any fiber selection program should accommodate the system utilized to acquire cotton bales from the warehouse. Two main methods of bale acquisition may be used: (1) storage and retrieval by group and (2) storage and retrieval by bale identification number. The first method is the most common one and is discussed next.

"By-Group" Fiber Selection

Selection of fibers by group and categories is based on knowledge of fiber characteristics of each individual bale in the warehouse population. In a typical bale warehouse, the population of cotton bales may be represented by one group, or it may be divided into several groups according to growth area, grade, cotton type, certain fiber characteristics, spinning systems, end products and so forth. A single-group system is hardly found in today's technology because of the increasing trend toward buying different cotton types for economical reasons and for satisfying different product specifications.

In a by-group selection system, bales belonging to each group are mainly identified by their corresponding group. Cotton bales are then picked from each group in quantities that are more or less proportional to their quantity in the group population. Figure 12.9 illustrates the general principle of the by-group system.

The method by which bales are picked from each group population may be divided into two types: (1) random picking and (2) category picking. In random picking, each bale in a group population has virtually an equal chance of being selected in the fiber mix, or laydown. In category picking, each group population of a fiber characteristic is divided into a number of categories, and bales are then picked by both

FIGURE 12.9. Selection by Cotton Group

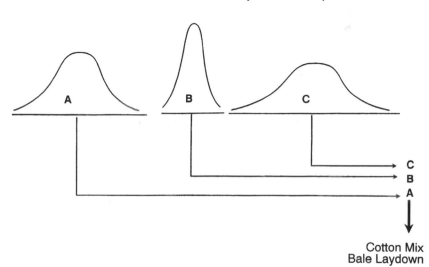

Cotton Mix
Bale Laydown

group and category within the group. Figure 12.10 illustrates the general principle of fiber selection by group and category.

As the number of fiber properties considered in a category system increases, the number of category combinations will also increase. This is a result of the increase in the joint occurrence of values of fiber properties. Mathematically, the number of category combinations in a single-group scheme is as follows:

$$Number\ of\ Category\ Combinations\ (k) = x^j$$

where x = the number of levels used for a single fiber property and j = the number of fiber properties considered for selection.

The previous equation indicates that when several fiber properties are considered in the category system, or when several levels are utilized, the number of category combinations increases rapidly. For example, if the number of categories per a single fiber property equals 3 and the number of fiber properties equals 3 (e.g., micronaire, fiber length, and fiber strength), the corresponding number of

FIGURE 12.10. Selection by Cotton Group and Category

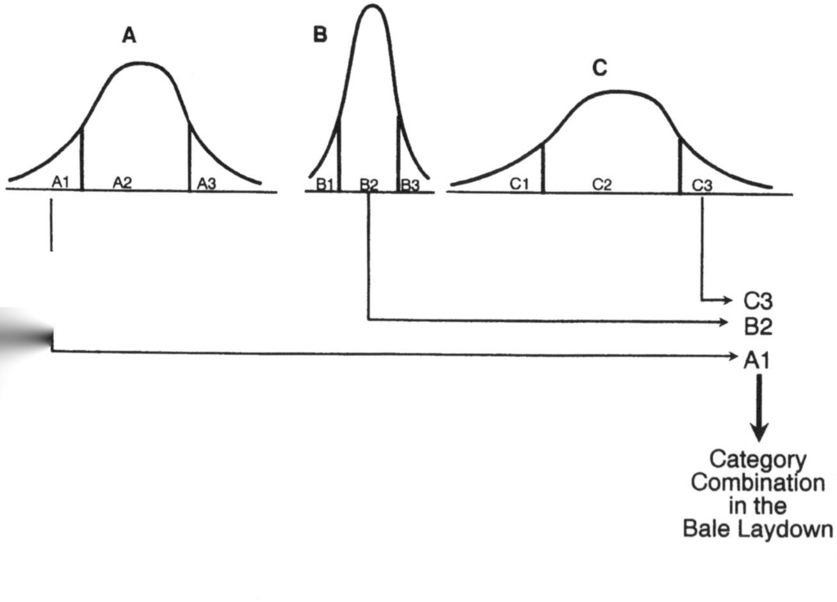

category combinations will be 27. The addition of one more fiber property (e.g., length uniformity) to this category system will result in 81 combinations, and so on. In a multigroup system, the number of category combinations is multiplied by the number of groups (El Mogahzy and Gowayed, 1995a,b). It is therefore important to consider a selection process that can accommodate the bale management scheme of the textile mill.

CONTROL OF OUTPUT CHARACTERISTICS

To control output characteristics, relationships between fiber properties and desirable output characteristics should be developed. Several methods for developing such relationships are discussed in the literature. The most common approach is multiple regression analysis (El Mogahzy, 1988; Kleinbaum, Kupper, and Muller, 1988; El Mogahzy and Broughton, 1989; El Mogahzy and Gowayed, 1995a,b).

In a typical textile mill, the following steps should be taken to perform reliable regression analysis:

1. From the warehouse bale population, frequency distributions of important fiber characteristics are constructed. From these distributions, typical average values and ranges of fiber properties of the bale population are determined.
2. From the bale population, a statistically sound database is then selected for regression analysis. This database should consist of special bale laydowns representing a number of experimentally designed combinations of fiber characteristics. The purpose of this design is to create a data set involving independent fiber parameters that is a basic regression assumption.
3. The special regression laydowns are then processed and corresponding yarns are tested for desirable characteristics.
4. From the data of fiber properties and their corresponding values of yarn characteristics, multiple regression equations are developed.
5. The developed regression equations can then be used in the fiber selection program in two different ways: (a) as a prediction tool that acts as an additional constraint on the range of

fiber characteristics selected in the bale mix and (b) as a selection tool that uses the predicted output characteristic (e.g., yarn strength) as the primary quality parameter.

BALE-PICKING SCHEMES

Bale-picking schemes represent the methods by which bales can be picked from a warehouse to meet fiber selection criteria. Two methods of bale-picking may be utilized: (1) the random-picking scheme and (2) the category-picking scheme. In the following sections, I will discuss the concepts underlying these two schemes.

The Random-Picking Scheme

In a random-picking scheme, laydown bales are picked randomly from the parent bale population (the warehouse). By definition, any value of the fiber characteristic (or of any bale) in the population will have the same opportunity to be represented in the laydown. If complete randomization can be achieved, this would result in an ideal mixing. This hypothetical situation largely resembles the old approach of massive cotton blending.

Despite the fact that pure random picking is not widely used, its algorithms serve as the basis for other picking schemes (Stuart, 1962; El Mogahzy and Gowayed, 1995a). Furthermore, any other type of picking scheme involves limited random picking from a particular group or category.

The principle utilized in the random-picking scheme is selection of a random sample from a finite population. If the population is of size N, and a simple random sample of size n is required (i.e., the sampling fraction $f = n/N$), this sample will be chosen at random from the distinct possible samples, with each including no population member more than once. Such a sample can be obtained sequentially by drawing members from the population one at a time without replacement, so that at each stage, every remaining member of the population has the same probability of being chosen.

In the random-picking scheme, between-laydown variance depends on two main factors: (1) the overall variance of the bale population and (2) the ratio between the number of bales in the bale laydown and the total number of bales in the warehouse population.

As expected, the larger the variability in the bale population, the larger the between-laydown variability. For a given bale population size, N, the larger the number of bales in the laydown, n, the lower the between-laydown variances. In practice, however, the size of a bale laydown may be constrained by technological criteria, making it difficult to increase. For a fixed laydown size, picking from a smaller bale population will result in smaller variance. Obviously, the overall population size depends on several factors, most of which are independent of laydown variability. Nevertheless, it is this concept that serves as the basis for the category-picking scheme, discussed next, which aims at minimizing variability by dividing the overall population into smaller populations, each representing a certain category.

Category-Picking Schemes

As noted earlier, a category system is based on dividing the total range of the population of a fiber characteristic into a number of categories from which bales are selected. Statistically speaking, a categorized population may be divided into k nonoverlapping categories of sizes N_1, N_2, . . . , N_k and corresponding weights, W_1, W_2, . . . , W_k, where $W_i = N_i/N$. Each category is considered a subpopulation. Thus, means and variance values of different categories will be as follows:

Means: μ_1, μ_2, . . . , μ_i, . . . , μ_k

Variances: σ^2_1, σ^2_2, . . . , σ^2_i, . . . , σ^2_k

Several category-picking schemes may be utilized (Stuart, 1962; El Mogahzy and Gowayed, 1995a). The most common one is the proportional weight category-picking (PWC) scheme discussed next.

Proportional Weight Category-Picking Scheme

Any type of category picking should account for inventory constraints. In other words, cotton bales belonging to a certain category should be represented in the mix proportionally to their population amount. This simple concept is the basis for the proportional weight category-picking (PWC) scheme. Within a given category, bales should be picked randomly. Accordingly, a proportional weight category-picking scheme should satisfy the condition that the probability of the presence of a cotton bale from a certain category in the mix or laydown is equivalent to N_i/N. In other words, picking fractions, $f_i = n_i/N_i$, should be identical for all categories. Accordingly, the number of bales, n_i, belonging to a certain category, i, is given by the following equation:

$$n_{ij} = n \left(\frac{N_{ij}}{N} \right)$$

where n = the total number of bales per laydown; N_i = the total number of bales in the population belonging to category i; and N = the total number of bales in the overall population.

In a category-picking scheme, the between-laydown variance may be reduced by increasing the number of bales per laydown. As mentioned in the case of random-picking, changes in these values are limited by other technological considerations.

Another approach to reducing the variance in the PWC scheme is to reduce the category variance. This reduction is systematically achieved when the population variance is small. In populations exhibiting high variability, the category variance may be reduced by increasing the number of categories. However, an increase in the number of categories requires more strict bale management, which may not be possible in some warehouses.

APPLICATIONS: CASE STUDY

In this chapter, two critical phases of FFE were discussed: (1) determining the technological value of cotton and (2) fiber selection techniques. In practice, these two phases are closely associated. In this section, I will discuss some applications in which the two phases are implemented. Due to the limited space, one case study representing a U.S. cotton denim mill will be used.

The denim mill under examination produces open-end spun yarn of English counts ranging from 5s to 6s. Using the procedures discussed in phase one, premium/discount scales were set for the mill. These are shown in Figures 12.11 through 12.14. These premium/discount values were used for three applications:

1. Comparative analysis between the market value and the technological value of cotton
2. Proper decision making regarding fiber selection in view of the technological value model
3. Establishment of objective methods for determining the relative contribution of cotton cost to the total manufacturing cost

Comparative Analysis of Market Value and Technological Value

Because this is a U.S. mill, cottons were purchased on the basis of the USDA classing system represented by the CCC loan chart. Thus, market premium/discount (CCC PDI) values can be compared with technological premium/discount values (TPDI). Figure 12.15 shows correlation coefficients between premium/discount values and some yarn properties. As can be seen, the technological premium/discount values had superior correlations to those of the

FIGURE 12.11. Fiber Defect Premium/Discount Values (Denim Mill; Ne = 5s-6s)

	SFC	Neps/g	Trash/g	Sugar
300	5	50	100	0.1
0	8	300	200	0.3
-400	9	375	621	0.4
-1000	25	700	600	0.6

FIGURE 12.12. Processing Performance Premium/Discount Values (Denim Mill, Ne = 5s-6s)

	FL	LU	Mic	FS	FE	Rd	+b
300 ◀	1.25	86	4.1	36.0	9.0	82	6
0 ◀	1.06	80	3.0	24.5	6.0	72	9
0 ◀	1.06	80	4.6	24.5	6.0	72	9
-300 ◀	0.85	75	2.6	18.5	4.5	60	16

FIGURE 12.13. Typical Target Mean Quality Premium/Discount Values for Open-End Spun Yarns, Coarse 5s-6s/Denim Fabric

OE-Coarse Denim Yarn	FL	LU	Mic	FS	FE	Rd	+b
+230 ◀	1.10	84	4.3	32.0	9.0	78	7
+100 ◀	1.09	82	4.6	29.0	8.0	72	8
0 ◀	1.05	80	3.0	27.0	6.5	70	8
0 ◀	1.05	80	4.6	27.0	6.5	70	8
-90 ◀	1.00	75	3.4	25.0	4.8	78	10
-235 ◀	0.95	76	5.5	23.0	5.0	70	10

market values. The impact of this observation can be reflected in making proper decisions on what cottons can be purchased by the mill and at what values. Quite often, a cotton that is sold at a discount may be of premium value to the particular end product under consideration.

FIGURE 12.14. Typical Variability Premium/Discount Values for Open-End Spun Yarns, Coarse 5s-6s/Denim Fabric

OE-Coarse Denim Yarn Variability PDI	C.V [FL]	C.V [Mic]	C.V [FS]	C.V [FE]	C.V [Rd]	C.V [+b]
+223 ◄	2.00	5.0	5.0	4.0	3.0	6.0
+16 ◄	3.02	7.7	6.7	10.4	4.5	10.0
0 ◄	3.00	8.0	7.0	7.0	4.0	9.0
-60 ◄	3.50	8.5	7.8	8.0	4.0	9.0
-100 ◄	2.80	9.5	8.3	9.0	3.3	7.1
-300 ◄	5.00	12.0	10.0	9.0	6.0	10.0

FIGURE 12.15. Correlations Between PDIs and Yarn Quality Characteristics (Denim Mill—Open-End 5s-6s)

Yarn Characteristic	TPDI	CCCPDI
CSP	0.540	0.013
Single-End Strength	0.600	0.078
C.V Percent Irregularity	−0.444	−0.256

Fiber Selection on the Basis
of the Technological Value Model

The techniques of cotton fiber selection discussed earlier aim at feeding a uniform mix to the textile process. Examples of uniform fiber profiles based on proportional weight category-picking schemes are shown in Figure 12.16. As can be seen in this figure, Micronaire values are consistent from one mix to another. Furthermore, the variability within a mix (C.V percent) is maintained below a maximum allowable threshold.

In recent years, major developments have been made in the area of bale management. In the United States, traditional large bale inventory and huge warehousing has given way to just-in-time and short bale inventory. This approach has necessitated the use of cotton shipments almost as they come to the mill. In this case, cotton shipments, typically a truckload of about ninety bales, should resemble the cotton mix or the bale laydown to be immediately selected by the mill. Figure 12.17 shows a case in which cotton

FIGURE 12.16. Example of Fiber Profile in the Textile Process

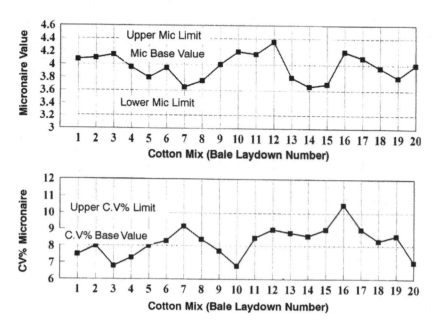

FIGURE 12.17. Shipment Fiber Profiles and Corresponding Premium/Discount Indices

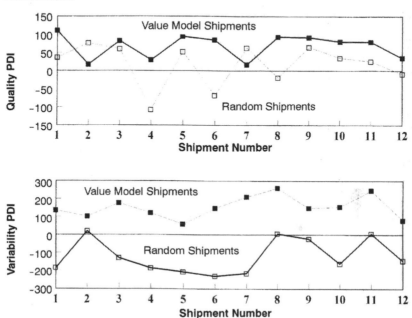

shipments are selected based on premium/discount values of both target mean and variability, as compared to purchasing random shipments. As can be seen in this figure, shipments purchased on the basis of the technological value model exhibit premiums of both target means and variability measures. The corresponding fiber profiles (using micronaire) are shown in Figure 12.18. Note the superior uniformity associated with bale laydowns selected from shipments purchased on the basis of the technological value model.

Objective Relationships Between Cotton Cost and Manufacturing Cost

Objective relationships between cotton cost and manufacturing cost can be established using the technological premium/discount values of cotton and corresponding premium/discount values of manufacturing parameters. One of the areas in which the cotton

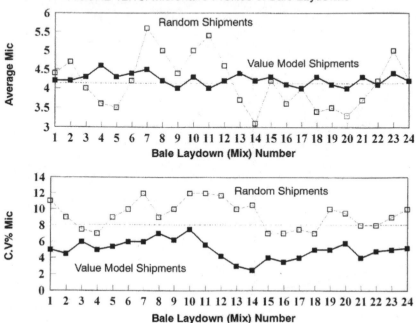

FIGURE 12.18. Micronaire Profiles of Bale Laydowns

value directly affects manufacturing cost involves fiber waste that is encountered due to excessive fiber defects in the incoming cotton (i.e., high trash and short-fiber contents). Figure 12.19 shows a direct relationship between the technological premium/discount value (fiber defect PDI) and discount values associated with manufacturing waste. As can be seen in this figure, discount cottons result in higher manufacturing costs associated with waste.

CLOSING REMARKS

This chapter has reviewed some of the major developments in fiber-to-fabric engineering. Obviously, other elements of this important issue were not covered. However, my intention was to encourage cotton users to implement more scientific/practical applications that can result in substantial savings and significant optimization of cotton quality/cost. I should also point out that there are powerful software

FIGURE 12.19. Values of Fiber Defects Premium/Discount Index and Corresponding Loss Points Associated with Opening and Cleaning Waste for Typical Bale Laydowns

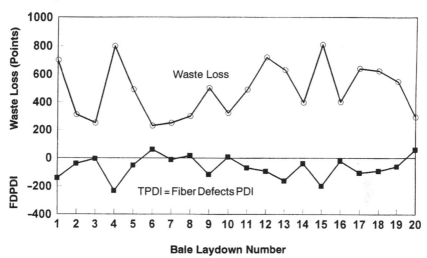

programs available today that can perform the analysis described in this chapter. In the United States, the Engineered-Fiber-Selection (EFS®) program, developed by Cotton Incorporated, has been used for selection of over 90 percent of the total bales produced in the United States. The program is also used in Europe and South America.

REFERENCES

Chen, C. and D. Ethridge (1996). Valuation of cotton characteristics by U.S. textile manufacturers. Beltwide Cotton Conference Proceedings, Cotton Economic and Marketing Conference, pp. 427-434.

Deussen, H. and C. Farber (1994). *Cotton Valuation Model.* Schlafhorst Publications.

Earnest, D. (1993). Field-to-bale distribution study. Paper presented at the Beltwide Cotton Conference, New Orleans, LA.

El Mogahzy, Y. (1988). Selecting cotton fiber properties for fitting reliable equations to HVI data. *Textile Research Journal* 58: 392-397.

El Mogahzy, Y.E. (1992a). Optimizing cotton blend cost with respect to quality using HVI fiber properties and linear programming. Part I: Fundamentals and advanced techniques of linear programming. *Textile Research Journal* 62: 1-8.

El Mogahzy, Y.E. (1992b). Optimizing cotton blend costs with respect to quality using HVI fiber properties and linear programming. Part II: Combined effects of fiber properties and variability constraints. *Textile Research Journal* 62: 108-114.

El Mogahzy, Y. and R. Broughton (1989). Diagnostic procedures for multicolinearity between HVI cotton fiber properties. *Textile Research Journal* 59: 440-447.

El Mogahzy, Y., R. Broughton, and W.K. Lynch (1990). A statistical approach for determining the technological value of cotton using HVI fiber properties. *Textile Research Journal* 60: 440-447.

El Mogahzy, Y. and Y. Gowayed (1995a). Theory and practice of cotton fiber selection. Part I: Fiber selection techniques and algorithms. *Textile Research Journal* 65: 32-40.

El Mogahzy, Y.E. and Y. Gowayed (1995b). Theory and practice of cotton fiber selection. Part II: Sources of cotton mix variability and critical factors affecting it. *Textile Research Journal* 65: 75-84.

El Mogahzy, Y. and R. Kearny (1992). Effects of gin/mill lint cleaning on processing performance and yarn quality. A technical report submitted to USDA, SRRC, New Orleans, LA.

Hudson, D. and D. Ethridge (1996). Texas-Oklahoma producer cotton market. Beltwide Cotton Conference Proceedings, Cotton Economic and Marketing Conference, pp. 380-384.

Kleinbaum, D.G., L.L. Kupper, and K.E. Muller (1988). *Applied Regression Analysis and Other Multi-Variable Methods*. Boston, MA: PWS-KENT Publication Company.

Leifeld, F. (1993). The secret of perfect blending from bale to sliver. Paper presented at the Beltwide Cotton Conference, New Orleans, LA.

Leifeld, F. (1994). New features of a high-tech card. Paper presented at the Beltwide Cotton Conference, San Diego, CA.

Meredith, W. Jr. (1995). Biotechnology, yield, and fiber quality. 8th Engineered Fiber Selection System Conference, Cotton Incorporated.

Selker, H. (1992). A new Trutzschler blow-room generation. A presentation at Trutzschler Indonesia customer's day, Bandung, Indonesia.

Stuart, A. (1962). *Basic Ideas of Scientific Sampling*. London: Griffin.

U.S. Cotton Fiber Chart for 1995, Textile World.

Uster Statistics and Uster Report (1993), No. 39.

Index

Page numbers followed by the letter "i" indicate illustrations; those followed by the letter "t" indicate tables.

Printed and bound by CPI Group (UK) Ltd, Croydon, CR0 4YY

23/10/2024

01777672-0004